ANTARCTICA

The Frozen Continent's Environment, Changing Logistics and Relevance to India

For

Meher and Ananya

ANTARCTICA

The Frozen Continent's Environment, Changing Logistics and Relevance to India

Dr Jagadish P. Khadilkar

BLOOMSBURY
LONDON • NEW DELHI • NEW YORK • SYDNEY

First Published, 2017

BLOOMSBURY PUBLISHING INDIA PVT. LTD.
London New Delhi New York Sydney

ISBN: 978-93-86432-00-1

10 9 8 7 6 5 4 3 2 1

Published by Bloomsbury Publishing India Pvt. Ltd.
Vishrut Building, DDA Complex, Building No. 3, Pocket C-6 & 7
Vasant Kunj, New Delhi 110 070

Laser Typeset by SHINU
A-5/56, Moti Nagar, New Delhi 110 015

Printed and bound in India by Thomson Press India Ltd.

SYMBIOSIS
INTERNATIONAL UNIVERSITY
(Established under Section 3 of the UGC Act 1956 vide notification No. F9-12/2001-U.3 of the Government of India)
Prof. Dr. S. B. Mujumdar M.Sc. Ph.D.
CHANCELLOR
(Awarded 'Padma Bhushan' by President of India)

Foreword

Since the start of the global calendar two millennia ago, I consider the last century and a half to be the most significant – in both productive and destructive terms. During this time, on the productive side, we have seen the transformation of human lifestyle from near-primitive to ultra-modern, with tremendous technological advances in every field, industrial and digital revolutions, the advent and explosion of Internet and information technology, to name a few. Landing on the moon and Mars no longer remain in the realm of the imagination of sci-fiction authors or poets. On the destructive side, however, the same period saw human greed at its worst, with colonisation and exploitation of resources, two world wars, the spectre of atomic and other weapons of mass annihilation, global warming posing a grave threat to the earth's eco-system stable for thousands of years – all within a few decades.

Interestingly, the period also saw the unravelling of various branches of science like geography, geology and atmospheric sciences, and of life in the continent of Antarctica, explaining the scientific theories and associated evolutionary processes. In Antarctica, the enormous ice cap, with a thickness ranging from a minimum of 800 m to a maximum of 4.8 km, covers over 98 per cent of its land surface and doubles the original area in the process. It makes the continent bigger than the continents of Europe and Australia or India and China put together. Even the scientists, using explosives and sound-wave

Senapati Bapat Road, Pune - 411 004 | Tel: 020-2565 2444 | Fax: 020-2565 9209.
E-mail: sbmujumdar@symbiosis.ac.in | website: www.symbiosis.ac.in

techniques during the International Geophysical Year (IGY) in the 1950s, could not believe the startling results when this was recorded for the first time. Atmospheric research revealed how hundreds of thousands of years' of the world's climatic history and records are preserved in the identifiable yearly rings in the enormous ice cap. But for the research in Antarctica, the world would have never known about the menace of the ozone hole, let alone take effective measures to contain it. Oceanographic research threw light on the self-sustaining giant food web in the surrounding Southern Ocean, developed over thousands of years, a web which the world may have to turn to as a possible food source in an uncertain future caused by global warming. While research on this and other branches of science was going on in Antarctica, geologists the world over, including the likes of German Alfred Wegener remained busy solving the mystery of the origin of the continents and oceans. Still at the hypothesis stage, with Antarctica at the centre and Africa, South America, Australia and the Indian subcontinent conjoined from all sides, Austrian Edward Suess coined the term Gondwanaland (Gondwana) to name the supercontinent after a tribe in central India. Another scientist, South African du Toit, called Antarctica the keystone of Gondwana to underscore its importance in the geological evolution. With further evidence, scientists could finally explain the mechanism about evolution through a geological timeline and develop the now widely accepted plate-tectonic theory in the 1960s. Though commercial exploitation is banned by the Madrid Protocol till 2048, the potential for natural mineral resources found in the other Gondwana counterparts and the hydrocarbon deposits in the surrounding Southern Ocean will always be matters of interest for the global community.

The uniqueness of all the environmental aspects of Antarctica was also reflected on the geopolitical front. Not knowing the environmental situation, territorial claims from seven countries, their refutation from others and minor conflict situations arose in the initial period. But soon it was clear that conventional colonisation was impossible in Antarctica, due to the surrounding fierce Southern Ocean, isolating Antarctica from the rest of the world and the continent's harsh environmental conditions precluding any permanent human or animal inhabitation. This realisation led to intense negotiations and the enactment of the Antarctic Treaty and its subsequent instruments called the Antarctic Treaty System (ATS). Coming into force in the thirty years during the 1960s to the 1990s, they include stringent stipulations like suspension of all territorial claims, a prohibition on military (though use of military personnel for support is permissible) or nuclear activity as well as any dumping, a ban on any commercial mining activity, and a total protection of the environment while conducting scientific research with transparency.

As if all these complications were not enough, the spectre of global warming made its appearance on the stage and became a major cause for concern for the world community. Success or otherwise of the implementation of the recently concluded Paris Agreement on climate change, actual environmental situation on ground, environmental interdependence between Antarctica and rest of the world and resultant consequences will be watched, keenly and anxiously, in the near future. As far as India is concerned, the effective implementation of the Paris Agreement by all countries is vital, and failure is not an option.

As of now, however, Antarctica acts as the world's unique scientific laboratory through the research stations of various countries under the umbrella of the ATS. Factors mentioned above in connection with the consequences of global warming and the situation in the Arctic Ocean caused by the possibility of exploitation of natural resources in the near future, too, has the potential to severely strain the ATS. However, unless disturbed, today's seemingly tranquil situation may continue till the year 2041 when the entire gamut of Antarctic Treaty System (ATS) will be up for review. All the leading countries in the world have, over the years, developed effective logistics networks, based on self-reliance, vitally needed for an influential and effective presence in Antarctica. As of now, they are well and truly entrenched with viable scientific programmes. As a result, the present system of international governance through ATS suits these countries in their wait and watch policy till 2041 and beyond, or till the situational change warrants a global rethink.

As far as India is concerned, things are different. Though a late entrant in matters Antarctic, India enjoys a prominent position in Antarctica, as one of the few countries with two or more permanent stations and viable scientific programmes. However, in terms of logistics, India is totally dependent on other countries and international tie-ups, a situation which can put the country at peril at the time of reckoning as and when it may come. China, on the other hand, though a latecomer like India, has well understood the writing on the wall and has taken great strides to develop self-reliant Antarctic logistics and may catch up with leading Antarctic countries in this regard.

If one has to choose just a handful of subjects of great relevance to India's strategic interests in the twenty-first century, the polar areas would surely be one of them. I am very happy that Dr Jagadish Khadilkar has chosen to write on this important subject. With excellent qualifications and a first-hand experience in Antarctica and other countries, he has analysed all of Antarctica's natural and anthropogenic environmental aspects, their impact, and evaluated what India needs to do to establish an effective and influential presence in the continent to safeguard its strategical interests. His advocacy for a national dialogue to create a polar vision and doctrine leading to a long-term plan speaks well in India's national interests. He urges the creation of national awareness on this important subject through a variety of ways, which is appealing. Seeing the conditions around, I also feel that the emphasis on polar awareness, studies, research and adventure projects may provide an effective and creative outlet for the energy of the Indian youth. Books like this can work as a catalyst in such national efforts.

I wish to conclude this foreword on a personal note. Way back in the 1940s, when I was a student at M.R. High School in the small city of Gadhinglaj, Maharashtra, we had a very knowledgeable and dedicated teacher of history, Mr P.L. Khadilkar. During an interaction while writing this foreword, I was delighted to learn that the author of this book is his son. I am sure that our old world-revered late teacher would be very happy that his son has written such a scholarly book on a subject of national importance with deep global significance.

I hope this book receives the attention it truly deserves.

Pune
November, 2016

Padmabhushan Dr S.B. Mujumdar
Founder and President, Symbiosis

In Praise of the Book

'Antarctica, the world's final frontier, represents a unique outdoor laboratory to study planetary processes, including the impacts of climate change. Though preliminary explorations in this icy continent revealed large coal and mineral reserves, Antarctica still contains many geological secrets. Hence Antarctic logistics assume great importance. Skilfully written, this book analyses the continent's environmental aspects, both natural and anthropogenic, their global significance and the Indian relevance. The observations made by Dr Khadilkar are interesting, far-reaching and new to science.

Professor M.S. Srinivasan
Professor Emeritus, Banaras Hindu University

I am convinced that, someday, there will be a conflict situation in Antarctica. When and how it comes about, only time and the changing Antarctic kaleidoscope, described in the book, will tell. With his intimate Antarctic knowledge, experience and thorough research, the author has made a strong case as to what India needs to do to establish an effective and influential presence in the continent to safeguard its strategic interests. There are many books on the subject of Antarctica from Indian authors, but to me, Dr Khadilkar's book is the most comprehensive, analytical and appealing. I think,the dispassionate analysis of observations and recommendations made in the book without parochial consideration is the need of the hour to forge a national polar doctrine. At a personal level, I felt like visiting Dakshin Gangotri once again.

***Major General S.S. Sharma,** KC, VSM (AV) (Retd.)*
Leader of the first Indian wintering team in Antarctica (1983–84)

Antarctica is of great importance for the ecological survival of Earth. It impacts meteorological phenomena and, so far as India is concerned, the efficacy of the monsoon. In addition its impact on marine life and deep sea-based minerals is of great importance for mankind. In the harsh environment surrounded by hostile Southern Ocean, effectiveness or global influence of any country's Antarctic presence can truly be judged on its capability to support scientific programmes with self-reliant logistics network. This book points out that India, in stark contrast to the other prominent countries in Antarctica, woefully lacks a self-reliant logistics network which needs to be created. If or when India decides its polar vision and doctrine on this premise as advocated by Dr Khadilkar, a great challenge will be presented to the armed forces and, in particular, to the Indian Navy and the Indian Air Force, to turn this vision into reality through the execution of long-term plans. India may ignore this to the peril to its polar and strategic interests in the continent. This is a well-researched, insightful and analytical book which deserves a close look by our policymakers.

Air Marshal B.N. Gokhale*, PVSM, AVSM, VM (Retd.)*
Former Vice-Chief of the Air Staff and currently Director,
Centre for Advanced Strategic Studies, Pune

Antarctica will attract major world attention in the twenty first century, thanks to the vast potential for natural resources, global warming and environmental interdependence. Col Jagadish Khadilkar, in his book makes a strong pitch for a national dialogue for polar doctrine and strong and influential Indian Antarctic presence with self-reliant logistics network to safe guard India's strategic interests while continuing to contribute with the scientific research programmes as a prominent member of the Antarctic Treaty System (ATS). He laments very poor polar awareness in India in stark contrast to leading countries in the world. Our educational institutions can play a significant role in this regards and stress the importance of polar areas in to national conscience. This will make up for India's late entry in to matters Antarctic. In the process, Indian youth may also get a challenging avenue to perform on a global arena. The author's detailed analysis and thoughtful projections need serious national consideration.

Lt. Gen. Zameer Uddin Shah (Retd.)*, PVSM, SM, VSM*
Former Deputy Chief of the Army Staff and member Armed Forces Tribunal
Currently Vice Chancellor, Aligarh Muslim University (AMU)

Acknowledgements

This book is the culmination of my fascination for the continent of Antarctica and a long-cherished dream to write a book about it.

Two and half decades ago, I spent two summers and a winter in the first Indian Antarctic station, Dakshin Gangotri, on the ice shelf. During this period, the station, in its sixth winter and also its last, was buried under snow much above the level of its roof. Having become dangerous to inhabit and maintain, the country decided to abandon it and build another station, about eighty-five kilometres away, on a snow-free, hilly area, known as an oasis in Antarctica. I was the soldier-in-charge, station commander and leader of a composite team of military and civilian personnel. In every capacity, I was introduced to a variety of experiences on the frozen continent, and also came in contact with many persons of eminence from other countries. It was far from the dangerous but protected world of an army officer. It is said that anybody who visits Antarctica is smitten by it and I was no exception. I could never go back, but the spirit of 'once an Antarctican, always an Antarctican' remained. The fascination only grew with time, through the changing scenarios of my life, varying from an initial sense of achievement at the unbelievable adventure, to a lacklustre feeling upon return, premature retirement from the army, a long and challenging overseas corporate service, a sense of disappointment at the total lack of awareness about the continent in India, a profound realisation of the catastrophic effects of global warming and the environmental interdependence between the polar areas and the rest of the world and, finally, a concern for the future of Antarctica from a global and Indian perspective. Simply put, the passage of time had changed the individual in me from a proud Antarctic winterer to a strategic thinker. Years passed but the urge to write about my Antarctic experience and perspective remained.

Thanks to the introduction from Air Marshal Bhushan Gokhale, executive director, Centre for Advanced Strategic Studies in Pune (CASS), I started interacting with Mr. Suresh Gopal, Publisher, Special Projects, Bloomsbury Publishing India Pvt. Ltd. He showed keen interest and we started working on the book. His sudden and untimely demise truly saddened me. But his replacement, Mr. Paul Vinay Kumar, took over the project with the same consideration. I thank both of them, Ms. Jyoti Mehrotra, Mr. Shinu and other team members.

I am fortunate to have Aradhana Bisht as my editor. Patient perusal back and forth, an eye for details, clarity and queries for logical progression and wide-ranging general remarks characterised her style. Thank you, Aradhana, for using your critical pen with skill and elegance, while maintaining encouragement. I am pleased to appreciate and thank Ms. Pallavi Kapur, Co-founder, Graficus Design for impressive jacket cover reflecting the soul of the book.

The recent studies started through personal encouragement from the Hon. Dr D.Y. Patil, former governor, Bihar. Prof Dr S.B. Mujumar, founder chancellor of Symbiosis International University (SIA), consented to my request to pen a foreword to this book. Prof Dr M.S. Srinivasan, Lt Gen Zameer Uddin Shah, PVSM, AVSM, SM, VSM (Retd), Air Marshal Bhushan Gokhale, PVSM, AVSM V.M. (Retd.) and Maj Gen S.S. Sharma, KC, VSM (AV) (Retd) went through the details of the book and provided expert comments. I am grateful to all of them.

I am indebted to the late Prof Dr K.B. Powar, renowned professor of geology, former secretary general, Association of Indian Universities, New Delhi (AIU) and Chancellor Dr D.Y. Patil, University, Pune and Dr Rasik Ravindra, former director, National Centre for Antarctic and Ocean Research (NCAOR), Goa, and presently serving as a member of the International Commission on the Limits of Continental Shelf (CLCS) at the United Nations in New York. The inspiring guidance of Prof. Powar and personal and institutional help from Dr Ravindra and NCAOR were of great help during my study.

I would like to mention two people with whom I struck an Antarctic bond – Dr Eugeny Uranov of the Arctic and Antarctic Research Institute (AARI), St Petersburg, the chief of the Russian Antarctic Station, Novolazarevskaya, and Mr. Bruno Zehnder, a famous Swiss photographer settled in New York, best known for his striking images of penguins. While Eugeny retired from AARI some time ago, Bruno died in a blizzard whilst photographing breeding emperor penguins near the Russian Antarctic station, Mirny, a few years ago. He wished and was buried in the nearby oasis area in their company. Both used to urge me to write about the Antarctic and Bruno would offer his photographs. I remember them both with respect.

I visited NCAOR, Goa, many times while conducting research for this book, and would like to thank Dr M. Ravichandran, director (NCAOR) and the central Ministry of Earth Sciences, for granting me blanket permission to use material from any NCAOR publications. I am also individually thankful to the many scientists and staff members at NCAOR, including Dr Alvarinho Luis, Dr A.K. Tiwari, Dr Ravi Mishra, Dr Shubhrata Mishra, Mr. M. Oturkar and Ms. Kaveri Kumbar, who were particularly helpful.

A number of publishing houses and authors gave individual and collective permissions to use their material and I have given them due acknowledgements in the book. I am thankful to the following: Cambridge University Press (UK), Diamond Publications (Pune), New Age International (New Delhi), Dronning Maud Land Air Network (DROMLAN) (Cape Town), Prof. David W.H. Walton, Dr C.S.M. Doake, Dr. J.R. Dudney, Dr I. Everson, Dr Hartwig Gernandt, Maj. Gen. S.S. Sharma, Brig Dr R.C. Pathak and Mr. D. Pashte. I wish to also thank Ms. Clair Taylor of Cambridge University Press (UK) who was very helpful in the matter of obtaining permissions.

Visits to Victoria State, Melbourne City, Melbourne University, and some regional libraries in Australia were very useful. I was able to find in them a vast number of books that were of great value to my research. I am truly grateful to the staff there for their kind help on innumerable occasions and in a variety of ways. My visits to the Antarctic Centre in Christchurch and *Fram* Museum, named after Amundsen's ship, in Oslo, were very pleasant and beneficial.

I greatly appreciate my friends, Shekhar Satpute, Mahesh Khadilkar, Jagdish Kelkar, Kaustubh Mudgal, Harshada Kshirsagar Sarma, Shrikant Athavale and Suraj Lotlikar for sparing the time to provide competent help in finalising the scripts, appendices and figures despite their busy schedules.

I thank the following for their encouragement, suggestions, friendship and other forms of kindness in the making of this book:

Lt Gen V.A. Bhat (Retd.), Brig V.V. Prabhudasai, VSM (Retd), Dr Vishwas Mehendale, Prakash Misal, Dr A. Koppar, S. Sivan, Suhas Kane, Ajay Dhar, C. Sivadasan Avinash Phatak and Ganesh Mhaiskar.

I wish to remember following close friends and relatives who have departed for their kind association:

> Dr S.D. Gokhale, M.R. Joshi, R.S. Gangadhar, Vijay and Ashok Patwardhan, Mukund Khadilkar, Shridhar and Anil Sahasrabuddhe, Rahul Khadilkar, Sharnappa Waghmare, Col Shrikant Yeri, Wg Cdr Vinayak Patankar, Cdr Narendra Albal, Dr M.V. Kale and Mrs Usha Grover.

I would like to mention my late mother here, too, as she was very fond of travelling and I feel all my travels are for her. Both my late father and father-in-law were from the old school of world history and geography lovers and would have been delighted to see this book. My mother-in-law who always appreciated me, passed away a few months ago, leaving a secret gift to claim after the publication of this book. The blessings and moral support of my family, including my sisters, brothers and others were invaluable.

My daughter, Amrita, and son-in-law, Mohit, were of great help in my numerous visits to Australia and New Zealand for sourcing books and references. Still active in his nineties, my daughter's grandfather-in-law, Mr. Roshan Lal Grover's enthusiasm for life is truly inspiring. I have dedicated this book to my granddaughters, Meher and Ananya, who are a constant source of joy and motivation for me.

Losing our younger daughter, Anu, in her prime, a few years ago, was a devastating blow to my wife, Rajashree, and me. With the support of many and the fortitude of my wife and elder daughter, I somehow recovered the will to live, keeping Anu close to my heart. My wife is my pillar of strength. I can never appreciate or thank her enough for her lone struggle to raise our children in my frequent absences as a career soldier, for taking care of my terminally ill mother while I was in Antarctica, for keeping all my records intact and transferring them to digital mode during my long overseas stints, and being there for me all the time. I prefer to remain indebted to her.

Before going to Antarctica, I was a senior member of an army project, headed by a general. He was very reluctant to permit me to volunteer for the Antarctic expedition due to his fear of my possible selection and, thereby, loss of participation in the project. He would say that I would gain nothing by going to Antarctica, except creating a few stories for telling my grandchildren one day. He was proven right on one count – I did gather a lifetime of tales. Also, if my grandchildren are ever inspired to visit Antarctica some day, may this book provide them with a perspective about the continent from a generation gone by.

Pune
November, 2016

Lt Col Dr Jagadish P. Khadilkar
(jpkhadilkar@hotmail.com)

Contents

List of Abbreviations

AAD	Australian Antarctic Division
AARI	Arctic and Antarctic Research Institute, St Petersburg (Russia)
AAT	Australian Antarctic Territory
ALCI	Antarctic Logistics Company International
AMCAFF	Agreed Measures for Conservation of Antarctic Fauna and Flora
ANARE	Australian National Antarctic Research Expedition
ASOC	Association of Antarctic and Southern Ocean Coalition
AT	Antarctic Treaty/Treaty
ATCM	Antarctic Treaty Consultative Meetings
ATS	Antarctic Treaty System
AWS	Automatic Weather Station
BANZARE	British Australian New Zealand Antarctic Research Expedition
BIOMASS	Biological Investigation of Marine Antarctic Systems and Stocks
CCAMLR	Convention for the Conservation of Antarctic Marine Living Resources
CCAS	Convention for the Conservation of Antarctic Seals
CDML	Central Dronning Maud Land
CEP	Committee for Environmental Protection
CFC	Chlorofluorocarbons
COMNAP	Committee of Managers of National Antarctic Programmes

COP	Conference of Parties
CP	Consultative Party
CRAMRA	Convention on the Regulation of Antarctic Mineral Research Activities
CSAGI	Comite Speciale de l'Annee Geophysique
DG	Dakshin Gangotri
DRDO	Defence Research and Development Organisation
DROMLAN	Dronning Maud Land Air Network
GI	Galvanised Iron
GDR	German Democratic Republic then East Germany
GOI	Government of India
Gondwana	Gondwanaland
GPS	Global Positioning System
IAA	Instituto Antarctito Argentino
IAP	Indian Antarctic Programme
ICJ	International Court of Justice
ICSU	International Council of Scientific Union
IGY	International Geophysical year 1957–58
IHO	International Hydrographic Organisation
IIGM	Indian Institute of Geomagnetism
IISD	International Institute of Sustainable Development
IITM	Indian Institute of Tropical Meteorology
IMD	Indian Meteorological Department
IPCC	Intergovernmental Panel on Climate Change
IPY	International Polar Year
ISC	International Seismological Centre
ITRF	International Terrestrial Reference Frame
IUCN	International Union for Conservation of Nature
IWC	International Whaling Commission

Ma	Millions of years ago
Madrid Protocol	Protocol for Environmental Protection to the Antarctic Treaty
NAP	National Antarctic Programmes
NCAOR	National Centre for Antarctic and Ocean Research
NGO	Non-Governmental Organisation
Novo	Novolazarevskaya (Russian Antarctic station)
OPP	Office of Polar Programmes (US)
PGE	Platinum Group Elements
NSF	National Science Foundation of the United States
PSCs	Polar Stratospheric Clouds
SCALOP	Special Committee on Antarctic Logistics and Operations
SCAR	Scientific Committee on Antarctic Research
SPRI	Scott Polar Research Institute
UNFCCC	United Nations Framework Convention on Climate Change
USAP	United States Antarctic Programme
USGS	United States Geological Survey
WMO	World Meteorological Organisation
°E	Degrees East (Longitude)
°N	Degrees North (Latitude)
°S	Degrees South (Latitude)
°W	Degrees West (Longitude)

Disclaimer

In the case of some sources, where I tried to contact copyright holders for permission, my letters and emails were returned undelivered or returned with failure notices. In one case, publishers were found but the author was not traceable. The books for which these efforts were made are *The Green Peace Book of Antarctica: A New View of the Continent*, *Antarctica: Great Stories from the Frozen Continent*, *Antarctica: Secrets of the Southern Continent* and *Antarctica: Exploring the Extreme, 400 Years of Adventure*. My emails, to the Australian Antarctic Division, remained unanswered. At a few places, I have used blank Antarctica maps from Google for placing my information on them. Though most of the information I used is available on multiple public domain sites, I have given due references to authors and sources thus utilised and thank all concerned.

Glossary of Common Terms used in Polar Areas

Ablation: The process by which ice is carried to the sea; or the loss of ice from a glacier or an ice sheet caused by ice melting at the surface or base, evaporation and calving.

Albedo: Derived from the Latin word *albus*, meaning white, it denotes the proportion or percentage of incident light or solar radiation reflected by ground or ice surface.

Antarctic Convergence or Polar Front: Roughly a circular belt of water about 40 m wide lying between latitudes 48°S and 60°S in the Southern Ocean, where the cold Antarctic water meets the ocean's warm waters flowing south, producing a sharp temperature increase of up to 2°C.

Antarctic Divergence: A zone around Antarctica, at approximately 65°S latitude, where opposing winds and water masses collide, separating the east and west wind drifts.

Aurora Australia: Also referred to as the Southern Lights, it is a phenomenon of the upper atmosphere, caused by complex processes between solar wind and the earth's magnetic field, which gives intermittent flashes of light in various colours. The Maoris of New Zealand called it *Tahu-Nui-A-Rangi*, meaning the great burning of the sky. In the Arctic, the phenomena is known as Aurora Borealis or the Northern Lights.

Blizzard: Blowing snow with a thickness of two to 100 m and a wind speed of 20 m per second.

By-catch: Incidental but very cruel killing of thousands of sea birds and fish species in the course of commercial fishing for tuna, swordfish and other species in the Southern Ocean by ships using the long-line technique.

Calving: Breaking off of icebergs from glaciers and other ice masses along the Antarctic coast.

Crevasses: Cracks or splits on the surface of a glacier or an ice shelf. These are formed by stress in ice when it is stretched over ridges and flows through valleys. They can vary between a few centimetres to several metres and pose great danger to movements of vehicles and explorers.

East Wind Drift or Antarctic Coastal Current: Polar easterly winds that drive the east-west flowing surface currents in the ocean caused due to strong airflows off the Antarctic continent. They are shallower than the West Wind Drift and are separated from the latter by Antarctic Divergence.

Fast Ice: A region of multi-year sea ice around the margins of Antarctica within pack ice, with a thickness of several metres, the extent of which is around four million square kilometres.

Glaciers: A flow of ice which moves from the mountain to the sea. Glaciers are extremely powerful and slow moving. They crush great boulders and reshape the land over which they move. A classic example of this movement is at the South Pole. A special marker to denote the exact position of the South Pole has to be moved back by about nine metres every year to offset its movement.

Icebergs: Masses of floating ice, varying in size from very small to gigantic, detached or calved from a glacier or ice sheet and carried to the ocean. Two million tons of ice calves off from the ice sheet and flows into the ocean every year in thousands of pieces. They were first sited in 1774 by Captain James Cook.

Ice Sheet: A massive ice cap with thickness ranging from 800 m to 4.8 km on about ninety-eight per cent of Antarctica. Ninety per cent of ice on earth is made up of the Antarctic ice sheet. The ice sheet has nearly doubled the original area of the continent.

Ice Shelf: This occurs when a vast mass of ice sheet meets the ocean in the form of a floating slab of ice. It rises and falls as per the ocean tide and forms an exoskeleton that shapes, binds and protects the icy continent. Ice shelves finally disintegrate into the ocean in the form of icebergs. They slow the flow of ice from the glaciers to the ocean, called the doorstop effect by some scientists.

Ice Streams: Rapidly flowing rivers of ice from the interior of the Antarctic ice sheet. Though accounting for only ten per cent of the total ice sheet, their size is up to 50 km wide, 200 km thick and hundreds of kilometres long.

Katabatic Winds: Originated from the Greek word *katabatikos*, meaning 'go down' or 'downwards'. Cold, dense air rolls down from the polar plateau after gaining speed due to gravity. After travelling for about 1600 km, the wind speed can reach up to 300 km per hour, creating severe blizzards, particularly in coastal areas.

Oases: Ice-free areas which form less than two per cent area of Antarctica. There are twenty such areas in the continent. Exposed rocks absorb heat which evaporates any snow falling on oases. This, along with katabatic winds, allow oases to remain ice free.

Overwintering: People who spend a winter in Antarctica refer to it fondly, or with pride, as overwintering.

Pack Ice: Plates of thin, frozen sea ice that surround Antarctica in winter months with the maximum extent of 20 million sq km. Though it generally appears to be a solid mass, wind and ocean currents keep it in continuous motion. It is a generic term with many subtle variations.

Pole of Relative Inaccessibility: The centre of the Antarctic continent, this point is the farthest from all the coasts. The Antarctic Pole of Relative Inaccessibility is located at a latitude of 85° S and longitude of 65° E, and at an altitude of 3718 m.

Polynyas Areas: Areas of open water amid pack ice in the Southern Ocean. Most Emperor Penguin colonies are located close to recurring polynyas. Early explorers got an indication of polynyas due to occurrence of grey skies.

Rookery: A generic term for the nesting areas for seals, penguins and birds.

Satstrugi: An Antarctic snow formation, caused by high winds resembling giant ripples or sharp ridges. They may be small-scale, similar to ripples in sand or like rough, hard ridges several meters high. They pose difficulties for vehicular movement.

South Geographic Pole: The southern extremity of the earth's axis of rotation. A special marker is placed to indicate it at the US Antarctic station, Amundsen-Scott, located at the South Pole. Due to the movement of the glacier on which it is located, every year the marker has to be moved back by about nine meters.

South Magnetic Pole: The southern extremity of the axis of earth's magnetic forces. This point, which is generated by electric currents flowing both inside the molten core of the earth and externally, is vertically upwards, and a compass points straight down.

Stratigraph: A method of determining earth's age by counting the successive layers of ice formed by annual snowfall in Antarctica.

West Wind Drift or Circumpolar Current: One of the six major ocean circulation systems in the world, the West Wind Drift is a surface ocean current that flows 24,000 km around the Antarctic continent from west to east. Inside the West Wind Drift, the East Wind Drift is a shallower, slower-moving current that flows off the Antarctic land mass. The two are separated by Antarctic Divergence.

World Park: An idea to establish Antarctica as a World Park was mooted by New Zealand in the 1970s but did not go ahead due to lack of support. The Antarctica and Southern Ocean Coalition (ASOC) established a World Park base in Antarctica in support of the idea between 1987 and 1993. Signing of the Madrid Protocol in 1991 and the adoption of a minimum fifty years' prohibition on all mineral resources exploitation was essentially due to efforts in this regard.

List of Figures

List of Tables

List of Appendices

List of Illustrations

CHAPTER 1

The Beginning

"The land looks like a fairytale and we must always remember with gratitude and admiration the first sailors who steered their vessels through storms and mists, and increased our knowledge of the lands of ice in the south."

—***Roald Amundsen (1872–1928)***
(The first person to reach the South Pole in 1911)

1.1 Sand Leads to Snow

Well before dawn of a warm summer day in 1985, loud cheers broke out as the Corps Commander, after a rousing speech, flagged off an army expedition dubbed '*Desert Safari*', at a grand function in Bikaner, Rajasthan. Driving the first of the eight vehicles, as the leader of a team of thirty officers and men, I felt both elated and anxious. As we drove out of the town through arches and banners wishing us success, I missed my wife and young daughters who were not present for this early morning event. However, as we veered off the road to go cross country, I saw my dear wife standing by the side with her bicycle and, though the team members looking on amusedly, it felt great to stop for a moment and exchange a few words of endearment before heading out for the tough days ahead of us. The next four weeks would turn our skins dark, our bones weary and test our endurance in the interiors of the Thar Desert. Later, with the extensive video material shot there, I would edit a short film on the expedition, which, hailed as a success by the army, would pave my way for a truly global and unique experience in a few years' time – a dream that all explorers cherish – of going to the continent of Antarctica.

1.2 Early Days of India's Antarctic Quest

Due to the foresight of the then Prime Minister, Mrs. Indira Gandhi, the country embarked on Antarctic expeditions in 1981. In the early days of India's quest for a physical presence in the continent, two summer-only expeditions lasting a few days had been undertaken. The first permanent Antarctic station was built during the third summer expedition between December 1983 and February 1984. This also commenced a round-the-year national presence, with special winter teams. The country named this station Dakshin Gangotri (DG), taking the flavour of our cultural icon to the bottom of the earth. But the station was on an ice shelf and, after serving as home for five winter teams, it was nearly submerged under rising ice levels by end-1988. Even during summer, only about two per cent of the total expanse of Antarctica, which is roughly four and half times India's size, is free of ice. This ice-free land is scattered all over and known as oases. It was decided to build a new station, to be named Maitri, on one such oasis, existing about 85 km from DG and in the vicinity of two other stations, Novolazarevskaya (Novo) and Georg Forster of the Soviet Union (now Russia) and then East Germany respectively. Once built, it would be safe from ice burial and long lasting like its neighbours. However, if Maitri was to be properly built and commissioned, it was necessary to extend the already precarious life of DG so that the latter could provide support to the new station for a year before it was decommissioned and abandoned. The risk was worth taking in pursuance of national interests.

Destiny called me and my team to serve the country in Antarctica in the dangerously ice-submerged DG before it merged with the enormous ice cap.

1.3 Calling from the Seventh Continent

Except for school lessons about penguins, seals and whales in the Southern Ocean and the absorbing story of the race between Norway's Roald Amundsen and Britain's Robert Scott, to become the first human to reach the South Pole, I had very little knowledge about Antarctica. But when the army called out volunteers for the Antarctic expedition, I was drawn to the challenge and responded. My obvious drawback was that I had never served in areas with snow and this was quickly seized upon by my interviewers. A very senior officer on the other side of the table asked what made me aspire for the expedition. Thinking for a moment, I said that I had faced temperatures of 50°C, severe sand storms and other adverse conditions in the desert, so, I added, living on ice at minus 50°C with blizzards seemed striking and appealed as a balancing experience. But for the usual army bravado, I knew it would really be tough. After my selection as the last wintering leader and station commander of DG, I read whatever books and material I could lay my hands on about the continent and its early explorers and I also met a few people who had been there earlier. A sense of anticipation and expectancy hung in the air. By the time we boarded the expedition ship, *MS Thuleland,* in Goa for a month-long voyage in November 1988, I was physically and mentally prepared and able to confidently communicate with my team members, preparing them for the task ahead. The ship captain, the very experienced Hedlund Lennart, became a friend and he and his crew helped in the learning curve.

1.4 The Voyage

The voyage was educative but tough. Crossing the Southern Ocean with its roaring forties, furious fifties and screaming sixties (as the latitudes there have been named by old sailors and navigators in accordance with the sea behaviour) was exciting, physically tough and mentally taxing. Many aboard the ship were unable to look at food, let alone eat it, for days. The ship had to cut through thick pack or sea ice to be able to move forward. Thousands of north-moving icebergs of all shapes and sizes with penguins or seals resting on them provided majestic glimpses of ocean life. Globetrotting birds in transit flying to Antarctica for their summer stay regularly visited the ship. They would return after disappearing for long periods. Stories about them abounded since early days. One such tale talked of them as guiding and helping fellow travellers to navigate the white wilderness. Probably, they had not realised that now, with satellite navigation, man no longer needed their help.

One day, three weeks into the voyage in severe sea, winds and weather conditions, the ship passed by a remote island. I was chatting with the ship captain at the controls when his radio came alive with a call from the garrison commander of Prince Edward Island of South Africa. When told of my presence, he asked to speak to me. I politely conveyed to him through Lennart that since India is opposed to his country's inhuman policy of apartheid, I did not wish to talk. He gracefully accepted the refusal and asked Lennart to convey their good wishes to the Indian expedition team. Many years later, while on a corporate assignment in Nairobi, I met the South African High Commissioner in Kenya, and narrated the incident to him. He embraced me with affection and became a friend.

Though the voyage was memorable, the spectacle on reaching the continent surpassed everything else, virtually rendering everyone speechless. We had reached the floating ice shelf – a gigantic vertical wall of ice about 80–90 m high, a table top, stretching from one side of the horizon to the other, standing sentinel, protecting the continent behind, reminding the ship that it could go no further.

1.5 Our Unusual Stay in Antarctica and the Decommissioning of Dakshin Gangotri (DG)

On landing, what followed was an eventful fifteen months. I will not narrate the daily routine or common occurrences, but only a few that may merit the reader's attention and help to appreciate the uniqueness of life on the continent. They will also highlight the fast-changing logistics scenario in the continent as never before. I will reflect on few other personal experiences in later chapters with specific contexts.

For the first two weeks before we took over DG and before the earlier wintering team moved to the ship, we stayed on snow outside the station in improvised, ice-buried containers. While going to the main station one day in full blizzard, one of my team members, a scientist, asked me if I could send him back to India as he was not that adventurous. His request was barely audible due to strong, hissing winds but I could hardly suppress a chuckle at his plight.

One day, returning from work, I saw a lone penguin sitting atop the steps of the survival container where I stayed alone. As it ran away at my approach, I noticed that it was injured. The next morning, coming out to go for work, the same scene was repeated. This happened daily for a few days and I could sense the injury aggravating. Sometimes, on those twenty-four-hours-sunlit days, I would, without making a sound inside my container, see the poor penguin quietly sitting outside, obviously unable to travel a long distance. One morning I found it had vanished and was never to return again. I felt sad and missed my companion for quite a while.

Another small incident in the early days would teach me an important lesson about surviving blizzards. As it so happened, several crates of juice were stored in my survival container. One night, as a raging blizzard blew swathes of snow into the container through a hairline crack in the window, I stacked all the juice tins against the window. Feeling secure, I slept peacefully. The next morning, one of my teammates, Hav Gawade (later retired as Sub Maj), came with a cup of tea and was startled to find me sound a sleep in my sleeping bag, surrounded by snow much above the level of my camp cot. I would keep this incident in mind when going out for convoys and never forget to take more than adequate sealant with us.

A few weeks after the summer team completed its work, the expedition ship *MS Thuleland* left the ice shelf for India, leaving sixteen of us at DG for a year of solitary wintering.

We quickly got used to the snow-buried station and the need for near-constant snow clearance. There was a 30 ft high shaft with an inside ladder and long, slanting ladders joined together to enter and exit from the station. The slanting ladders swayed with the wind and the whole arrangement was daunting. Optical illusions and mirages played amazing mind games on us. Icebergs permanently grounded in the ocean miles away looked like nearby housing colonies. As summer faded and winter approached, conditions on ground quickly changed. Soft ice on the surface hardened by the day. We no longer walked in knee-deep soft snow nor were we able to easily shovel it into the snow-melt plant, but had to hack hard ice with axes with great effort for water. A short time later, we would not require mirrors to look at ourselves as our beards grew, and falling on glassy surface while walking became common. One had to stand over the heaters for several minutes to free one's beard from sticky hard snow after every outside venture. The regular release of balloons for atmospheric and ozone research (ozone sonde), hourly recording of weather data and other experiments occupied the scientists and other researchers there. It was no comfort to know that we were staying in the thick of the ozone hole which the world had discovered only a few years earlier.

That year, for the first time, soft drinks were brought in plastic bottles. Like all items, except those that met a few days' requirements, they had to be stored in containers placed on barrels outside the station. While fetching replenishments one day, we found that all the bottles had been distorted as the gas had escaped, rendering them useless. In a short while, amusement gave way to distress.

Ice accumulation in containers creates considerable air pressure. One day while struggling to open the door of a container after a blizzard, it suddenly burst open due to inside pressure and violently struck the sub-team leader Captain Rao on the head. He suffered from severe concussion. Fortunately, with constant attention from the team doctor, Lt Col Chaturvedi, he fully recovered in a few days. An important lesson as to what trapped ice can do in Antarctica if proper precautions are not taken, even in very simple activities, was learned.

As winter began, the Russian icebreaker, *Mikhail Somov*, berthed at the shelf (known as the Russian Bay) for a day to pick up the last winter teams from Novo and Georg Forster. Colonel Chillingarov had rescued *Mikhail Somov,* stuck in thick winter pack ice just a couple of years earlier, from a hopeless situation and the saviour and the ship had both become polar celebrities. Being a clear day, I was able to go to the shelf for a goodwill visit with a few bottles of liquor. When asked if I wish to send anything home, I wrote a letter to my wife and gave the original and carbon copies to the returning Russian and East German team leaders. The letters, after travelling all over the globe were received in Pune by my wife through the Indian embassies in Moscow and Leipzig after five and seven months respectively.

On India's Independence Day, the Russians and Germans came to the shelf in a convoy. In the thick of winter with temperatures at about minus 40°C but with a tolerable snowdrift that day, we played a football match with the visitors. On such a hard and glassy snow ground there is no friction and unless the ball is stopped, it just vanishes. In ten minutes, we had lost three footballs and one volleyball and the match ended in a goal-less draw before we moved in for the party. For quite a while, we wondered how far the balls had gone.

Months earlier, I had won a chess tournament held on the ship in which I beat the summer expedition leader in the finals. It was amateur standard but the Russians seemed impressed and asked if I could play a correspondence game with the chess champion at their Antarctic and Arctic Research Centre (AARI) in Leningrad (in a couple of years it would revert to its old name of St Petersburg after the fall of communism in Russia). The moves between us would travel in the form of Morse code telegraphic messages from DG to Novo to Molodoznia (Russia's coordinating station in Antarctica) to St Petersburg and back in the same sequence. Mighty pleased with this unusual game between continents, my naval sub-team leader and DG communications officer Lt Cdr Soman and his assistant POEL (R) Kottol used to make grand announcements after the arrival of every move. In four months, we exchanged about fifteen moves. I had a better position but the Russian player was a pawn up. But then it was time to return and an honourable draw was agreed. I still treasure a fine chess set presented as a memento by the Russians. Many years later, I would fondly remember the game during my visit to St Petersburg.

Running the convoys was a very important activity. It also afforded opportunities to examine the routes, see Aurora Australis in different perspective and gain valuable experience about the area conditions. Once, before running a large convoy on a different and longer route, I travelled with a Russian convoy with two of my colleagues for a route acquaintance and assessment of ice conditions, crevasse proneness and our snow vehicles' suitability for the summer route. At the first meal, I found that the rice, cooked in a huge utensil, had in it every type of meat. When I expressed my inability to eat it on account of being a vegetarian, the cook, who also happened to be their surgeon, took out all the meat pieces and offered me plain rice as a vegetarian dish. On learning that I could not eat even that, they gave me a large bottle of Hungarian cucumbers as food for the next three days. It was fun the first day but I starved for the next two.

Another heart-thumping incident occurred while running a convoy on a glacier. I was driving the vehicle in front, while my team member, Captain Rao, was following in another vehicle. As he told me later, he was worried that my vehicle, instead of going forward had been going sideways on the glacier slope for quite a while. Suddenly, my vehicle stopped in a patch of soft snow at the stiff edge of the glacier. Had there not been that soft patch, my vehicle would have toppled from the edge of the glacier and fallen down below.

It is famously said with reference to blizzards and crevasses that one never knows when conditions in Antarctica may turn from comfort to survival. This sentiment can be applied to an incident concerning my spectacles. I had packed four sets for my Antarctic sojourn. For nearly nine months all were intact, until one day, one pair broke while I was working in the morning, a second pair fell soon after and was crushed under my feet and the third was trampled upon by one of the team members the same evening. Within a day, I was left with my last pair of spectacles for the remaining seven months of the expedition.

Every day before dinner, sub-team leaders and scientists briefed us on their next day's planned activities followed by my briefing and task allocation. Even with all the activities, the twenty-four hours of dark nights stretching for months made time crawl and one needed to come up with creative ways to keep the team busy and away from negative thoughts. Sports competitions and social functions with special prizes and certificates were of great help. As a novelty, I also used to put a question on the notice board every week. The first to give a correct reply used to be awarded the station commander's dollar of the week.

Days passed and the first penguin came to visit DG at the end of October, signalling the end of winter. This was cause for great celebration. December arrived with a new team from India and hectic summer work. Only this time there would be no new winter team for DG as the station was to be handed over to the snow god, as we used to say.

One of my army team mates, Subedar Garlyal (later retired as Honourary Captain), was approved for promotion to the coveted rank of Subedar Major. After the flag hoisting ceremony on 26 January 1990, I had the privilege to pip him, as his station commander, in front of the Indians, Russians, Germans and Swedish nationals. For the first time in months, both of us shaved, wore our uniforms and braved both low temperature and wind in front of the fluttering tri-colour. Only the two of us faced the elemental fury and by the time the ceremony was over, we were half frozen.

After transferring all the possible stores and equipment to the new station Maitri, we closed DG with prayers on 25 February 1990. It was very emotional to say goodbye to the abode which had protected the sixteen of us for sixteen months. The return voyage was uneventful and we returned to India by end-March.

With the decommissioning of DG, an important chapter of the Indian Antarctic Programme (IAP) had successfully concluded and a new one had started at Maitri.

1.6 Passage of Time and the Antarctic Label

It is said that for those returning after spending a winter in Antarctica, things are never the same. It was no different for me. Many people suffer from a condition known as the 'winter effect' which negatively impacts the psyche then and for years after return from the expedition. Fortunately, I did not. For me, the environment had changed in other ways. The realisation of gains and losses followed the early days of submission and presentation of reports, some praise, felicitations and the odd award. While the main gain was an enduring feeling of achievement, the losses were many. We had not received much news of the outside world for a period of sixteen months, including all of 1989, and, subsequently, were unaware of the fall of communism, the taking down of the Berlin Wall, pieces of which were sold as mementos, pre-democracy protests and massacre at Tiananmen Square in Beijing, Aung San Suu Kyi's victory and arrest in neighbouring Myanmar and many other major historical events.

In India, we were to learn that the economy was preparing to go into ICU mode, that a prodigy in cricket named Sachin Tendulkar and a new sensational *'ek, do, teen'* actress named Madhuri Dixit had arrived on the scene, a few prominent leaders had died, and so on. Till today, I often learn something new about that period that I had missed. On the personal front, my wife had kept away the news of my mother's terminal illness from me, braving it alone at military hospitals, so as to not affect my performance. Daily office routine seemed very different and I felt as if I had come from another planet. With all these factors and other obligations, life had become very tough. Soon after, I took premature retirement from the army to seek outside pastures.

The next decade and a half took me to Saudi Arabia, USA and Kenya on various corporate assignments and to many other countries for training, visits and travel. Everywhere, in interviews or meetings or other interactions, people would ask me about my Antarctic trip. There was hardly any place I went where I did not speak about the continent and my experiences there. The response from the young and old was equally enthusiastic. Many eminent people from all over took out time to learn about my Antarctic association with great interest. So, when one such interaction back home with a state governor and a university chancellor created the opportunity for doctoral studies, I decided to pursue it.

Time would pass but the Antarctic label I now had would not fade.

1.7 Changing Times and Aim of the Book

In the course of my study and research, I realised that times had changed enormously and much water had passed under the bridge since my days as a college student. In Antarctic parlance, we would say that an umpteenth number of icebergs had been calved off the shelf and gone into the ocean. With the advent of Internet and other technological advances, the extent of changes in Antarctica, in the last few years, have surpassed all traditional perceptions. Some of my own experiences of just a few decades ago seem prehistoric and dated today.

As my research on the subject got more detailed and deep, it cemented my views about the strategic importance of Antarctica to the world in general and to India in particular in the years to come. This book is the culmination of this thinking. Scientists have estimated the age of our earth at about 4.53 billion years, a time span in which the changes in the natural environmental aspects will be minimal but profound and will have an impact that is global and equal to all. The area where the extent and rate of change experienced has been most dramatic is within the anthropogenic aspects of the Antarctic environment. This will have a bearing on other countries in proportion to their influential and effective presence in Antarctica. In such a scenario, I am deeply concerned about India's role in the shape of things to come, independently and vis-a-vis other countries.

The aim of this book is to present a view of both the natural and anthropogenic aspects of the Antarctic environment and their kaleidoscopic scenarios with reference to the IAP. Irrespective of technological advances, even today, about ninety per cent of all efforts, time and resources are directed towards logistics and the remaining ten per cent towards other aspects. Consequently, the major emphasis, while analysing India's presence in Antarctica, would be on the competency of its logistics systems based on self-reliance. Antarctic logistics may need to be viewed from this angle rather than any other. And that is the strength of the argument in this book.

1.8 Organisation of the Book

A basic idea of how material in this book has been organised will help readers understand the logic behind its presentation and arguments. To keep as aim the highlighting of India's relevance in this region, the work done by Indian expeditions in the last thirty years or so is described in every chapter, along with the general global picture.

To start with, the book will take readers through the development of Antarctic literature. It is interesting to see how this literature has evolved over the years, keeping pace with geographical discoveries, scientific revelations and technological advancements in logistics. As in all other fields, the dissemination of information about Antarctica, too, has undergone a dramatic change since the advent of the Internet. In discussing the unfolding of Antarctic literature, this chapter charts the chronological flow of events, right from early discovery to subsequent reflections that various aspects have generated. India's perspective towards Antarctic literature and the necessity of its presence in the continent are discussed in this chapter as well.

Studies on the natural environmental characteristics go through regular paces, tracing changes over millions of years. The next chapter deals with the evolutionary process of the geographical and geological perspectives of the continent. It follows the sequence of hypothesis, scientific theories based on evidence gathered over the years, and firm conclusions and maps to show continental divisions based on physical features and natural resources. The discovery of an unbelievably enormous ice cap over the entire land mass, the surrounding floating ice shelf, the calving of icebergs, the close balance between ice gain and loss, the huge variations in summer and winter sea ice in yearly cycles over thousands of years, all make for fascinating study among the global scientific community. Research also points out why, while discussing the continent, the Southern Ocean has to be taken as an integral part.

The chapter on Antarctic climate acquaints the readers with how it has evolved over millions of years, storing global climatic history in the permanent ice cap through yearly differentiable rings which can be seen in ice cores, and how the structure of the atmosphere and the effect of various winds make Antarctica the inhospitable continent we know, where no conventional human or animal habitation is possible. This chapter also touches on important aspects concerning us in recent times, like the greenhouse effect and the ozone hole.

The next chapter on Antarctic life winds up the discussion on natural environmental aspects. It discusses the existence of the interdependent Giant Food Web, its individual components and their present status. It also chronicles the spectre of exploitation against seals, whales and certain fish witnessed in the past century, necessitating many of the present Antarctic Treaty (AT/Treaty) and outside instruments, and the disturbing yet likely scenario if climatic changes are not kept in check. The successful implementation of the Paris Agreement at the climate summit in December 2015, therefore assumes enormous importance in the next few decades of earth's environmental journey which is at crossroads today.

The anthropogenic environmental aspects cover human involvement in Antarctica. The chapter on international presence, the Antarctic Treaty System (ATS) and world governance traces the discovery of the continent, national interests leading to territorial claims by seven countries, their refutation by others and concerns in the likelihood of a conflict situation. It further describes the observance of the International Geophysical Year (IGY), leading to the Antarctic Treaty and its instruments, including the Madrid Protocol, for environmental protection. The chapter also talks about some countries with a prominent presence in Antarctica, their proximity and rivalry with one another, their long-existing research stations and activities, highlighting their perspectives and policies with clear reference to their position on the continent. Furthermore, it briefly discusses the present situation in the Arctic Ocean, mode of governance in the form of the Arctic Council and the effects of probable changes in Arctic on Antarctica. We will certainly hear much more on this subject in the coming years.

The book enters its final and important phase with chapters on Antarctic logistics and the complete gamut of the IAP. In fact, this particular section and ensuing arguments form the backbone of this book. Though preliminary exploration has revealed large coal and mineral reserves, Antarctica still contains many geological secrets. Hence, Antarctic logistics have assumed great importance. Once that is understood, the author expounds his views and findings from the Indian perspective.

The chapter on Antarctic logistics describes developments in major logistics instruments like ships, aircrafts, vehicles, dogs, shelters, station structures, harbours and airports over the years. It traces the modifications over the years in food, medicines, communications and clothing. How humans have transformed everything that was primitive and dangerous about a mere two hundred years ago into safe and modern methods of today makes for a compelling read. These transformations occurred as technological advances elsewhere got transferred to Antarctic expeditions which went from being individual ventures to extensions of national policies. The chapter discusses a couple of major organisations engaged in the Antarctic logistics and programmes of countries like USA, UK, Australia, Argentina and China in detail, and of other countries in passing, to understand the perspectives and underlying intentions when the time of reckoning comes.

The chapter on the Indian Antarctic Programme (IAP) covers the background, organisation, expeditions and progress made by India over the years. In a threadbare manner, it discusses all three Indian Antarctic stations with respect to their location, structure, safety, environmental protection, efficacy of life-support systems and longevity. It discusses the entire logistics gamut of how expeditions are taken to Antarctica, how they are maintained to endure and fulfil the vital long-term need for self-reliance and how the mission mandate proclaimed by IAP's apex body, the National Centre for Antarctic and Ocean Research (NCAOR), Goa, compares with the conditions on ground.

1.9 Looking Ahead

After much study, interactions with people from various countries and personal reflections, I am of the considered view that the polar areas will engage world attention in the future in a way like never before. Perhaps, a few decades from today, the possibility of a conflict situation in Antarctica may not come as a surprise. How and when that may manifest itself, only time will tell. Some indications, though, may come from future developments in the Arctic and from the review of the binding Antarctic Treaty in 2041 and the Madrid Protocol in 2048. Countries like USA, Russia, UK, Australia and Argentina have developed logistics systems and are ready for any situation at any time. Some countries are very developed but with natural limitations like size and population. China is a late entrant like India but, unlike India, has taken tremendous strides to catch up with the leading countries as regards its readiness for the future. These things, probably never announced, are only deduced by actions on ground.

I was in my sprightly forties when I visited Antarctica as a soldier but soon thereafter started a career outside of the army. After serving and extensively travelling in many parts of the world, I am now past seventy. So a long reflection precedes the book which aims and suggests as to how India should progress in its Antarctic journey in preparation for future global situations. My suggestions also have a by-product. If India can develop a self-reliant logistics system to take care of the needs of its Antarctic programme, this skill may be effective anywhere in the world and used as and when required to great benefit. This is, of course, easier said than done and requires sustained investment and effort over a long period of time. But then the protection of a country's strategic interests does not come easy or cheap. As far as Antarctica is concerned, we have time till 2041 and it is sufficient for any plan to show realistic results. What we need, I argue, is for India to set its own strategic blueprint for long-term unaffected execution. Examples of Roosevelt's thinking and Reagan's presidential proclamation after a gap of forty years, as given in the book later, are worth national attention as examples in consistency. Governments come and governments go but national interests remain.

I do not claim the book to be all-encompassing. Any intellectual failings or mistakes in perception are all mine. I am also aware that a work of this nature raises more questions than it answers. In fact, it should. If this happens, my efforts will be well rewarded.

CHAPTER 2

Antarctic Literature Comes of Age Keeping Pace with Physical Discoveries, Scientific Developments and Technological Advances in Logistics

"If Antarctica were music, it would be Mozart. Art and it would be Michelangelo; **Literature and it would be Shakespeare**; the only place on earth that should is still as it should be. May we never tame it."

—***Andrew Denton***
(Australian Television Producer)

2.1 Introduction

The history of Antarctica is rich in adventure, yet human activity in the region extends only to about two hundred years ago. The surrounding Southern Ocean, an inaccessible location and a totally inhospitable environment presented difficulties in physical and scientific exploration, reporting and exchange of information. Also, the objective of most early expeditions was hunting and exploitation of marine resources like seals and whales. This effectively precluded any other exploration work.

As technological advances eased hardships in logistics, individual quests gave way to government-sanctioned expeditions with national interests and aspirations. This led to organised efforts for a physical presence on the continent and, thereby, the establishment of permanent stations with a view to create evidence for territorial claims. The realisation that the harsh environmental conditions would preclude territorial domination and colonisation in the conventional sense, led to efforts to create a world order through entities like the International Geophysical Year (IGY) 1957–58, permanent stations for scientific research and, finally, the creation of the Antarctic Treaty (AT/Treaty) and its additional instruments.

It is seen that explorations, national interests, scientific pursuits and the development of logistics all grew simultaneously as interdependent aspects of the evolving process. These activities also played a role in transforming earlier forms of literature such as basic record-keeping, diaries, memoirs, personal journals and adventure stories into formal reports, books, autobiographies, international seminars and other forms of individual, official and institutional literature. They also inspired poets and writers of fiction to produce memorable novels and verse based on polar adventures. Though this is not related and, therefore, not attended to in the present book, they also played a significant part in raising public awareness and interest through cinema, documentaries, media coverage, tourism, travel books and other channels.

Chronicling all this material gives an insight into the development of non-fiction Antarctic literature, which has roughly followed the timeline of physical exploration till the advent of Internet. Though most of what is known about the continent has been discovered only over a century or so, its foundation was laid by the early explorers. A representative timeline of the Antarctic exploration is shown in *Appendix I.*

2.2 A Review of Literature

The earliest publications about Antarctica were mainly personal diaries and journals, ship logs and accounts of scientific observations like weather data collected during voyages made mainly for sealing or whaling, as also for exploring new areas for national domination. They include such works as James Cook's *Journals* (1772–75), James Weddell's famous book *Voyage Towards the South Pole* (1825), five volumes of the narrative of the US expedition with twenty volumes of scientific reports (1845), Joseph Hooker's *The Botany of the Antarctic Voyage* (1847), and some others. Most of the early works have now been republished by modern-day publishers, but the original documents still inspire and hold a fascination for scholars who study them and provide new interpretations. Even new books that throw light on early heroes' personal and public lives often hit the stands frequently as a result of modern-day investigative journalism. Amongst these, *An Empire on Ice: Scott, Shackleton and the Heroic Age of Antarctic Science* by a Pulitzer Prize winner, stands out with its compelling objective analysis of all the facts over a century of exploration with his final observation as to how science gave meaning to the death of Scott and his men in a manner that a failed dash to the pole could not (Larson, 2011).

Earlier, the publication of reports took a long time as information was guarded and its exchange limited. Secrecy and non-sharing of information were seen as key to securing and safeguarding national interests. For example, the Russian explorer, Admiral Thaddeus von Bellingshausen, returned to Russia with a huge quantity of specimens and multi-location scientific data on ice, weather, temperature, pressure, volcanic activity, oceanic currents and other aspects. But his report was published in 1831, ten years after he had become, unknowingly, the first person to sight the continent in one of the greatest voyages of all times. The first book published in Antarctica was *Aurora Australis* in 1908. Members of the British Antarctic Expedition wrote and produced it during a wintering at Cape Royds (Trewby, 2002).

However, since the earliest papers, dating from the seventeenth century, the contemporary nature of Antarctic literature can be seen from the fact that, of the world's recorded 61,000 non-fiction papers and books published about Antarctica till then, about ninety-one per cent were after 1951 (*NSF, Antarctica, Past and Present*, 1997). Organisations like the *Reader's Digest* (*Great Stories from the Frozen Continent)* and *National Geographic* (Issues of April 1980 and April 1987) gave attention to bringing out special publications or issues to serve as excellent references.

Though India showed an interest in matters related to the Antarctic in the 1950s at the United Nations (UN), its active involvement started only in 1981 after it launched its first expedition. Indian literature on the continent is still quite limited to expedition reports, reports from participating organisations, sporadic articles in journals and newspapers by expedition members and a few books and memoirs by, for example, Sharma (1997 and 2001), Sehra (2003) and Raina (2006). India had also published over 300 research papers based on Antarctic studies by 2007 (Wikipedia, 2012, Indian Antarctic Programme). More clinical reviews and articles based on the Indian experience on various aspects of Antarctic research like geosciences, glaciology and environment, geomagnetism, atmospheric science, oceanography, ocean forecasting and air-sea interaction, paleoclimatology, etc, are available from seminars (for example, S Rajan, 2004) and will be useful for study and reference. A number of general books have also been published in recent years by the National Centre for Antarctic and Ocean Research (NCAOR) and other institutions (Tiwari, 2008, Chaturvedi, 2009), which may just introduce people to the continent and demonstrate its importance from the Indian point of view. Concerning Antarctic logistics, apart from some scanty reports as part of expeditions, very little literature on the subject is published. A book in Hindi entitled *Bharatiya Antarctic Sambhar Yantra* (trs. *Indian Antarctic Logistics*) (Mishra, 2012), gives statistical and factual details on some aspects of the logistics of Indian expeditions from 1981 to 2010. All put together, one can say that the Indian body of Antarctic literature is still in its elementary stage commensurate with the overall poor state of awareness about the continent in this country.

As in all spheres, the availability of information about Antarctica, too, on search engines like Google, websites of national Antarctic programmes and other organisations and other sites in the public domain, has proliferated in recent times. However, the quantity and quality of the information available varies hugely from one website to another, depending on the organisation and country. For example, the information available about USA's Antarctic Programme (USAP) is different from that of the Russian Antarctic programme, though the intensity of their involvement in matters Antarctic is the same. This can be seen from ensuing discussions in later chapters. Technology has also helped take the polar areas into drawing rooms all over the world in the form of unique documentaries and clips by media and travel organisations. One such brilliant example is the BBC's 2011 multi-episode documentary, *The Frozen Planet* with an oral narration by Sir David Attenborough, a personality hailed as a national treasure in the UK. It gives viewers a close and in-depth experience of the polar areas and their inhabitants, the penguins, seals, whales and polar bears.

While reviewing a broad spectrum of Antarctic literature, my preconceptions about early explorers were also cleared. I had always looked upon them as some sort of supermen. A close study of their personal lives and expeditions revealed, however, that while some could not boast of physical fitness, others did not know how to pitch a tent or walk on ice, and some had hidden heart problems, etc. When one realises that they were frail, like most of us, the admiration for their pioneering and inspirational deeds in most trying circumstances with primitive technology, goes up many notches. We can never admire or thank them enough.

2.3 Antarctic Exploration and Understanding

It can be seen that the exploration time line stretches through many phases since ancient times, starting from the concept of Terra Australis (the great southern land), to present governance through the Antarctic Treaty System (ATS). This timeline can be divided into seven phases based on major chronological development landmarks as described below.

2.3.1 *Historical Perceptions (upto 1500 AD)*

Pythagoras and Aristotle (in the sixth and fourth century BC respectively) argued that the earth is sphere-shaped, hypothetically filling this new world with land and sea, which led to the concept of Terra Australis. In 150 AD, Claudius Ptolemy published his eight-volume guide on geography and named the imaginary continent Terra Australis Incognito or the Unknown Southern Land, as opposed to the known northern land. European maps continued to show this hypothetical land for centuries. However, in medieval Europe, flat-earth orthodoxy also held sway until the fifteenth century till expeditions to the southern hemisphere commenced. Once, the Portuguese explorer Ferdinand Magellan (1480–1521) said, 'The church says earth is flat, but I know it is round, for I have seen its shadow on the moon, and I have more faith in a shadow than in the church.'

2.3.2 *End of the Concept of Terra Australis (1501–1819)*

As expeditions of European sailor-explorers started, knowledge about the southern waters slowly increased. Bartholomew Diaz (1488) and Vasco de Gama (1497) went around Cape of Good Hope. Vasco de Gama took the expedition further to India by sea route and, in the process, also marked the extent of Africa's landmass. Magellan (1520) entered the Pacific Ocean from the east, proving that South America was not connected to a southern continent. Francis Drake (1578) went round Cape Horn, showing that no visible land existed south of it. Despite this, the myth of the southern land persisted. French explorers Jean-Baptiste Bouvet de Lozier (1739) and Yves Joseph de Kergulan (1772) discovered the sub-Antarctic islands, jumping to the conclusion that they had reached the edge of the southern continent.

Captain James Cook finally cleared the various myths. During his voyage of 1772–76, he crisscrossed the Indian, Pacific and Southern Oceans without discovering any new continent. Heavy pack ice forced him to withdraw after reaching a point just seventy kilometres from the the Antarctic ice shelf. He concluded that if any land mass existed beyond, and he believed it did, it lay well south of the pack ice and was of no use to the rest of the world. He published his two-volume narrative, *Resolution,* in 1779 and demolished the myth of a great temperate southern land (*Reader's Digest*, 1985). By 1818, many outposts or sub-Antarctic islands had been discovered, but what lay behind the pack ice stretching away till the ice shelf we know now as Antarctica, was still unknown to the human mind.

2.3.3 *Physical Discovery and Shaping of the Coastline (1820–1894)*

On 27 and 30 January 1820, in two remarkable events involving two explorers, the Russian Bellingshausen and British Edward Bransfield, the first two sightings of the continent were made 2000 km apart, unaware of each other and of the significance of their achievements. While Bellingshausen did not even record it, Bransfield, in his diary, called the spectacle very gloomy and depressing. Instead, the discovery of the island, South Shetlands, a year earlier in 1819, with its rich fur seal colonies seemed more important and alluring from a commercial angle. But within a few years these sealing grounds were exhausted and new search began. John Biscoe (1831), Peter Kemp (1833) and John Balleny (1839) all sighted new coastline locations on the other side of the continent (Indian Ocean sector). By this time, the interest of many countries was aroused and nationally organised expeditions took over from private sealing and whaling ventures. In 1840, French Dumont d'Urville and American Charles Wilkes discovered separate sections of the coastline south of Australia. Dumont d'Urville named the beautiful and petite species of penguins found there after his wife Adelie and the name has stuck since then. In 1841, James Clarke Ross discovered Victoria Land and named the sea after him. For the next fifty years or so, till nearly the end of the century, apart from sealing and whaling, expeditions made visible the outline of a continent, but there was no inland exploration worth the name.

2.3.4 *The Age of Heroes (1895–1922)*

Till 1895, no one had ventured beyond the coastline of Antarctica, but the next two decades witnessed intensive exploration activities. Ships were now steam-powered and stronger and could penetrate the pack ice around the continent. In 1895, the Norwegian whaler, *Henryk Bull*, made a first landing outside the Antarctic Peninsula at Cape Adare in Victoria Land. In 1898, the ship *Belgica* got stuck in pack ice and forced scientists and sailors to endure the Antarctic winter, the first party to do so. A year later, Carsten Borchgrevink spent a scheduled winter in a hut at Cape Adare. In 1902, Swedish Otto

Nordenskjold made the first major Antarctic sledge journey using dog sledges. The techniques of transport in Antarctica were developing as rapidly as in the Arctic. A year later, Briton Robert Scott made the first ascent of the polar plateau in a balloon to survey the progress of his sledging party. Wilhelm II and Coats Land were discovered by Erick von Dragalski and W S Bruce respectively. Douglas Mawson led the 1911–14 Australian expedition to set up a base in Commonwealth Bay, made the first venture into George V Land and used a dysfunctional aeroplane as a mechanised tractor. By transmitting a message to Australia (relayed though Macquarie Island), he also became the first person to use radio in Antarctica. He lost one colleague and the sledge he was hauling with most of the supplies in a huge crevasse and the remaining one through scurvy. Himself deeply affected, he however, endured one of the most hazardous expeditions in heavy blizzards and survived to tell the story in his famous book, *The Home of the Blizzard* (1915). Earnest Shackleton nearly reached the South Pole in 1908 but retreated from a point just about a hundred miles from the pole due to harsh conditions. He, however, lived to fight another battle, famously saying, 'Better a live donkey than a dead lion'.

But it was the epic battle between Robert Scott and the Norwegian Roald Amundsen to reach the South Pole that captured everyone's imagination. Amundsen reached the South Pole in December 1911, a month before Scott, and became the first person to do so. He triumphantly returned to tell the story of his superbly organised expedition. Scott perished on his return journey with four colleagues but set an example of fortitude, and his last days' account and diary entries have become an integral part of Antarctic history and literature. His story of the courageous deeds and the team spirit of his colleagues resounded all over the world for decades and still lives on to inspire.

Earnest Shackleton tried to cross Antarctica in 1914–15. The expedition failed but it became one of the greatest sagas of courage, endurance and survival. Sir Raymond Priestly (1886–1974) who was part of the expeditions led by Shackleton in 1907–09 and by Scott in 1910–13 came across Roald Amundsen and his team on their ship, *Fram.* He may be among the very few to have met all three heroes of that age. Many years later, he summed up their relative merits in a remarkable quote, 'For scientific leadership, give me Scott; for swift and efficient travel, Amundsen; but when you are in a hopeless situation, when there seems to be no way out, get on your knees and pray for Shackleton.'

With the death of Shackleton on the island of South Georgia during his third expedition in 1921, the inspiring Age of Heroes of Antarctic exploration came to an end.

2.3.5 *From Mechanised Age to International Geophysical Year (IGY) (1923–58)*

It was only natural that rapid developments in transport systems in other parts of the world would reach Antarctica. Shackleton and Scott had both experimented with motor vehicles without much success, but their potential was never in doubt. Within a short time, vehicular transport in the form of snow scooters, tractors, cranes, snow cutters, sledges and other equipment, now common, became an integral part of any Antarctic activity. As motor vehicles started to govern movement, the use of dogs, the predominant mode of transport until then, gradually waned. Finally, the use of dogs was officially discontinued from April 1994. The ban was introduced due to concerns that the dogs might transfer diseases such as canine distemper to the seals' population or disturb the wildlife (Mcgonigal, 2008 and wissard.org, 'No dogs allowed', 2016).

In the case of aviation, the first flight was made by Hubert Wilkins in 1928. It marked a revolution in the method of Antarctic exploration. In the next decade, more of the coastline was identified and mapped than in all previous expeditions put together. American Admiral Richard Byrd systematically developed the use of aircrafts. In 1928, he controversially claimed to be the first to fly over South Pole. He used a two-way radio communication with the outside world and drove tracked vehicles in Antarctica. His radio communication was accidently picked up by someone in New York's Times Square and the world virtually went mad with excitement. American Lincoln Ellsworth became the first to fly across the continent in 1935.

This period was also marked by territorial claims from seven nations, assertions by some, refutation by others and occasional localised conflicts. Operation High Jump (1946–47), a massive US exercise, involving over 4700 people and scores of vehicles and aircrafts, followed by Operation Windmill (1947–48), was organised with the primary aim to establish a US Antarctic research base. It also marked the beginning of international interest in Antarctica after World War II.

The international community was also getting alarmed over resources' exploitation with the stocks of some whale species reaching near extinction. This led to the creation of the International Whaling Commission (IWC) in 1946. It introduced a quota system for hunting endangered species of whales.

Unexpectedly, the International Geophysical Year (IGY), observed from July 1957 to December 1958, became a masterpiece of international scientific cooperation despite political differences. Under the IGY, twelve countries established about sixty research stations for eighteen months. The concept of a permanent research station in Antarctica took root. Over forty-eight tomes of documents in various branches of science, astrophysics, oceanography, glaciology, meteorology and zoology were produced. Many important firsts were achieved, including recording the unbelievable depth of the Antarctic ice cap through detonations and sound waves and making the first trans-Antarctic crossing from Weddell Sea to Ross Sea via South Pole by Sir Vivien Fuchs, covering 2180 miles in 99 days. Sir Edmund Hillary was part of Fuchs' team responsible for laying depots for Fuchs. But after doing that from Ross Sea, Hillary pressed on and reached South Pole without Fuchs' knowledge or permission. He became the first to reach South Pole after Scott's ill-fated expedition of 1912, also creating a controversy in the process.

Up until 1819, no human being had seen Antarctica. In just 139 years, by the end of IGY, the continent had been fully explored and mapped. A handful of men in about fifty expeditions had unlocked most of the secrets of the world's last wilderness. The IGY succeeded because of its political neutrality and focus on science. It was, indeed,

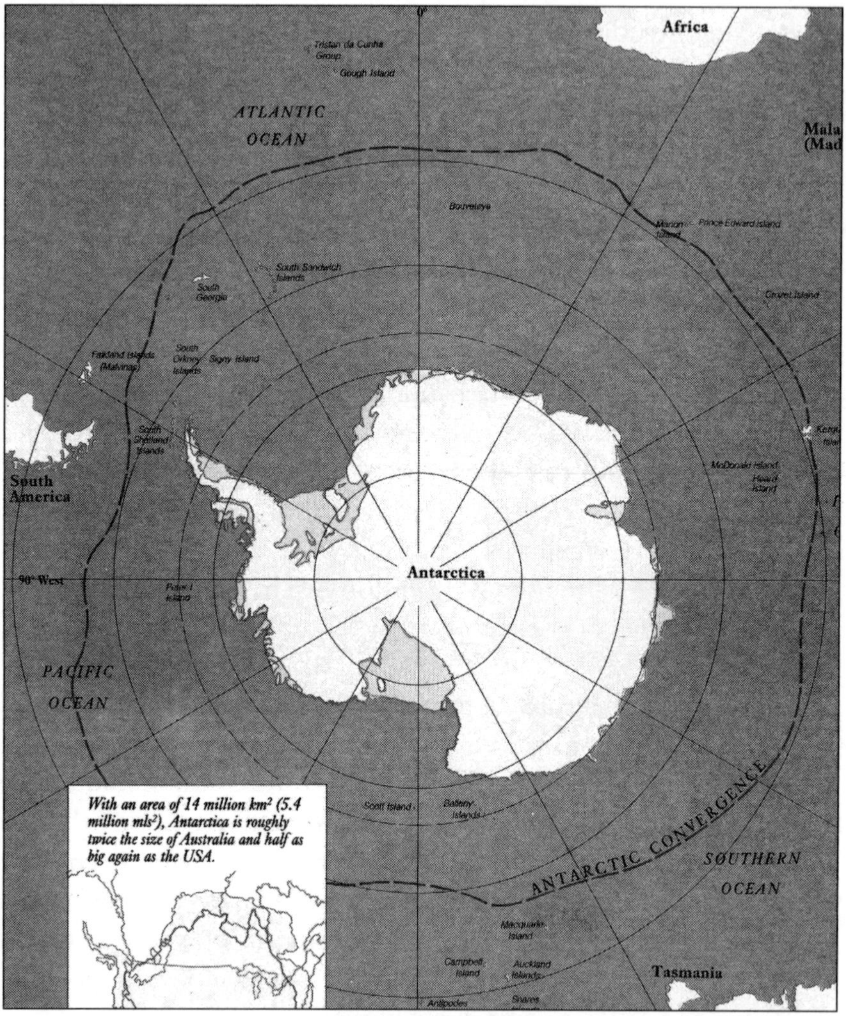

Fig. 2.1: Antarctica and the Surrounding Southern Ocean (In terms of area, Antarctica is over four times the size of India's 3.3 million sq km)

a watershed in Antarctic logistics and records, paving the way for the current system of international governance through the ATS.

Another significant development in this period pertains to seven countries searching and collecting evidence and laying formal territorial claims in Antarctica to protect their strategic interests. This added an entirely new dimension to theirs' as also others' Antarctic perspectives. A detailed analysis of this aspect is discussed in the relevant chapter in the book.

The hitherto mostly unknown continent on the earth, represented by a few wavy lines in the south not so long ago, had finally taken its due place in geography books as shown in Fig. 2.1 (May, 1989).

2.3.6 *The ATS and Governance till the Madrid Protocol (1959–98)*

After intense negotiations, the twelve participating nations of the IGY concluded a treaty, known as the Antarctic Treaty (also referred to as AT or Treaty in the book) in December 1959, encompassing the continent proper and all areas south of 60°S latitude. The Treaty came into force in June 1961. It effectively barred military operations of any kind (but not use of military personnel for peaceful purposes), nuclear testing and dumping of nuclear waste, and declared that the continent could be used only to promote peaceful and scientific pursuits. As of now, about thirty countries, including India, maintain about seventy permanent stations. There are a total of fifty-three member nations out of which twenty-nine are known as Consultative Parties (CP) who enjoy decision-making and veto rights. The period also saw the signing and coming into force of many agreements as parts of the Treaty. The Treaty and all these agreements are collectively called the Antarctic Treaty System or ATS. There are many agreements within the ATS as also outside it, protecting the environment and wildlife of Antarctica. They pertain to the protection and conservation of Antarctic flora and fauna, seals, whales, issues relating ozone depletion, waste management, dumping at sea and other environmental concerns. The Convention on the Regulation of Antarctic Mineral Resources Activities (CRAMRA) was signed in 1988 but never came into force due to opposition from non-governmental organisations (NGOs), like the Association of Antarctic and Southern Ocean Coalition (ASOC) and the Green Peace, and the refusal from Australia, Belgium and France to ratify. Though CRAMRA never came into force, it triggered negotiations for a comprehensive agreement covering all aspects of Antarctic environment and not just mineral resources, leading to the Madrid Protocol.

2.3.7 *The ATS and Governance since the Madrid Protocol (1998 onwards)*

After the failure of CRAMRA to come into force, intense negotiations paved way for the signing of an important agreement known as the Madrid Protocol, as part of the ATS, in 1991 which, after due ratification by members, came into force in 1998. The Protocol prohibits all mining activities for fifty years and designates Antarctica as a 'natural reserve, devoted to peace and science'. It has six annexures which make mandatory the requirement of an environmental impact assessment (EIA) on all activities, including scientific and tourist operations, and an approval from Treaty instruments, overseeing all aspects of governance for any new stations. However, the sixth annexure with penalty clause is not in force due to non-ratification by all members.

A few stations established after 1998, including India's third station Bharati, have been built in accordance with the requirements of the Madrid Protocol. Most of the stations existing from earlier times do not conform to these. It is hoped that the situation will gradually change, as all Antarctic stations have to be rebuilt or replaced after every few years. For example, India's second station Maitri, operational since 1989, is ageing and India plans to rebuild or replace it in the near future. This and all future stations from other Treaty countries will have to meet stringent stipulations of the Protocol.

The Antarctic Treaty and Madrid Protocol will be up for review in 2041 and 2048 respectively. Conditions then may depend upon various factors like the situation of renewable and non-renewable resources, implementation of the recently concluded Paris Agreement and global warming effects, the geopolitical situation in the Arctic and other factors as discussed in details later.

2.4 The Natural Environmental Aspects of Antarctica

Physical research on the continent started only after 1895, during the Age of the Heroes, when explorers ventured beyond the coastline. The pace of research was slow and limited to areas which were accessible by the explorers. The exploration and study of environmental aspects, however, picked up once the use of motor vehicles and aircrafts became common after the end of WW II (1939–45) followed by the IGY in the late 1950s.

Though we will be discussing all environmental aspects in details in separate chapters, it would be pertinent to see their relation with respect to the physical discovery of Antarctica and the chronological sequence of information build-up through various forms of literature.

The geological record of Antarctica is mostly hidden beneath a permanent ice cap covering over ninety-eight percent of its area. Modern techniques such as deep drilling for ice cores, ice-stream base samples and remote-sensing techniques, including seismic and radar soundings, allow research below the ice sheet. The first geological investigations were made by Hartley Ferrar during the National Antarctic Expedition (1901–04) by mapping mountains of Victoria Land in West Antarctica. The large number of fossil wood, marine plants, fish, penguins, whales and reptiles indicated a semi-tropical climate, deep forests and animals. Edward Wilson's discovery of a piece of coal with impressions of stem and leaves in 1912 were the first of many samples of the glossopteris fossils that provided an important link to similar fossils recovered from other continents, consolidating proof of Gondwanaland. Through long sledging journeys, Vivien Fuchs and R.J. Adie, after WW II, confirmed that East and West Antarctica were once separate land masses and the latter experienced more tectonic activity.

Alfred Wegener's theory of a continental drift in his masterly book, *On the Origin of Continents and Oceans,* in 1915, later led to the now widely accepted theory of plate tectonics. In 2001, nine seismic stations were set up to measure tremors; those recorded are caused usually by volcanic activity or fracturing of the ice shelf. A massive earthquake that occurred off the Balleney Islands in 1998, measured 8.1 on the Richter scale. At present, however, the continent is the most earthquake-free area in the world. The probability of mineral resources similar to those in other Gondwana parts of the world gave rise to concerns, culminating in the signing of CRAMRA in 1988. However, though CRAMRA never came into force, it gave way to the signing and coming into force of the comprehensive Madrid Protocol on environmental protection, banning all commerçial mining activities till 2048.

Early geographical research focused on coastline exploration and discovery of new regions. Aerial surveys pioneered by Byrd in 1928 dramatically widened this scope.

As far back as 1700, Edmond Halley was the first to measure the magnetic variations in the southern waters and to realise that a magnetic field controlled Aurora Australis. Two centuries later, Scott and Mawson tried to photograph the same, but were unsuccessful. Geophysics got a boost only from the IGY. Data gathered was collated with simultaneous information from different locations to produce a new understanding of the earth's environment. A number of unmanned stations played their part in this. Today, with direct links to satellites that orbit the South Pole, geophysical data can be transmitted all year round from Antarctica back to laboratories around the world.

The first detailed climatological records in Antarctica were kept by William Colbeck and Lois Bernacchi in the 1898–1900 Southern Cross Expedition. In 1950, there were only twenty meteorological stations in Antarctica and some of the data on weather till the 1960s was from whaling fleets. Only after Automatic Weather Stations (AWS) were established during IGY, could multi-locational data on temperature, humidity, air pressure, wind speed and direction be available for comparative analysis. Now every Antarctic base records weather conditions and this information along with AWS data is transmitted to supercomputers all over the world for global weather predictions. Balloons are tracked across the continent and satellites are used most commonly to show cloud formations.

Ozone Hole is a term used to describe areas of the atmosphere where the ozone has been reduced by at least fifty per cent. This condition and resulting scenario occurs during every polar winter in Antarctica. Monitoring of the Antarctic ozone began at Halley Station during the IGY in 1957. By 1984, it was observed that ozone values had gone down by thirty per cent more than those observed a decade earlier. This depletion was first termed as an ozone hole in the findings published in *Nature Magazine* in 1985 and has since become a widely accepted scientific term. Today, it is a major science project in Antarctica. Since its discovery, the hole has continued to grow. By 1993, the ozone reached a record springtime low.The volcanic acid particles from the 1991 eruption of Mount Pinatubo in Philippines had accelerated the ozone loss to nearly seventy per cent of the ozone level existing in the year 1960. The hole in the year 2000 measured 29 million sq km, twice the size of the continent. International agreements such as the Montreal Protocol have limited the use of ozone-destroying chemicals. However, it is unlikely that ozone concentrations will recover by 2050 as was once estimated (Mcgonigal, 2008).

Even in the forbidding and hostile environment, there are a few areas, like Dry Valleys, cold deserts, saline lakes and exposed rock surfaces on isolated mountains where snow melts in summer, that support plant life. The flora includes lichens and mosses and, at a microscopic level, algae, fungi and different forms of bacteria. There are over 350 types of lichens alone. Out of a total of the 120 new microbes discovered as a result of international efforts till 2005, twenty have been discovered by Indian scientists (Pande, 2007).

The fauna, non-existent on the continent proper, abounds in ocean waters. All the organisms live in a cycle of interdependence, with the Southern Ocean as its centre. Life begins and ends in a process called the giant food cycle or web. It goes from phytoplankton to zooplankton to krill to sea birds, penguins, seals, whales to dead plants and animals settling at the bottom of the ocean to plant nutrients to phytoplankton and so on, endlessly.

The blooming of the pink algae at the base of pack ice and its sinking to the ocean's bottom when that ice melts in summer, creating food for the elements underneath in a yearly cycle, is a very special and unique feature of the giant food web. As of now, pack ice extends to about twenty million square kilometres at the peak of winter. Any reduction in this extent due to global warming will have far-reaching consequences on the functioning of the food web.

With krill at the centre, the food web has evolved over millions of years. All predators, including seals, whales, penguins, birds and fish together consume hundreds of millions of tons of krill every year. Overfishing of krill, therefore, will have a highly adverse effect on all components of the food web.

In fact, in the earlier period of discovery, human greed and the lack of understanding of the damage it can cause to the balance of the global ecosystem, led to sustained attacks on the ocean's animal life by sealers, whalers and commercial fishing enterprises. Some species of seals and whales, due to indiscriminate and thoughtless harvesting, were driven to near extinction. Unintentionally though, thousands of birds were also killed due to the long-line technique used to harvest whales. But the situation has been retrieved to an extent with the AT and its various instruments created for conservation of the ecosystem.

All this highlights the fragility of Antarctica and demonstrates why humans, whether they are harvesters, visitors, scientists or tourists alike, must abide by international conventions, measures and regulations to limit the degradation of the world's last great wilderness.

2.5 Antarctic Science

In the early phases of Antarctic exploration, the sole emphasis was on sealing and whaling which either neglected or relegated any work concerning science. Many times, desirous scientists had to join expeditions as helpers on a ship to enable them to carry out some scientific work at least. However, scientists the world over knew the importance of polar areas to science. The pioneering efforts of German explorer Karl Weyprecht led to first International Polar Year (IPY) in 1882–83 wherein ten countries established stations, three of which were in the southern hemisphere, with the most southerly on the sub-Antarctic island of South Georgia. Detailed and synchronised measurements of meteorology, magnetism, aurora, and ice were taken for the first time in both polar regions. The same conditions of sun spot min (minimum activity of the Sun) again prevailed fifty years later and the second IPY was observed under the aegis of the International Council of Scientific Union (ICSU) and the Comite Speciale de l'Annee Geophysique (CSAGI) in 1932–33. With this background, the IGY was successfully conducted during 1957–58, as mentioned earlier.

On the global organisational front, a number of new supporting instruments were created. An interdisciplinary body of ICSU, called the Scientific Committee for Antarctic Research (SCAR), was formed in 1958 to coordinate the scientific activities during IGY 1957–58. It has permanent working groups for various disciplines like upper atmosphere, biology, human biology and medicine, geodesy and cartography, geology, solid-earth geophysics, meteorology, oceanography and logistics. It provides scientific advice to the Antarctic Treaty Consultative Meetings (ATCM) and other organisations on issues concerning the management of Antarctica and the Southern Ocean. Membership of SCAR, however, is not limited to acceding countries of the AT. Likewise, AT members are not necessarily members of SCAR. Due to the rigorous and hazardous environment and very expensive logistics, the promotion and development of Antarctic science can only be done effectively under the umbrella of the AT and within the framework of the ICSU committee, especially SCAR.

The scientific and objective nature of SCAR and the political nature and vulnerability of ATCM pose a number of difficulties. Within limitations, however, SCAR, has played a significant role in the formation and working of some ATS instruments like Agreed Measures for Antarctic Flora and Fauna (AMCAFF), Convention for Conservation of Antarctic Seals (CCAS) and Convention for Conservation of Antarctic Marine Living Resources (CCAMLR).

The role of scientific research in Antarctica will be very significant in the fields of biology, and its applications in the continent's ecology, conservation and pollution studies and earth sciences, to further demystify past events, and in various aspects of atmospheric sciences. The AWS in Antarctica has hugely contributed towards automatic data collection. In addition to SCAR, a number of international organisations are involved in Antarctic research. SCAR, however, retains the leading role, being closely associated as adviser to ATCM.

2.6 Interdependence and Impact of the Environment in Antarctica with the Rest of the World

If one topic is to be singled out as occupying world attention in recent times, it is global warming. Various climate change conferences under the aegis of the United Nations Framework Convention on Climate Change (UNFCCC) have addressed this issue in the last few years. The conference at Copenhagen in December 2009, though otherwise unsuccessful, underscored the grave dangers of the rising average temperature of the earth's atmosphere, popularly referred to as global warming. It has been concluded that global warming must not exceed 2°C from the nineteenth century pre-industrial age level. At this point, termed tipping point, massive amounts of warming gases trapped in the Siberian permafrost and in the world's humid rain forests are likely to be released into the atmosphere (D Singh, 2010). It would result in, as per scientific analysis, a likely loss of the world's many low-lying lands. In the past hundred years, the earth's average surface temperature has increased by about 0.8°C, with about two thirds of it in just the last three decades. It is caused by increasing concentrations of greenhouse gases produced by human activities such as deforesting and burning of fossil fuels (Wikipedia, 2012, Global Warming). Subsequent climate change conferences at Cancun, Durban, Doha, Warsaw (every year from 2010 to 2013) and the climate summit in the UN in 2014 have discussed issues threadbare and taken few positive steps, including resolutions like securing the second phase of the Kyoto Protocol (1991) and finalising a new climate deal by end-2015 at the Paris climate summit, its implementation from 2020 and the establishment of the Green Climate Fund (*Climate Summit Bulletin*, Volume 172; No. 18, 26 September 2014, published by the International Institute of Sustainable Development [IISD] and the UN). At the just concluded climate summit in Paris, known as

Conference of Parties (COP 21), from 30 November to 11 December 2015, over 190 countries have concluded an agreement on this vital issue, covering all aspects. Vague language giving leeway for different interpretations marks this agreement too. Implementation of this agreement would be keenly observed in the coming years.

The major concern due to global warming in Antarctica would be its effect on the stability of the ice cap. Recent studies indicate that the average temperature rise is an increase in surface temperature of about 0.05°C per decade for the last five decades in East Antarctica, whereas the corresponding figure in the West is 0.1°C. Most of the studies on the subject are ongoing but, as of now, what is seen on ground is the collapse of large portions of Larsen B and Wilkes ice shelves in West Antarctica into icebergs in the last few years and likely such occurrences in the near future. For the first time, isolated cases of these have been noted in East Antarctica also (Mcgonigal, 2008).

Some of the serious concerns are the effect a large number of icebergs may have on the global weather pattern and the increase in sea levels due to the calving of icebergs in the peninsula area. The reduction in the area covered by pack ice and in the production of phytoplankton may have an adverse effect on all elements of the giant food web in the Southern Ocean. It may also seriously affect the vertical mixing of ocean waters, posing dangers to animal life in other oceans. All these may form burning issues, demanding immediate attention, study, analysis and establishment of corrective actions. The Intergovernmental Panel on Climate Change (IPCC) in its report in 2007 estimated that the sea level rose globally from 1993 to 2003 at the rate of 3 mm per year and about ten per cent of this was from the Antarctic ice sheet. If fully melted, the total volume of the Antarctic ice sheet would be equivalent to a sea rising of about 57 m. The IPCC and the former US Vice President Al Gore jointly received the Nobel Peace Prize in 2007 for 'their efforts to build up and disseminate greater knowledge about man-made climate change and to lay the foundations for the measures that are needed to counteract such changes'. IPCC's latest and fifth report was published in four parts. The first three are working group (WG) reports – WG I (Physical science basis), WG II (Impacts, adaptations and vulnerability) and WG III (Mitigation of climate change) and were published in September 2013, March 2014 and April 2014 respectively. The fourth part, AR 5 Synthesis Report, was published in November 2014. More than 800 authors from 3000 nominations were involved in its preparation. Advancing the argument further, AR 5 says, among other things, that a global surface temperature increase by the end of the twenty-first century is likely to exceed 2.7°C or much more in some scenarios. The Paris Summit has tackled all these issues as discussed in later chapters.

Along with the above, issues which will influence the Antarctic environment are changes in the extent of winter pack ice, growth and effects of the ozone hole, caused by chlorofluorocarbons or CFCs, overfishing of krill, non-observance of instruments like the IWC by taking advantage of exceptions made for scientific reasons, human footprint in the continent due to permanent research stations, visiting personnel for various reasons and Antarctic tourism. Widely varying policies of world governments and disagreements, too, pose a danger to the fragile Antarctic ecosystem. An open conflict between Australia and Japan over the alleged killing of whales by Japan, causing Australia to appeal to the International Court of Justice (ICJ) at Hague is a case in point (cnn.com, 2012, 'Australia heads to court over whaling disputes with Japan'). Finally, the ICJ backed the landmark case in April 2014 and demanded that Japan stop its whaling programme in the Southern Ocean with immediate effect. Japan said it would abide by the verdict. However, it may continue the same programme in the North Pacific which underscores the limitations of international agreements and institutions.

Global warming's more pronounced impact is lately felt at the other end of the earth, that is, the Arctic Ocean. Here, shipping has become a real possibility in the near future due to ice melting in summer. This may have far-reaching consequences in terms of drastic distance-reduction in shipping routes between Asia and Europe, likely exploitation of natural resources like oil and natural gas and potential conflict situations between countries of the Arctic Council and others. It may affect the present serenity and smooth functioning in the ATS (Shyam Saran, 2012 and Wikipedia, 2012 Arctic Council).

2.7 Antarctic Logistics

In earlier days, the art of movement and supply of goods to a body of troops in military operations was referred to as logistics. Volume, range, preservation, distance, modes of transportation, communication and other aspects emphasised the need for a skilful use of the art. As the world became more complex and technologically advanced, logistics too changed its form from art to a combination of art and science. As an important tool of management, logistics became the management of the flow of goods, information and other resources, including energy and people between points of origin to points of application in order to achieve desired results.

Separated from the rest of the world by the Southern Ocean with its pack ice and harsh environment precluding any human inhabitation, Antarctica is a place which tests the limits of logistics skills. Tremendous progress in Antarctic exploration was possible only due to the advancement of logistics, with IGY 1957–58 as a watershed. A changing logistics scene saw modes of transportation shifting from sledges hauled by either men or dogs, to modern ice-breaking ships, aircrafts, helicopters and tractors/sledges trains, greatly reducing travel time from months to hours. Antarctica came closer to other continents and the Southern Ocean lost its formidable inaccessibility. Human and equipment safety took great strides. Instant communication and data transfer over Internet is a far cry from the conditions just a few years ago. Today, sacrifices of the early explorers have become a thing of the past.

Despite all this, maintaining permanent stations for scientific research and extending logistics support still remain a big organisational and logistics challenge due to long distances and harsh environment, as can be seen from the detailed discussion later. For example, for the three main stations and a number of field camps of the United States Antarctic Programme (USAP), Mcmurdo Station serves as a central supply depot. About 58 million pounds of fuel and 14 million pounds of cargo have to be supplied every year. The last heavily ice-clogged stretch, 135 km long and 55 km wide, connects Ross Sea in the north with the Ross ice-shelf, where Mcmurdo Station is located, is called Mcmurdo Sound. Over the years it has become very difficult for ships and escort icebreakers to navigate in Mcmurdo Sound due to the increased number of calved icebergs from the ice-shelf. A revision of the entire set-up is under consideration as there is no back-up at present.

Most of the Antarctic stations are on oases areas at the periphery of the continent. Those built on the ice shelf or inland ice plateau are bound to be buried in snow at some stage. However, modern stations last longer due to the techniques used to build them above ground on adjustable stilts or on skis, to facilitate shifting of their locations when threatened by ice accumulation. All this is a far cry as compared to conditions in the not-so-distant-past.

Issues related to logistics were handled by the ATCM with various SCAR subcommittees. The first symposium on Antarctic Logistics and Operations was held in 1962 at Boulder in the US. After its inception, the Committee of Managers of National Antarctic Programmes (COMNAP) has been organising them since 1988 through its Standing Committee on Antarctic Logistics and Operations (SCALOP). From 2010, the symposium has widened its scope to cover all activities relevant to the National Antarctic Programmes (NAP) in addition to logistics and operations.

International organisations and consortiums like the Antarctic Logistics Company International (ALCI) and Droning Maud Land Air Network (DROMLAN) over the last few years have also revolutionised the entire gamut of Antarctic logistics with reference to transportation of personnel and cargo. Constantly changing logistics scenario will always pose new challenges.

2.8 India's Antarctic Presence and Programme (IAP)

As per the established geological timeline, India was part of the supercontinent but is a late entrant to Antarctica, unlike its Gondwana counterparts – Australia, New Zealand, South Africa and countries of South America. In over thirty national expeditions since 1981, about 2000 Indians have been to Antarctica *(Appendix VII)*. Two Indians, Mr. G.S. Sirohi and Dr. P.S. Sehra, were part of the US and the then-

USSR expeditions in 1961 and 1971 respectively, in their individual capacity (Sehra P.S., 2003). After two initial summer expeditions on the ice shelf off the Princess Astrid Coast in Central Dronning Maud Land in East Antarctica (*NCAOR Technical Publications*, No 1 and 2), India took the decisive step to establish a permanent presence in Antarctica. A permanent station, Dakshin Gangotri (DG), was built by the third expedition team during December 1983–February 1984. Wintering started with a team of twelve members before the summer team returned to India (*NCAOR Technical Publications,* No 3). Within a short time of signing the Antarctic Treaty (19 August 1983), India was granted Consultative Party (CP) status on 12 September 1983. Since then, without any break, India has maintained its presence in Antarctica with its National Antarctic Programme (NAP).

The structure and systems for DG were imported in a knocked-down condition from a firm in the UK. An Indian Army team, trained in the UK, constructed the station in one austral summer during 1983–84. Though it was supposed to last for two to three years, it survived for six years due to the excellent work and ingenuity by successive winter teams in snow clearance, periodical structure strengthening and modifications and maintenance of life-support systems. It became very hazardous during the sixth and last wintering team's stay and had to be decommissioned in February 1990 due to snow burial.

In all these years, many support structures were also erected around DG, like a hanger for vehicle repairs, indigenous container accommodation for emergencies as also for summer teams (popularly known as the Defence Research and Development Organisation [DRDO] complex), portable cabins, fuel dumps, etc. Despite snow-clearing efforts and repeated shifting of containers on barrels from time to time, everything was buried due to rising snow levels.

As the fate of DG, on the ice shelf, was evident from day one, a location about 85 km away from DG, in a hilly area known as Schirmacher Oasis, was selected to build the second Indian Antarctic station, Maitri, a few km away from two other stations, Novolazarevskaya (Novo) of Russia and Georg Forster of the GDR or East Germany (GDR, 1984). With no danger of snow burial and survival conditions vastly better than on the ice shelf, the location assured less danger, longevity and better continuation with a greater range of scientific research activities, situated as it was in the largest Wolthot mountain range in East Antarctica.

Maitri, built by the seventh and eighth expeditions, in the austral summers of 1987–88 and 1988–89 was totally designed and fabricated by DRDO and erected by the Indian Army. The station consists of a main block and two side blocks at right angles at both ends. The structure stands on stilts and is partially on relatively level ground. Over all these years, considerable scientific research work has been carried out in the fields of geology, ice-core analysis, upper-atmosphere studies including ozone and greenhouse gases, meteorology, geomagnetism, biology and environmental science. Though designed to last ten years, Maitri is still fully functional after twenty-five years. The ageing station is likely to be rebuilt in the near future.

With over thirty years' experience and an urge to expand its sphere of activities in quantum and range in Antarctic research, India decided to establish its third research station, Bharati. After detailed reconnaissance expeditions, due consideration of various logistics and scientific aspects, a location in the Lars Mann Hills of Prydz Bay Area of East Antarctica was selected. There are four stations (about 7 km away) namely, Progress I and II (Russia), Law-Racovita (Romania) and Zhongshan (China) in the same area. The Australian station, Davis, is located about 70 km from these stations.

A design from a German firm was approved for the station, conforming to the exacting needs of logistics, operational requirements, life-support systems and environmental safety in accordance with the Madrid Protocol on Environmental Protection. The station was completed and functional as of 2013–14.

It may be by coincidence that DG, Maitri and Bharati represent three of the four main areas of Antarctica with very different logistics parameters, one each on shelf, oasis and promontory respectively. Bharati is about 3000 km away from India Bay, the disembarking point for Maitri, while the area around the buried DG serves as a supply base for Maitri. As for Bharati, the approach for berthing and transportation of stores on rocky area is over a short distance and the creation of a road and sophisticated systems have created great logistics ease and convenience. The need to maintain two stations widely spaced apart on Antarctic coast, with different locations, approaches and terrains certainly add a new dimension and complexity to the logistics.

In meeting these challenges, maintaining the two stations, any future stations and expansion of activities, lies the success of the IAP.

2.9 Antarctic Links

The limited spread of knowledge about Antarctica underwent a sea change once Internet arrived on the scene. Over the years, a number of websites have been developed. Most of them have the advantage of the centres and libraries established a long time ago and the original material from early expeditions. Some of the libraries in the UK, Australia, US, New Zealand and Russia are known to have an excellent collection of varied and valuable literature in terms of books, artefacts, photos and other forms. The sites provide information on exploration history, science, logistics, current news and all other aspects of Antarctica. Some of the principle websites are described below.

2.9.1 *National Antarctic Programmes (NAPs)*

Each country that has an Antarctic presence has a website but most are sparse, containing limited information. Some sites give detailed information useful in research work. The COMNAP, established in 1988, maintains an excellent site, www.comnap.aq. It provides information on all research stations, other facilities and activities of all NAP. The Australian Antarctic Division (AAD) maintains a site, www.aad.gov.au, which gives detailed information on stations, maps and facilities in Antarctica. Sites for British Antarctica Survey (BAS), www.antarctica.ac.uk, United States National Science Foundation (NSF), www.nsf.gov, and New Zealand Antarctic Institute, www.antarcticanz.govt.nz, produce good information about their countries' NAPs and other information.

2.9.2 *Antarctic Centres and Other Prominent Websites*

Some of the sites that give excellent academic information on polar expeditions, ice, glaciers, etc, are from Byrd Polar Research Centre in USA http://bprc.osu.edu, Antarctic Cooperative Research Centre in Australia, www.acecrc.org.au, Rice university in Texas, www.glacier.rice.edu, International Antarctic Centre, Christchurch, New Zealand, www.antarcticanz.govt.nz, and some others. Here are some principal websites:

Scott Polar Research Institute (SPRI): The world's leading academic polar organisation based in Birmingham, UK, maintains the website www.spri.cam.ac.uk. It contains 120,000 bibliographic records, picture library database, special index to Antarctic expeditions and directories for worldwide polar museums and polar and cold regions organisations. The institute also maintains the site www.scar.org which contains facts and statistics concerning NAPs.

Antarctic History and Philately: The site www.south-pole.com contains information on stamps and postal history, south polar expeditions and a vast collection of photo illustrations.

Gateway Antarctica: The site www.anta.canterbury.ac.nz originates from the University of Canterbury in New Zealand. It strives for increased understanding of Antarctica and the Southern Ocean and attracts international participation in collaborative research, analysis, learning and networking. Very detailed sections on history, environment, life, logistics, education, images, news, tourism, etc, are a source for enormous material.

Antarctic News: The site www.antarctican.com is produced in Tasmania (Australia) delivers the latest news, happenings and comment about Antarctic life, South Polar endeavours, the world of ice and the Southern Ocean.

Many other sites give information on tourism, virtual tours of some stations and history. Search engines like Google and Wikipedia as also websites of various ATS organs also have a good reservoir of information.

Antarctic Tourism: There has been a surge of tourists, from a few hundred in 1969 to an average of about 30,000 as of now. Presently, tourism is largely regulated by the International Association of Antarctic Tourists Operators (IAATO) formed in 1991 with a present membership of 111. It maintains its website, http://iaato.org, and abides by all regulations for protecting the pristine Antarctic environment. It advocates that all fortunate travellers to Antarctica return as ambassadors for the continent and support its continued protection for future generations. Many organisations from the tourism industry, for example, http://www.quarkexpeditions.com, too, provide excellent information as part of their marketing strategy.

The above are just a few representatives Antarctic sites for scholars to refer.

2.10 In Conclusion

Antarctic literature has followed in the footsteps of exploration. In the initial stages, exploration was undertaken by individuals and its recording was localised in the form of ship logs, personal diaries, reports to the kings or rulers and declaration of sovereignty. As elements of logistics like ships, vehicles, aircrafts, helicopters and communication started developing, state sponsorship replaced individual quests and literature also changed from personalised accounts to detailed expedition and scientific reports, books and other sophisticated forms. IGY 1957–58 gave the continent, as also the literature on it, a boost and true international flavour. With IGY, SCAR, AT, ATS and other instruments, the generation of data and sharing of information and scientific progress was the logical outcome. As has been brought out earlier, over ninety per cent of Antarctic non-fiction literature has resulted only since the IGY.

As if all this galloping development in exploration, as also the resultant literature, were not enough, Internet made its appearance. Public domain sites, search engines, tourism, national, institutional and other websites proliferated. In just about a decade or so, a stage has been reached where, with a little patience and skilful use, one can access any information loaded from any source of any vintage. Here again, nations in the forefront of exploration and their institutions are far ahead of other countries and their organisations. SPRI is a case in point to illustrate the richness of information and its availability.

Since its physical entry in matters Antarctic since 1981, India's progress in its effective presence, participation in ATS instruments and other aspects is impressive but not as much when it comes to the development of Antarctic literature. India has a long way to go towards creating public awareness about the continent, its importance to science and the world's ecosystem. Concerted efforts on this score may be very useful, not only in the efforts towards effective and influential presence, but also towards enhancing public awareness on this subject of global significance. The establishment of libraries, information centres, special chairs at universities for doctoral and post-doctoral research, creating and making Antarctic literature easily available in regional languages, and other forms of research and development may have the desired effect. Wider participation alone will boost national efforts to maintain and enhance the influence and efficacy of the Indian presence in the Antarctic. With an influential and effective presence through self-reliance and by the creation and preservation of records, the development of all Antarctic aspects from the Indian perspective, including science, institutional competence, expertise in polar logistics and other complimentary areas will take place. The development of Antarctic literature will automatically follow.

CHAPTER 3

The Seventh Continent Evolves: Geological and Geographical Perspectives

"I felt as though I had been plumped upon another planet or into another geological horizon of which man had no knowledge or memory."

Admiral Richard E. Byrd (1888–1957)
(American Explorer, overwintering alone in a weather monitoring station in Antarctica in 1938)

A GEOLOGICAL PERSPECTIVE

3.1 Introduction

Geology is the branch of science relating to the history and development of the earth's crust. An understanding of the geological aspect of Antarctica, the history of which stretches back to millions of years (Ma) of geological timeline, is essential for a proper appreciation of all the environmental aspects of the continent. Until about 280 Ma, present-day Antarctica was part of the supercontinent Pangaea (meaning all lands). Pangaea subsequently broke and its resultant fragments were scattered in different directions, as per the widely accepted theory of plate tectonics which replaced the earlier hypothesis of a continental drift. To understand this, we need to have a look at the structure of the earth.

Detailed geophysical studies have revealed that the earth has a shell-like structure consisting of an inner core, a middle mantle and an upper crust. Within the inner core are two parts: the solid inner core and the liquid outer core. Likewise, the mantle, which is viscous, is divided into two parts, the lower and the upper mantle. The upper mantle flows more easily than the lower one. The outermost shell is known as the crust and is solid. The crust also consists of two layers – the upper Sima (composition of silicon and aluminum) and the lower Sima (composition of silicon and magnesium). The crust and the uppermost part of the mantle form the lithosphere. The lithosphere is separated from the mantle by the asthenosphere – a highly viscous, mechanically weak and ductile deforming layer. The movement of the lithosphere over the mantle along the asthenosphere gives rise to what is today known as plate tectonics.

If a pie-shaped cross-section of the earth is taken, the inner structure would look as shown in Figure 3.1 (Google, Structure of the Earth, 2012).

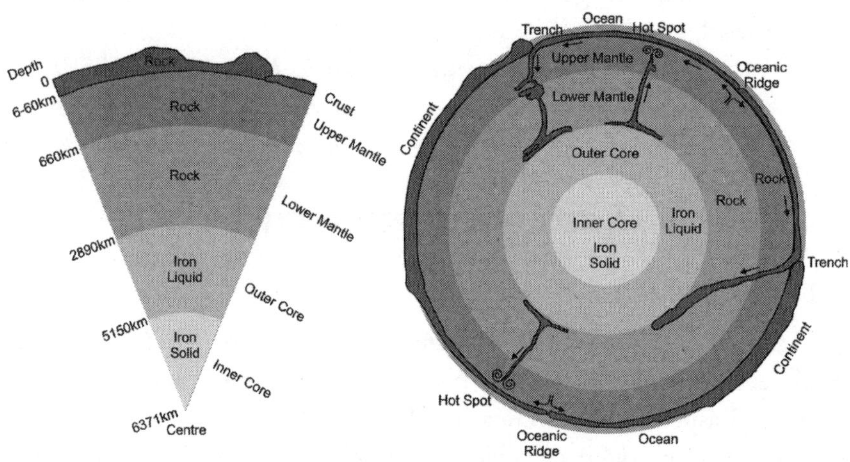

Fig. 3.1: Structure of the Earth

3.2 Lithosphere

Lithosphere is the strong, hard and rigid outer layer and consists of the crust and the uppermost mantle which is elastic. The lithosphere is underlain by asthenosphere which is the weaker, hotter and deeper part of earth's upper mantle. The lithosphere is broken into tectonic plates whose movements have resulted in the evolution and formation of various continents over a geological timeline of millions of years.

The concept of a strong lithosphere resting on a weak asthenosphere was developed by Barrel in 1914 and expanded by Daly in 1940 (Wikipedia 2010, Lithosphere). This concept is essential to understand the theory of plate tectonics developed later in the 1960s.

The thickness of the lithosphere is considered to be the depth to the isotherm or counterline of equal temperature, associated with the transition between brittle and vicious behaviour. The thicknesses of the continental lithosphere ranges from 40 to 200 km and that of oceanic lithosphere from 50 to 100 km. The thickness of the crust, however, is uniform for both and varies from 30 to 50 km.

3.3 Continental Drift and Plate Tectonics

3.3.1 *Continental Drift*

Around 220 Ma, the large supercontinent Pangaea, named so by German Alfred Wegener, comprising most of the land surface on earth until then, broke into two – Laurasia and Gondwanaland. Austrian Edward Suess first proposed the concept of a southern supercontinent in 1885 and named it Gondwanaland (now widely referred as Gondwana) after an area in central India inhabited by a tribe called Gonds. However, Suess could not explain the mechanism leading to the concept, explained later by Wegener. A striking similarity of coastlines on either side of the Atlantic Ocean prompted one thinker to observe that this could just not be a matter of chance and that fragmentation must have been the cause. However, this and the bathymetric contours of the surrounding 1000 m formed the principal evidence for Wegener to originate the hypothesis of the continental drift in 1912. He published his masterly book, *On the Origins of Continents and Oceans,* in 1915 with a superb graphic illustration of how Antarctica acted, in the words of du Toit, another early believer, as the 'keystone to Gondwana'. Further evolution saw Gondwana drift southwards, but later break up around 183 Ma, the separate pieces mainly forming what are now known as South America, Africa, India, Australia and Antarctica (Landis, 2001).

3.3.2 *Plate Tectonics*

The discovery of similar magnetic anomalies across the mid-Antarctic ridge further led to the widely accepted concept of plate tectonics in the 1960s which revolutionised the understanding of the earth.

According to this concept, continental plates make up the earth's outer shell, the lithosphere. They move slowly across the ocean floor in relation to one another, colliding with each other at times. When they are compressed together, they buckle in to form mountains and when they rift apart, they produce earthquakes and volcanoes.

Over geological timeline, the evolution of the earth from Pangaea to its present configuration is shown in Figure 3.2 (Walton and Doake, 1987).

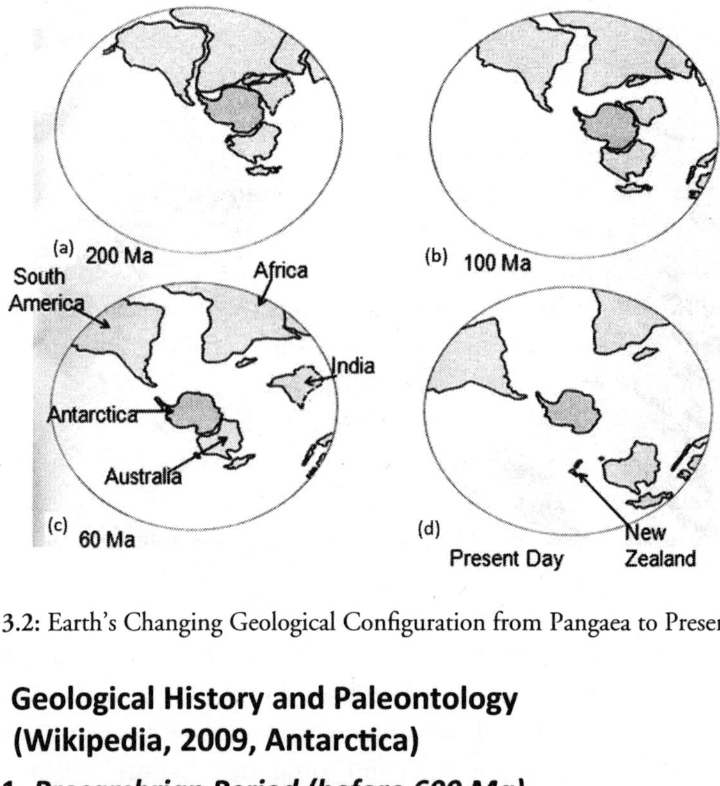

Fig. 3.2: Earth's Changing Geological Configuration from Pangaea to Present Day

3.4 Geological History and Paleontology (Wikipedia, 2009, Antarctica)

3.4.1 *Precambrian Period (before 600 Ma)*

Similarities between ancient crustal belts suggest that Antarctica may have been connected to southwestern North America more than 600 Ma ago in the Precambrian period. This theory has come to be known as the Southwest US-East Antarctica Hypothesis in a paper published in 2008 by an American researcher John Goodge and his team. The study funded by US National Science Foundation (NSF) significantly supports the theory of Southwest US-East Antarctica hypothesis. It says that around 600–800 Ma, a portion of the ancient continent, Rodinia, broke away from what is now Southwestern US and eventually drifted southwards to become East Antarctica and Australia as parts of Gondwana. A 1.4 billion-year-old chunk found recently has proven to be a piece of evidence linking Antarctica to western US (*The Antarctic Sun,* 2008).

3.4.2 *Paleozoic era (540–250 Ma)*

During the Cambrian periodic age, Gondwana had a mild climate. West Antarctica was partially in the Northern Hemisphere and during this period large amounts of sandstone, limestone and shale were deposited. East Antarctica was at the equator, where sea-floor invertebrates and trilobites flourished in the tropical seas.

By the start of the Devonian period (416 Ma), Gondwana was in more southern latitudes with a cooler climate. Though fossils of land plants are known from this time, sand and silts were laid down, forming the Ellsworth, Horlick and Pensacola Mountains.

By the end of the Devonian period (360 Ma), Gondwana was centered on the South Pole with cool climate and glaciations began. The flora, however, remained.

During the Permian period (250 Ma), Gondwana's plant life was dominated by fern-like plants such as glossopteris, which grew in swamps. Over time, these swamps became deposits of coal in the Transantarctic Mountains. Fossils of glossopteris plants found in various erstwhile Gondwana parts, form one of the principle evidences for establishing the existence of the supercontinent.

Continued warming witnessed dry and hot climate over much of Gondwana by the end of the Permian period.

3.4.3 *Mesozoic era (250– 65 Ma)*

Due to continued warming, the polar ice cap melted, rendering much of Gondwana a desert.

In East Antarctica, the seed fern became established and large amounts of sandstone and shale were laid down.

The Antarctic Peninsula began to form during the Jurassic period (65–46 Ma) and islands gradually rose out of the ocean. Ginko trees and cycads were plentiful as were reptiles such as lystrisaurus.

In West Antarctica, coniferous forests dominated through the entire Cretaceous period (146–65 Ma). Ammonites were common and dinosaurs were also present. It was during this period that Gondwana began to break up.

3.5 Breakup of Gondwana (160–23 Ma)

Africa separated from Antarctica around 160 Ma, followed by the Indian subcontinent around 125 Ma, the same time when Madagascar rifted from Africa. New Zealand, which was part of West Antarctica, rifted away around 85 Ma.

Around 65 Ma, Antarctica, still connected to Australia and South America, had a tropical to sub-tropical climate, complete with marsupial fauna. About 40 Ma, Australia-New Guinea separated from Antarctica, created latitudinal currents and isolated them. Around this time, the first ice began to appear in Antarctica.

Around 23 Ma, the Drake Passage was opened between the Antarctic Peninsula and South America. With this severance, geographically, Antarctica was totally isolated from rest of the world. This also resulted in circumpolar currents, accelerating ice formation. The ice began to spread, replacing the forests covering the continent.

It can be seen that the cooling of the continent occurred stepwise as the continental spread changed the oceanic currents from the longitudinal equator to pole temperature, equalising oceanic currents to latitudinal currents that preserved and accentuated the latitudinal temperature differences.

Since about 15 Ma, the continent has been mostly covered with ice, with the Antarctic ice cap reaching its present extension around 6 Ma.

3.5.1 *Fossil Evidence and Gondwana (Landis, 2001 and Trewby 2002)*

Fossil findings over the years have provided considerable evidence about Gondwana. Over 800 fossils have been found on Seymour Islands, off the northeast tip of the Antarctic Peninsula. They include a two-meter tall penguin weighing about 135 kg, a turtle the size of a Volkswagen Beetle car and abundant fossil remains of the glossopteris plant. A fossil of a small marsupial, Polydolops, proved that marsupials travelled from South America to Australia via the Antarctic land bridge during the Gondwana time.

The chronology of fossil evidence is very interesting as also clinching for scientific explanations. As early as 1829, American James Eighth found on South Shetlands, a piece of fossil wood, indicating that the islands had not always been cold and cheerless. Botanist Joseph Hooker was puzzled in 1840 by the similarities among plants in Isles Kergulen with those on Falkland Islands, Tierra del Fuego, Tasmania (Australia) and New Zealand. In 1851, he said he was becoming convinced that there was a great southern continent. Edward Suess wondered in 1885 as to how fossils of glossopteris plants in India, Africa and South America, looked so similar unless they were contiguous. But he was unable to offer the mechanism as pointed out earlier. In 1901, a scientist pointed out the striking similarities between rocks on Victoria Land and those in eastern Australia. He also pointed out the presence of marsupials both in Australia and South America, suggesting Gondwana. In 1912, Robert Scott found and a brought a fossil to his final camp where he perished. It was later identified as that of glossopteris, a legacy of Gondwana. Then came Wegener, his theory of continental drift, plate tectonics and a scientific explanation of the mechanism about geological evolution. Still, few felt that strong winds might have carried glossopteris and it necessarily did not indicate a single continent. Then, in 1967, American Peter Banet found in the Transantarctic Mountains a fossil jaw of a four-legged lizard-like amphibian. This amphibian was a shallow-water specialist incapable of long-distance swimming. This profound discovery was accepted as the final evidence of Gondwana.

In 1982, a rat-sized marsupial was found on Seymour islands, again proving that marsupials had indeed travelled from South America to Australia via the land bridge of Antarctica.

3.6 Present Day Geology of Antarctica

The geological study of Antarctica has been greatly hindered by the fact that nearly the entire continent is permanently covered with a thick layer of ice. However, new techniques such as remote sensing, ground-penetrating radars and satellite imagery have begun to reveal the structures beneath the ice.

From a geological evolution perspective, present-day Antarctica can be divided into two major areas, namely, greater or East Antarctica and lesser or West Antarctica.

3.6.1 *Greater or East Antarctica*

East Antarctica is the area bound roughly by the longitudes 30°W and 160°E and is geologically varied, dating from the Precambrian era, with some rocks formed more than three billion years ago. It is composed of a metamorphic and igneous platform which is the basis of the continental shield. On top of this base are various modern rocks such as sandstone, limestone, coal and shale laid down during the Devonian and the Jurassic periods to form the Transantarctic Mountains.

3.6.2 *Lesser or West Antarctica*

Geologically, West Antarctica closely resembles the Andes mountain range of South America. The Peninsula was formed by the uplift and metamorphism of sea-bed sediments during the late Paleozoic and the early Mesozoic eras. The most common rocks are andesite (a dark, fine-grained brown or greyish intermediate volcanic rock) and rhyolite (a fine-grained volcanic rock similar to granite with flow lines), formed during the Jurassic period. There is also evidence of volcanic activity, even after the formation of the ice sheet. The only anomalous area of West Antarctica is the Ellsworth Mountains region, where the stratigraphy is more similar to the eastern part of the continent. There are conflicting theories as to how it has moved from its likely original position on the margins of Greater Antarctica (near the West Coast of Coats Land or Droning Maud Land) to its present one (Walton 1987).

3.7 Potential for Natural Resources

The geological resources of primary interest are minerals and hydrocarbons. Based on the geological evolution and affinity with other continents that once formed the supercontinent Gondwana, predictions about the likely distribution of mineral resources are often made using the location of Antarctica in Gondwana as a guide, plus the traces found at various locations from time to time.

Precambrian rocks in the greater Antarctic shield have an affinity with those in South Africa, India and Australia where many rich metallic ores exist. In South Africa, the Proterozoic Witwatersrand System contains gold and uranium, while chromium, nickel, copper, platinum, iron and vanadium occur in the Lower Proterozoic Bushveld Complex, a layered gabbroic (a dark, coarse-grained rock from lava) intrusion. Iron and manganese are found in India in the rocks of the Achaean and Proterozoic Ages. The Australian shield contains nickel, gold and iron in the Achaean rocks and iron, copper, lead, zinc and silver in Proterozoic deposits. Paleozoic rocks in central and eastern Australia, comprising the Tasman belt, contain numerous but small mineral deposits. The Andean magmatic and deformational belt extends around the Pacific margin through New Zealand, Marie Byrd Land, the Peninsula and the Andes of South America. While the northern and central sections of the Andes are rich in mineral resources, the southern portion has few metallic ores. In New Zealand, most deposits are small and consist mainly of gold, tin and copper.

3.7.1 *Metallogenic Provinces*

Three metallogenic provinces – Greater Antarctic Iron Metallogenic Province, Transantarctic Metallogenic Province and Andean Metallogenic Province – are recognised in Antarctica as shown in Figure 3.3 (Walton and Doake, 1987).

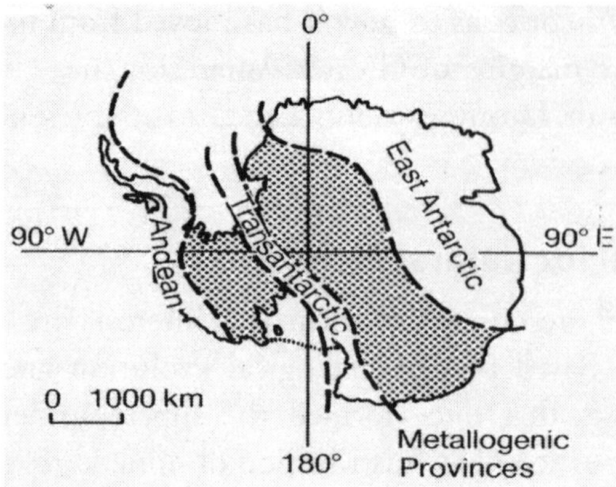

Fig. 3.3: Metallogenic Provinces of Antarctica

The Greater Antarctic Iron Metallogenic Province describes the extensive occurrence of iron minerals over the most exposed parts, from Wilkes Land to Droning Maud Land. The largest iron formation deposits occur in the Prince Charles Mountains. Aeromagnetic measurements suggest that the deposits extend up to 180 km under the ice. However, they are considered to be of lesser thickness and grade than the iron formations in Australia.

The Transantarctic Metallogenic Province has an affinity with the Bushveld complex in South Africa. Dufek Intrusion, mostly under the ice cap, spanning over an area of 50,000 sq km can be of interest due to the platinum group elements (PGE).

The Andean Metallogenic Province has an affinity with the Andean mountains in South America. Occurrences of copper and iron are most common, but minerals containing molybdenum, lead, zinc and silver are also found. The greatest likelihood of finding a commercial mineral deposit in Antarctica is thought to be in the Peninsula or Ellsworth Land, but the relative scarcity of minerals in the southern Andes may extend across Drake Passage.

3.7.2 *Traces of Mineral Findings*

The extent of minerals in Antarctica is not known and may not be known till proper exploration is done. However, this is an expensive exercise and does not seem feasible while the Madrid Protocol, banning all commercial mining, is in force up until 2048. However, numerous finds discovered so far in various parts of Antarctica can be summarised as shown in Figure 3.4 (May, 1989, and Google).

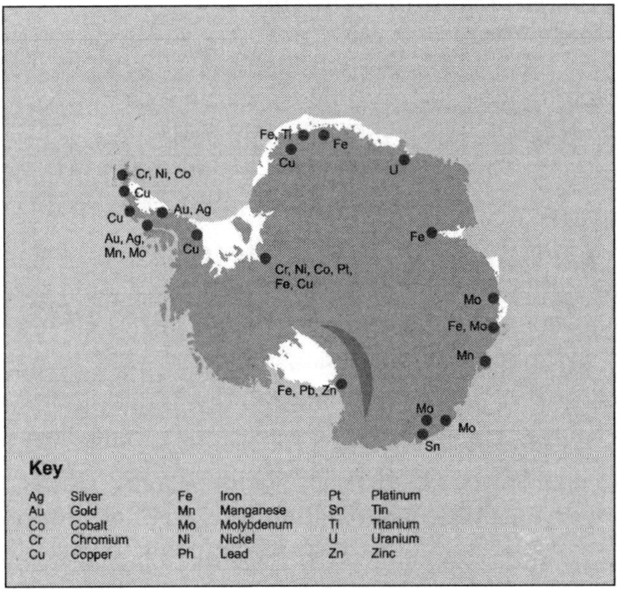

Fig. 3.4: Widespread Traces of Mineral Deposits in Antarctica (Low grade coal is present in the Trans Antarctic and shaded area in Prince Charles Mountains)

The main mineral resource definitely known on the continent is coal. Its presence was first recorded near the Beardmore Glacier by Frank Wild on Shackleton's Nimrod Expedition (1907–09), and now low-grade coal is known to be present across many parts of the Transantarctic Mountains. The Prince Charles Mountains, too, contain significant deposits of iron ore.

The most sought-after resources potentially lie offshore, in hydrocarbons (oil and gas). Considerable interest was generated in 1973 when they were reported in the Ross Sea area (Trewby, 2002). The extent of these, however, cannot be determined unless deep drilling is carried out. Since mining of all resources is banned as per the Madrid Protocol till 2048, except for scientific purposes, it may not be economical, as pointed out earlier, as in the case of other mineral resources.

3.7.3 *CRAMRA and the Protocol for Environmental Protection (Madrid Protocol)*

The thick ice cap and harsh environmental conditions have affected resource-exploitation issues and no efforts on a commercial scale have been made so far. In the 1980s, the Antarctic Treaty System (ATS) had prepared a Convention on Regulation of Antarctic Mineral Resources Activities (CRAMRA). It was signed in 1988, but could not come into force as mentioned earlier due to the refusal of some countriesto ratify it. Subsequently, the Protocol on Environmental Protection to ATS (Madrid Protocol) was signed in 1991 and came into force in 1998. It comprises a comprehensive protection plan for the Antarctic environment, specifically prohibiting commercial mining. The Protocol will be open for review in 2048. What happens then may depend on global population levels, effects of global warming, the availability or depletion of natural resources elsewhere in the world and the overall geopolitical situation at that time.

3.8 Geomorphology of Antarctica (Pathak, 2010)

3.8.1 *Soil Formation and Rock Patterns*

Due to the enormous ice cap, covering over ninety-eight per cent of the continent, very little soil is available for physical study. Soil formation is essentially an outcome of the type of rock and the effects of weathering. The soil patterns, therefore, emerge as a combined function of geology and geomorphology. It is influenced by various geomorphologic processes like climate, weathering and organic cycles. Rate of weathering, resultant physical and chemical characteristics and stability of the soil depend on the rock type. Proportion of fine and course material in the rock is largely determined by such factors as rock hardness, crystallinity and texture by virtue of rock-fracturing characteristics. Sedimentary rocks, for example, break down more easily into sand or silt-size particles than coarsely crystalline, hard, igneous rocks. Mineralogical differences between the rocks in Antarctica are also reflected in the soil through respective differences in their chemical properties.

3.8.2 *Land-Form Genesis*

The major land forms of Antarctica are glacial in origin, although they may have been eroded from initial fluvial topography. A good understanding of glacial history will be handy for understanding land-form development. The main Antarctic glacial influences have been the stable East Antarctic ice sheet, the fluctuating West Antarctic ice sheet and ice shelves responding to sea-level changes and independent Alpine glaciers.

Land forms are created or developed due to various processes such as glacial (covered with ice), fluvial (rivers and streams), lacustrine (inland wetlands or deep-water habitats from fresh-water lakes or reservoirs), aeolin (due to effects of wind), biotic (living organisms) and frost shattering (at freezing temperatures water enters cracks in exposed rocks during warmer day and freezes during colder nights). Along with this, tectonic activities also influence the pressure, volume and temperature. In the Transantarctic Mountain region when the uplift started, the continent was apparently not glaciated and, consequently, the mountainous topography was shaped by streams and rivers and the land form was essentially controlled by rock type.

Regions of relatively soft or complexly folded rocks, such as the North Victoria Land on Ellsworth Mountains, were finely textured and valleys followed a closely spaced dendrite (shrub or tree-like formations found on stones due to foreign material) pattern. Later, glaciations steepened the ridges and deepened the valley by the formation of curves before the topography was buried by a flood of ice where granite intrusive was present within the basement complex. They often stand out as mountain blocks, their peaks sculptured by shallow cirques (bowl-shaped, steep-walled mountain basins carved by glaciation often containing a small lake).

3.8.3 *Mountains*

In most parts of the Transantarctic Mountains, at Beacon Super Group rocks, the topography is one of steep slopes alternating with relatively level platforms. It becomes obvious once the ice cover is gone. In some ice-free areas, the topography is very reminiscent of South Africa and the land forms developed in an arid climate.

3.8.4 *Other Significant Landforms*

Volcanic landforms are exemplified by the McMurdo Sound region (40 Ma) and the inner core of mountain (Mt) Erebus on Ross Island which are still eruptive craters. Other cone-like land forms are Mt Byrd, Mt Terror and Mt Discovery. Some land forms are unmodified by glacial processes but some are modified by glacial erosion. Wright and Taylor Valleys remain unmodified and have identifiable craters, small cones formed of pyroclastic (fragments of rocks erupted by a volcano) bombs and flowing lava. In Marie Byrd Land, volcanic activity has destroyed the land forms and eroded hydroclastic (elastic due to water) or pyroclastic rocks.

When glaciations began, ice was formed in cirques (concave or circular basins carved by the base of a glacier), in small hollows and irregularities where ice could accumulate. A good example of cirque erosion occurs in Olympus Range, where the terrain appears to have gone through complete glacial erosion due to scarp retreat. Other examples are Asgard Range and the Hatherton Glacial area.

It is probable that rivers in the valleys gave way to ice due to glaciations and this overwhelming ice enlarged and deepened them. Many of them are ice-filled and it is only in the McMurdo Sound region that land forms become apparent.

The best example of the valley-form attained by glacial erosion is the Wright Valley, which is free of ice for 55 km of its length. Also a very broad U-shaped profile of the Upper Wright Valley can be recognised from a distance. Herein a special feature of labyrinth formation is also noticed, below which, two deep, cirque-headed canyons cut the valley floor. These canyons unite about 7 km down the valley, which then continues for a further 35 km as a broad valley, with an almost level floor. The lowest point on the valley floor (17 m above mean sea level) is on the bed of Lake Vanda and the valley rises gently to 277 m at the terminus of the Wright Lower Glacier.

Moraines are formed by rocks and debris carried down and deposited by glaciers. Many of the smaller alpine glaciers in ice-free areas are surrounded by loops, while ridges of lateral moraines often bind major glaciers. Moraines are made up of a variety of unsorted rock fragments in unbedded clay matrix. Some examples are found in Victoria and Taylor valleys also.

Drumlins are low, rounded hills, sometimes reaching a mile in length, found in the glaciated areas of Antarctica. Aligned in the direction of ice flow, their steeper, blunter ends point towards the direction from which ice gets collected. They produce a characteristic 'basket of eggs' topography.

Kames are isolated hills of stratified material formed from debris that fall into openings in retreating or stagnant ice. Kames terraces are benches of stratified material deposited between the edge of a valley glacier and wall of the valley.

Eskers are long, narrow and often breaching Simmons ridges of poorly sorted gravel and of sand formed by the deposition of former glacial streams.

In front of glaciers such as the Wright Lower or Victoria Upper, outwash plains or sanders are found surfaced with braided melt-water channels. The surface material is much sandier than the surrounding moraines, which usually contains a high proportion of coarser stones and boulders. In many places these features appear to be relict, the supply of debris or melt-water having ceased.

Kettleholes are depressions (sometimes filled by lakes) due to the melting of large blocks of stagnant ice, formed in any typical glacial deposit.

3.8.5 *Patterned Grounds*

Patterned grounds are essentially a pre-glacial environmental phenomenon. These are a very widespread feature in the Antarctic landscape in which the ground surface is covered by a more or less systematic pattern of polygonal features, generally of constant size in any locality usually found in a 3–4 m diameter. They have been studied in detail by Indian scientists around the Maitri station in the Schirmacher Oasis and Wolthot Mountains. One such patterned ground is shown in Figure 3.5 (Pathak, 2010).

Fig. 3.5: Patterned Ground on Schirmacher Oasis

3.8.6 *Continuing Research*

Continuing geophysical surveys reveal new discoveries from time to time. A gigantic chasm has been found under the ice sheet in the still largely unexplored Princess Elizabeth land. The chasm is 1000 km long, more than double the length of the Grand Canyon (450 km long). There is a possibility that this canyon network is connected to a previously undiscovered sub-glacial lake covering up over 1250 sq km of area (Jonathan, 2016).

A GEOGRAPHICAL PERSPECTIVE

3.9 General

The name Antarctica was derived as a fallout of the concept of Terra Australis (great southern land). It was believed, since the times of Ptolemy (first century AD) that a vast continent, in the far south of the globe, exists in balance to the known northern lands of Europe, Asia and Americas. The name Antarctica is also the Romanised version of the Greek compound word Ántarktike, feminine of Antarktikos, meaning 'Opposite to the North'. European maps continued to show this hypothetical land until James Cook crossed the Antarctic Circle on 17 January 1773 and came within 121 km of the Antarctic coast before retreating in the face of pack ice (Wikipedia, 2009, Antarctica).

Centered asymmetrically on the South Pole, Antarctica is an isolated land mass, surrounded by the Southern Ocean, unlike the Arctic which is an ocean basin surrounded by various land masses.

Its isolation from the rest of the world, a permanent ice cap and sea ice, circumpolar and katabatic winds, to name only a few aspects, have made Antarctica the coldest, windiest and highest continent with an entirely different environment from the one found elsewhere on earth. While the tip of South America is only about 1000 km from the Antarctic Peninsula, other parts of the erstwhile supercontinent Gondwana are much farther, with Australia, Africa and Indian sub-continent at about 2500, 4500 and 10,000 km respectively.

At 13.209 million sq km, area-wise, Antarctica is the fifth largest continent on earth after Asia (44.579), Africa (30.065), North America (24.265) and South America (17.819) and larger than Europe (9.938) and Australia (7.687).

The factor which makes Antarctica totally different, precluding any permanent human inhabitation, is the massive ice cap covering nearly its entire land mass. The enormity of the weight of the ice cap can be seen from the fact that it has nearly doubled the original area of the continent depressing the land mass below by about 600 m. Antarctica without an ice cap and with its present ice cap is seen is shown at Figure 3.6 (May, 1989).

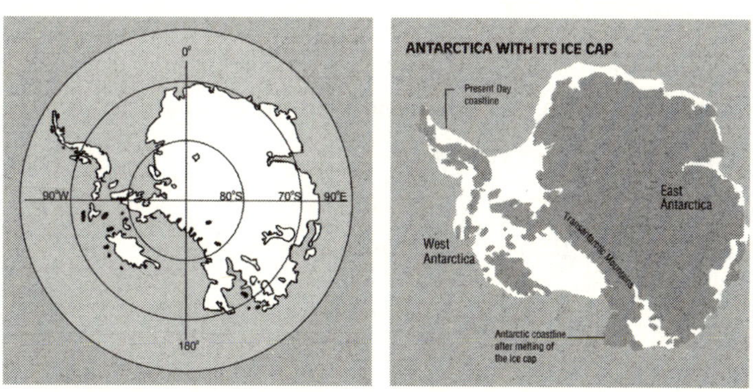

Fig. 3.6: Antarctica without Ice Cap*(L)* and with Ice Cap*(R)*

3.10 Major Geographical Features

The geographical features of Antarctica are wide-ranging and the environmental conditions at the coast, mountains, valleys, oases, lakes, polar plateau, vary considerably both location-wise as also depending on the seasons – summer or winter. All the features greatly influence the continental conditions as described below. Some of the features are as indicated in Figure 3.7 (Google).

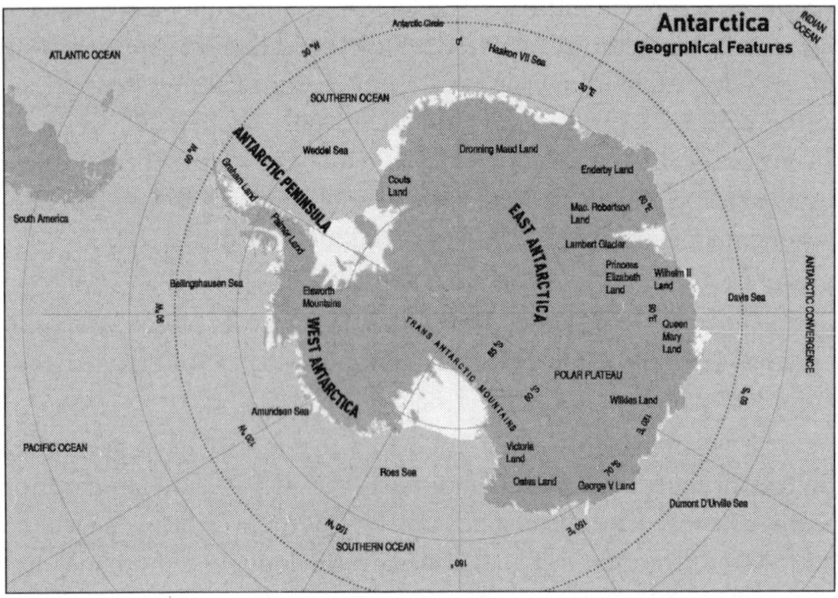

Fig. 3.7: Major Geographical Features in Antarctica

3.10.1 *Antarctic Coast*

The coastline of the continent measures about 17, 968 km and is mostly characterised by ice formations as under (Landis, 2001):

Ice shelves (floating ice front)	:	44%
Ice walls/cliffs (resting on ground)	:	38%
Ice streams/outlet glaciers (ice front or ice walls)	:	13%
Promontory (high areas of land or rock that stick out in the sea)	:	5%
Total	**:**	**100%**

3.10.2 *Regions of the Continent*

Antarctica is divided into two unequal parts by the Transantarctic Mountains, close to the neck between the Ross Sea and the Weddell Sea. The portion west of the Weddell Sea and east of Ross Sea is called West or Lesser Antarctica and the remainder is called East or Greater Antarctica. Roughly speaking, they correspond to the Western and Eastern hemispheres relative to the Greenwich Meridian. Ice sheets covering the East and West Antarctica regions occupy about eighty-eight per cent and twelve per cent of the total Antarctic ice respectively.

Whereas West Antarctica comprises of Palmer Land, Mary Byrd Land, Edward Land, to name the major areas, East Antarctica comprises of Coats Land, Queen Maud Land, Enderby Land, Mac Robertson Land, Wilkes Land, Victoria Land, to mention a few. Geographical and Magnetic South Poles are also located in East Antarctica.

3.10.3 *Mountains, Nunataks and Oasis*

Although much of Antarctica is overwhelmed by ice, mountainous outcrops of rocks, free of ice, protrude from the white surface cover in many parts of the continent. In some cases, only isolated spires of rock called nunataks appear above the ice, in others, long spines of rocks form the exposed backbone of partly hidden mountain ranges. The largest mountain chain, the Transantarctic Mountains, extends for about 3200 km, separating the two regions. The sentinel range of Ellsworth Mountains is the tallest in Antarctica, and includes Vinson Massif which reaches 4887 m above the sea level. Many smaller mountains are also spread all over the continent like Dry valleys, Horlick Mountains, Prince Charles Mountains, to name a few.

Exposed rock has a significant impact on the local climate. Over summer, rocks absorbs heat which evaporates any snow falling on the rock surface. This, along with strong katabatic winds, enables the rocks to remain ice-free. There are about twenty such ice-free areas in the continent called oases. The name is a misnomer as they are some of the coldest and driest places on earth, unlike the fertile oases of hot deserts else where in the world. Covering less than two per cent area of the continent, they are, however, very significant for scientific research which can only be carried out in an ice-free land.

3.10.4 *Dry Valleys*

Dry Valleys in Antarctica are areas that have a significant proportion of rock exposure. They are developed in the McMurdo area of East Antarctica and form part of the oases.

Three Valleys, Victoria, Wright and Taylor, are located in South Victoria Land with a combined area of about 2500 sq km. The Valleys run from and through the Transantarctic Mountains to the Ross Sea Coast. Known for strong winds, extreme cold temperatures (a mean annual floor temperature of minus 20°C) and dryness, the landscape is a mosaic of rock, gravel and ice. Snow that falls is invariably sublimated, evaporated by katabatic winds and the area receives less moisture than the Sahara Desert. This landscape is closest on earth to that of Mars. The US utilised the area to train astronauts and carried out experiments before launching probes to Mars and Venus (Trewby, 2002).

The ecosystem in Dry Valleys clings to existence in the rocks, lakes and small melt-water pools and streams. It represents life at its environmental limits. The microorganisms, mosses and lichens found here are among the oldest life forms on earth. Due to large-scale indiscriminate activities, a glaciologist from New Zealand, Mr T.J.H. Chinn, had warned in 1990 that the area has evolved from pristine desolation (complete emptiness) to scientific despoliation (actions to plunder). Such concerns have prompted the development of an environmental monitoring and scientific 'code of conduct' for the Dry Valleys (McGonigal, 2008).

3.10.5 *Lakes*

Both fresh-water and saline-water lakes occur mainly in the oases areas around the continent. It is here where most of the research stations are located. Many glaciers and glacial lakes, Lake Washburn being prominent amongst them, occupy significant area of the Taylor valley.

Lake depths vary from a few centimeters to over 60 m. The lakes are mostly capped by a permanent ice cover between 2 to 5 m thick. In summer, the ice melts at the shoreline, forming a moat. Lakes that freeze more slowly than the sediment around them have internal pools of water that eventually freeze and form ice domes on the lake surface.

Fresh-water lakes are fed by glacial melt water during summer, whereas saline lakes are located in closed basins. Both types support life which is thought to be one of the oldest life forms on earth and possibly one of the first sources of oxygen on the planet.

3.10.6 *Sub-Glacial Lakes*

Antarctica is home to more than 145 sub-glacial lakes which lie hidden below a thick layer of the Antarctic ice sheet. Lake Vostok, discovered beneath Russia's Vostok Station in 1996, is the largest of these sub-glacial fresh-water lakes. Study on these lakes is done using satellite surveying and radar probing of ice. A recent survey suggests that many lakes are interconnected and every so often, there are large flows of water from one lake to another which affects structure and movement of glaciers. A major breakthrough was achieved when after decades of ice-core drilling, Russian scientists accessed the waters of Lake Vostok on 6 February 2012, beneath 4 km of densely compacted ice. Considerable experiments in this regard from the UK, US and Russia are underway at Lakes Elseworth, Williams and Vostok with different drilling techniques to ensure that the lakes, deep down the glacier ice, are not contaminated in the process (Jones, 2012). Considerable research is likely to be done on this subject in coming years.

3.10.7 *Volcanoes*

Volcanoes form where hot mantle rises between rifts in the continental plates beneath Antarctica. Some of the magma is stored in chambers beneath the crust. When it reaches the surface, it erupts from volcanoes as lava or volcanic ash. There are about seventy extinct volcanoes in Antarctica.

Four volcanoes on mainland Antarctica are considered to be active on the basis of observed fumarolic (escaping of smoke and gas from a hole) activity or recent tephra (fresh-water fish) deposits. Whereas three of them, Mt Melbourne (2730 m), Mt Berlin (3500 m) and Mt Kauffmann (2365 m), are stratovolcanoes (alternate layers of lava and ash), Mt Hampton (3325 m) is a volcanic caldera (crater).

Several volcanoes on offshore islands have records of historic activities. The southernmost active volcano in the world, Mt Erebus, a stratovolcano on Ross Island is very active. It erupts ten times daily but it is rare for lava to be thrown beyond the crater. However, in 1984, gas trapped under the solid lava crust was released and a series of cannon-like explosions showered lava bombs 3 km beyond the crater rims. It also supports the only Antarctic algae that need high temperatures to survive. A volcanic caldera in Deception Island is also most active. Volcanoes on the Buckle Island in Balleny Islands, Penguin Island, Paulet Island and Lindenberg Island are also considered to be active in varied degrees.

3.11 Southern Ocean

Different organisations and nations have differing points of view over the existence and extent of the Southern Ocean, though mariners have long accepted the term 'Southern Ocean' as traditional and used it accordingly (Wikipedia, 2010, Southern Ocean).

3.11.1 *Definition and Extent*

The Southern Ocean, geologically the youngest of the oceans, was formed when Antarctica and South America moved apart, opening the Drake Passage, roughly around 23 Ma.

In 1937, the International Hydrographic Organisation (IHO), for the first time, defined an ocean around Antarctica, in its publication, *Limits of Oceans and Seas,* as Southern Ocean. Skipping the issue in 1953 (third edition), the IHO readdressed the question of the ocean's existence in 2000 (fourth edition). It defines the Southern Ocean to be the southernmost waters of the world, south of 60°S latitude and encircling Antarctica. Southern Ocean includes the Antarctic Circumpolar current, the Amundsen Sea, Bellingshausen Sea, parts of Drake Passage, Ross Sea, Cooperation Sea, a small part of the Scotia

Sea and Weddell Sea. The total area of the Southern Ocean is about 20.327 million sq km and it is the fourth largest of the five recognised world oceans, smaller than Pacific, Atlantic and Indian Oceans, and bigger than Arctic Ocean.

Though there was a broad consensus about the name and limits, the fourth edition was never published due to certain reservations from Argentina and Australia over the ocean's definition and extent respectively. Some geographers also do not regard the Southern Ocean as a separate ocean and consider its waters as part of the South Pacific, South Atlantic and Indian Oceans. Others regard Antarctic Convergence (a circular belt of water about 40 m wide between latitudes 48°S and 60°S, where the cold Antarctic water meets the oceans' warm waters flowing south) as the northern boundary instead of 60°S latitude or 60th parallel, as commonly called. The UK uses 55th parallel and Australia defines Southern Ocean as waters to include the entire body of water between Antarctica and the south coasts of Australia and New Zealand, although the latter does not follow suit.

Despite these differing perceptions and practices, the fourth edition definition has de facto usage by many organisations, scientists, nations and even by IHO itself.

3.11.2 *Features*

The depth of the Southern Ocean varies between 4000 m to 5000 m over most of its extent with some limited areas of shallow water. The Antarctic continental shelf appears generally narrow and unusually deep, its edge lying at depths up to 800 m, compared to a global mean of 133 m. Its greatest depth of 7235 m occurs at the southern end of South Sandwich Trench, at 60°S, 24°W.

The Antarctic circumpolar current, started as a consequence of opening of the Drake Passage, moves perpetually eastwards, chasing and joining itself, and at 24,000 km in length, it is the world's longest ocean current and transports about 130 million cubic m of water per second – a hundred times the flow of all the rivers in the world put together.

The area of Antarctic pack ice, fluctuates from an average minimum of 4 million sq km in March to about 20 million sq km in September, the increase being more than the area of the continent itself with total area equal to the area of the Southern Ocean.

3.11.3 *Climate and Storms*

Sea temperatures vary from about minus 2°C to 10°C. Cyclonic storms travel eastward around the continent and frequently become intense because of temperature contrast between the ice and open sea. The ocean area from 40°S to 67°S (the Antarctic Circle) has the strongest average winds found anywhere on earth. In winter, the ocean freezes till around 65°S in the Pacific sector and around 55°S in the Atlantic sector, lowering surface temperature to well below 0°C. At some coastal points, intense persistent drainage winds from the interior keep the shoreline ice-free throughout winter.

3.11.4 *Shipping and Natural Hazards*

Icebergs can occur at any time during the year. While the bigger ones may be up to several hundred meters, smaller icebergs, iceberg fragments and sea ice (generally 0.5 m to 1 m thick) pose problems for ships. The deep continental shelf has a floor of glacial deposits, varying widely over short distances.

As seen earlier, sailors famously knew the Southern latitudes from 40°S to 70°S as the 'Roaring Forties', 'Furious Fifties' and 'Screaming Sixties' due to changing sea and wind conditions. The absence of nearby land and the presence of pack ice, especially from May to October, make the area dangerous and sources of search and rescue scarce.

3.11.5 *Ports and Harbours*

Few ports or harbours exist on the southern coast of Antarctica, since ice conditions limit use of most shores to short periods in midsummer. Most of these ports are operated by government research stations and remain closed to commercial or private vessels.

Major operational ports include the Rothera Station (UK), Palmer Station (US), Esperanza Base (Argentina), Mawson Station (Australia) and McMurdo Station (US). However, many times, due to sea and wind conditions, ports are not available to most of the stations and they have to look for berthing places on the ice shelf or offshore anchorages. This aspect is discussed in detail in later chapters.

3.11.6 *Current Environmental Issues*

Increased solar ultraviolet radiation from the Antarctic Ozone Hole has reduced marine primary productivity (phytoplankton) by as much as fifteen per cent and started damaging the DNA of some fish. Illegal and unregulated fishing, especially the landing of an estimated five to six times more Patagonian toothfish, affects the sustainability of stock and high incidences of seabird mortality (McGonigal, 2008).

3.11.7 *International Agreements*

All international agreements regarding world oceans apply to the Southern Ocean. Antarctica, however, is subjected to certain agreements, specifically for its environment like the International Whaling Commission (IWC), prohibiting commercial whaling south of 40°S, the Convention for Conservation of Antarctic Seals (CCAS) and the Convention for Conservation of Antarctic Marine Living Resources (CCAMLR). The effects and efficacy of these are discussed in a later chapter discussing the Antarctic Treaty.

3.12 Antarctic Islands

In the Southern Hemisphere, oceans occupy most of the area, except for a few islands, from 40°S to 70°S where the Antarctic ice shelf begins. Thus, water hugely dominates the earth's surface in this area. Northwards from Antarctica's coast and based on air temperatures and type of vegetation, islands in this area can broadly be classified into four groups. The first and the second group are called coastal and maritime islands. They can be called true Antarctic islands. Their climates are dominated by their proximity to the continent and to the sea ice. The third and the fourth group are sub-Antarctic and temperate isothermal islands which are not directly affected by the continental conditions, but more so by the characteristics of the surrounding ocean waters and latitudes.

The islands located near the coasts of South America, Southern New Zealand and Tasmania (Australia) are not from any of the above category.

The islands and the four groups are shown in Figure 3.8 (Reader's Digest, 1985).

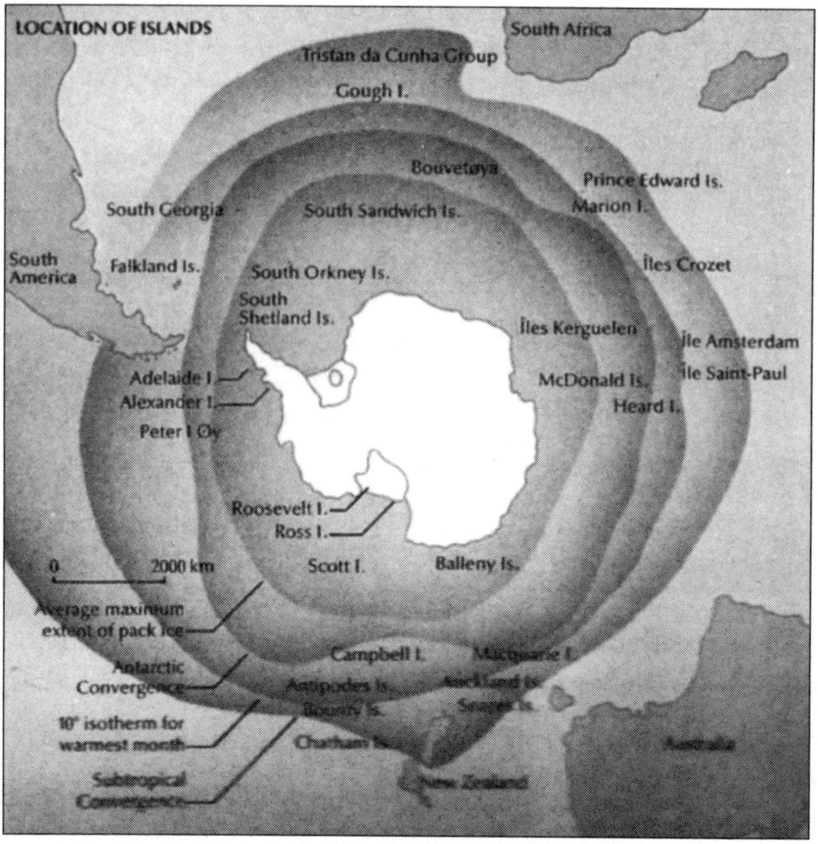

Fig. 3.8: Islands Surrounding the Continent of Antarctica

3.12.1 *Antarctic Coastal Islands*

This group includes Adelaide, Alexander, Ross, Roosevelt and a few other islands.

Under the direct influence of cold, dry air from the continent, they are more or less ice covered and beset by pack ice for much of the year. The mean annual air temperatures are minus 9°C or lower. Precipitation falls as snow and rain is unknown. Conditions are very similar to those on the nearby mainland.

Little rock is visible and there are very few plants. Ice-free, rocky areas form large summer breeding colonies of sea birds and several species of penguins.

3.12.2 *Antarctic Maritime Islands*

This group includes Bouvetoya, South Orkney, South Shetland, South Sandwich, Peter I, Scott and Balleny Islands.

These islands are subjected to the maritime influence of the Southern Ocean and are surrounded by pack ice in winter. The mean monthly air temperatures rise above freezing for only short periods in summer and mean monthly temperatures vary between minus 7°C to freezing point. Only two species of flowering plants are found here and deep peat banks are formed by slow moss growth.

The islands support a huge number of breeding penguins and petrels.

Scientific stations are situated on many of the islands near the peninsula.

3.12.3 *Sub-Antarctic Islands*

These include South Georgia, Heard and McDonald Islands.

These wet, windy islands lie near the Antarctic Convergence. The mean annual temperatures at sea level are between freezing point and 3°C. Coasts are seldom closed in by pack ice, but they have ice caps and glaciers that descend to sea level. Large parts are ice covered and furious katabatic gales are experienced.

The native land vegetation consists of coastal tussock grassland and wet, peaty herb fields. There are no woody plants and attempts to establish cold-tolerant trees and shrubs have been unsuccessful.

A number of elephant and fur seals breed ashore in summer.

3.12.4 *Temperate, Isothermal Islands*

Iles Kerguelen, Iles Crozet, Macquarie, Marion, Falklands, Diego Ramirez, Campbell, Auckland, Antipodes, Bounty and Snares Islands are included in this group.

All these lie between the Antarctic and Subtropical Convergences and are away from the influence of the dry, cold Antarctic air masses. They have extremely equable, wet, stormy and cloudy climate. The mean annual air temperatures range from 4°C to 10°C. Though Kerguelen has a small ice cap there is no permanent snow or ice on other islands. Most precipitation falls throughout the year as rain or sleet.

Most of the islands have tussock grass and treeless peaty uplands. The coasts are free of ice and more or less surrounded by kelp. Islands near New Zealand are warmer. Extensive thickets of low, woody vegetation are found on the Campbell, Snares and Auckland islands.

Breeding colonies of seals and penguins are found near beaches on most islands. The largest breeding population of the Great Albatross is found on Auckland Islands.

There are some unique features within the geographical environment of all these islands like the 'reverse waterfall' of Auckland Islands, deep fiords and lakes created by glacial action, geological origins like buckling of oceanic ridges (Prince Edward Island), volcanic hot spots resting on oceanic plateau (Kerguelen, Heard and McDonald Islands) and the only area in the world where rock from deep within the earth's mantle has been exposed by plate tectonics (Macquarie Islands) (K Scott, 1993).

On many islands, introduced animals have ravaged the native vegetation and wildlife. Rats devour birds' eggs, and rabbits, pigs and goats, brought as food for sailors and sealers, destroy fragile plant species. House cats were earlier introduced to control rodents or came through shipwrecks.

3.13 Antarctic Ice

Out of all the features of Antarctica, the massive ice cap covering the continent can be described as its single most unique and visible feature. According to astronauts, from space, Antarctica 'radiates light like a great white lantern across the bottom of the world'. It is also a repository of millions of years of Antarctica's and, to some extent, the world's environmental and climatic history.

3.13.1 *Ice Statistics*

How ice dominates the continent of Antarctica and its surroundings can be seen from the statistics given in Table 3.1 (Reader's Digest, 1985; Walton, 1987; IPCC, 2007; McGonigal 2008).

Table 3.1: Facts about the Antarctic Ice Cover

Area of continent without ice	:	7 million sq km
Area of continent with ice cap	:	13.209 million sq km
Percentage of area free of permanent ice cover	:	2%
Volume of total ice covering over 98% of the continent	:	32.4 million cu km
Out of the total ice as above, the amount of ice in East and West Antarctica	:	88% and 12% respectively
Percentage of Antarctic ice in total ice in the world	:	90%
Percentage of Antarctic ice in total fresh water in the world	:	61%
Depression of rock structure due to weight of massive ice cap	:	600 m
Average elevation of the continent over mean sea level	:	2300 m
Minimum thickness of ice over ground	:	800 m
Maximum thickness of ice over ground	:	4800 m
Average thickness of ice over ground	:	1600 m
Likely rise in sea level if all the Antarctic ice melts	:	57 m
Minimum area of sea ice during March	:	4 million sq km
Maximum area of sea ice during September	:	20 million sq km
Percentage of coastline free of ice (mostly in the Peninsula)	:	5%
Total ice calved per year in the form of icebergs	:	1450 cu km or 50% of the world's fresh-water usage.
Move of the pack ice edge from the coast or fast ice once the freezing starts	:	4.2 km per day
Increase of ice coverage per day in the ocean once the freezing starts	:	100,000 sq km per day

3.13.2 *Formation of Glaciers*

Out of all the Antarctic ice, ninety-five per cent of it is glacier ice. It is basically a densely packed ice which takes hundreds of years to form from snow. Glaciers move from mountains to sea. The movement is extremely slow but very powerful and crushing and it reshapes the landscape while moving, also taking detritus to the ocean. At the South Pole, the thickness of the ice sheet is about 3 km and it moves at the speed of 10 m per year. Therefore, a marker placed to indicate the exact location of the South Pole needs to be relocated every year as mentioned earlier.

Most of the ice is formed by the process of firnification. Initially, the snow falls in the form of hexagonal low-density crystals with a lot of air space between them. Gradual accumulation builds weight and pressure, which breaks the edges of the crystals, making them round. Consequently, the air space is closed off and the ice gets consolidated. This stage is known as 'neve' or 'firn'. As the ice still contains air bubbles, it looks white in colour. As it is further compressed, all the air is squeezed out and it turns into clear, hard impregnable ice known as glacier ice. It often appears blue as it is not able to filter out the blue section of the light spectrum.

The time taken for snow to become ice varies, as does the depth at which this process occurs. On the ice shelves, glacier ice forms relatively quickly, in about 200–300 years at a depth of about 50 m. In inland areas, the process occurs much more slowly because precipitation is considerably lower. At South Pole, glacier ice is formed at a depth of about 100 m over a period of thousand years or so.

3.13.3 *Polar and Sub-Polar Glaciers*

Most Antarctic glaciers are located on the edges of the ice sheets and in Dry Valleys. Mainly, they are of two types of glaciers: Polar or Dry-based and Sub-polar or Polythermal. Polar or Dry-based glaciers are small Dry Valley glaciers which occur when the total thickness of ice is below melting point throughout all seasons. In these, the base is welded to the bedrock and sliding is possible only through deformation, which occurs when ice grains shear past each other. Sub-polar or Polythermal-type glaciers are large outlet glaciers that drain the Antarctic ice sheet and the glaciers on the Peninsula area. The weight of this type of glacier is enough to generate sufficient shear stress or friction for melting point to be reached at its base. This creates a thin layer of water allowing the glacier to slide or move forward on its rocky basal bed. These glaciers move more quickly than the Dry-based ones.

The largest glacier in the world, known as Lambart, is 40 km wide and moves for 400 km through Prince Charles Mountains. At its lower reaches, it merges with the floating Amery Ice Shelf, and flows for a further 300 km to the seaward margin. It moves at about 230 m per year at the mountains and 1 km per year at the ice-shelf front, where it is nearly 200 km wide.

The West Antarctic ice sheet is more dynamic and responds more quickly to climate changes than its eastern counterpart. If this ice melts, the land beneath may become buoyant and rise up again over thousands of years.

3.13.4 *Crevasses*

As a glacier moves, the flow within it varies. The beginning and the end, generally, move faster than the centre and there are different flow rates between various levels of ice. This variation creates stresses in the ice which produce cracks or crevasses. The crevasses range in size from a few millimeters to over 30 m across and occur mainly in the brittle top crust. At a depth of 30 m, the ice is under so much pressure that it deforms plastically and does not crack. Crevasses form across the direction of the ice flow due to sudden changes in the slope of the underlying bedrock, which results in the glacier speeding up. Longitudinal ones occur wherever a glacier is advancing at more than one speed. Most glaciers contain both sorts of cracks to make a crisscross pattern. Traversing glaciers is, therefore, a very dangerous manoeuvre, especially as the cracks are concealed by often-fragile snow bridges.

3.13.5 *Ice Streams or Rivers*

Dynamic, rapidly flowing rivers of ice or ice streams pull ice from the interior of the Antarctic ice sheet into the ice shelves. However, the name 'river' is a misnomer here as the flow is not like water, but the speed is much faster than the speed of glaciers on both sides. Although accounting for only ten per cent of the ice sheet's volume, ice streams are sizeable features, up to 50 km wide, 2 km thick and hundreds of kilometres long. Nearly ninety per cent of ice flowing across West Antarctica converges into three main drainage systems – ice streams flowing into Ross and Ronne Ice Shelves and one that flows directly into Amundsen Sea. They move much more rapidly as compared to the flow of the surrounding ice and at a speed of 1 km per year.

3.13.6 *Ice-Tongues*

At the coast, the terminal of a glacier may form a tongue of ice that protrudes into the ocean. Periodically, parts of the tongue sheer off to form icebergs.

3.13.7 *Ice Sheet over the Continentand Continuing Research*

The ice sheet over the continent looks as shown in Figure 3.9 (Reader's Digest, 1985).

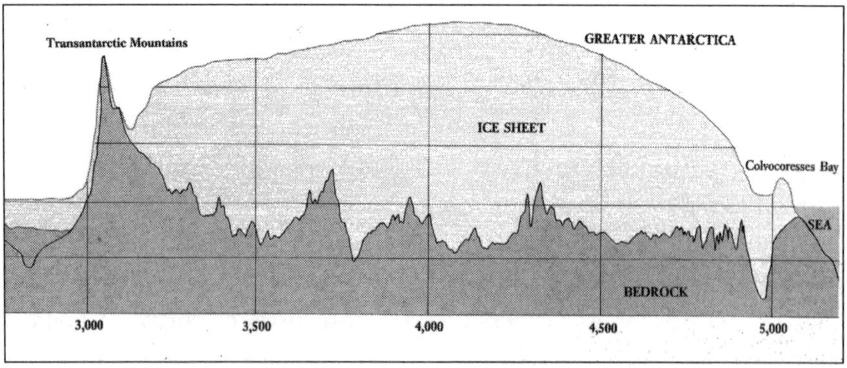

Fig. 3.9: Thickness of the Ice Sheet over East Antarctica

The massive Antarctic ice cap is a major area of attention for researchers. The study of the back and forth movements of glaciers during their journey, their eventual calving as icebergs and glacial earthquakes in the Greenland ice cap in the Arctic reveal that glacial earthquakes have increased seven-folds in last two decades with an increased mass of ice loss. It is of relevance to the ice cap in Antarctica (T Murray & others, 2015). Another study reveals that Antarctic snow accumulation, which began 10,000 years ago, is currently adding enough ice to outweigh the loss from glacier ice. The satellite data shows a net gain of 112 billion tons during the years 1992 to 2001 reduced to 82 billion tons during 2003 to 2008. As of now, this challenges the IPCC 2013 report that Antarctica is overall losing land ice. However, this indicates that the losses may catch up with the gains in the near future (Zwally & others, 2015). Another study points out that giant atmospheric rivers (long, narrow water vapour plumes stretching thousands of kilometres across the sky over vast ocean areas) add mass to the Antarctic ice sheet (Goridetskaya, 2014). This trend of new research findings will continue.

3.13.8 *Gravity or Katabatic Winds, Blizzards and Sastrugi*

At the South Pole, the total radiation received by the sun is much less as rays strike at a more oblique angle here than at other parts of the earth. Also, radiation reaching the surface is reflected back by the ice cap. Near the surface and in stratosphere, cold air is transported away from Antarctica and, in between, warm air is transported from the equator to the South Pole. Here it loses its latent heat as water vapours condense and freeze. This saves the ice cap from becoming progressively colder. This inversion layer traps cold air at the surface. Heavier than the air above, this layer flows downhill as a light breeze northwards from South Pole and picks up speed as it glides down. Therefore, this is known as a gravity or katabatic wind. The word is from the Greek *Katabatikos,* meaning, to go down or downwards. The cold air is dense and it flows down the sloping polar ice domain in all directions. As it reaches the coast 1600 km away, in a layer around 300 m, it gathers tremendous speed of up to 300 km an hour. The coastal pack ice is broken and in interaction with warmer, moister marine air, heavy low-lying clouds are formed at places. The winds also create moats and polynyas and impact the extent of pack ice, making their realistic studies difficult. Blizzards are a combination of this strong katabatic wind and falling or blowing snow, particularly in winter. They can reduce visibility to as low as one meter, bring all outside work to a halt and can cause considerable damage to unprotected machines and buildings.

Strong winter winds of Antarctica combine with heavy coastal snowfall to produce dramatic landscapes. These winds erode the snow surface, sculpting it into a corrugated sea of irregularly shaped forms like ridges and ripples. These grooves or ridges are known as satstrugi and they make movement very difficult across Antarctic ice.

3.13.9 *Ice Shelves*

The vast mass of the interior ice sheet meets the shore in the form of massive, awe-inspiring ice shelves and impregnable grounded ice cliffs. Ice shelves are floating slabs of land ice that rise and fall with the ocean tide and, in the words of Steven J. Pyne, form a 'kind of crystalline exoskeleton that shapes, binds and protects the icy continent' and slow the movement of glacier ice from inland polar plateau to the ocean.

The critical boundaries of an ice shelf are the grounding lines or hinges where a glacier or ice sheet begins to float and the ice front where the shelf disintegrates in the form of icebergs. There is a constant balance between ice flow from the continent, ice added by snowfall and bottom freezing, and ice lost by calving of icebergs and bottom melting. The stability of an ice shelf rests on maintaining this equilibrium. At present, the balance is slightly in favour of ice gain rather than ice loss.

Where an ice shelf passes over the rocky shoals below sea level, it forms dome-shaped ice rises or low ice rumples. These act as spot welds on a car, anchoring the ice shelf in place. Without these pinning points, the inland ice sheet can push some of the ice shelves out to sea. These anchors, however, also create instability as they create fractures and stresses within the shelves. Due to these anchors, the entire ice-shelf structure is riven by crevasses and its interior ice is infiltrated by brine.

The Ross Ice Shelf is the largest ice shelf of Antarctica. It is roughly the size of France and twice the size of New Zealand, and it alone accounts for thirty per cent of the Antarctic shelf ice area. It was first sighted by Sir James Clark Ross, who named it Great Barrier. In many books of old vintage, ice shelves were referred to as great barriers. It is continuously replenished by five ice streams and numerous glaciers.

Other major ice shelves include Amery, which is formed where the Lambart Glacier empties into an enormous fjord, Filchner and Ronne Ice Shelf.

3.13.10 *Ice Bergs*

When ice is calved from the Antarctic ice sheet or ice shelves or from glaciers meeting the ocean at the coastline, it is known as an iceberg. Nearly four fifths of icebergs are calved from ice shelves that make up for just over forty per cent of the coastline. About 1450 cu km of ice is calved off every year, equivalent to about half the world's fresh-water usage.

Thousands of icebergs are calved off every year in various shapes like tubular, rounded or irregular. Whereas tubular icebergs are big and calved off from ice shelves, round and irregular ones are smaller and are calved off from ice cliffs and floating glaciers. Usually, icebergs are large, measuring around 200–300 m in thickness and can tower ten storeys above the ocean surface. In other words, they can be gigantic. An iceberg sighted in November 1965 was 335 km long and 97 km wide, the size of Belgium, and another sighted after breaking off from the Ross Ice Shelf in March 2000 was 300 km by 40 km (McGonigal 2008).

Whereas most icebergs reflect blue light as they are from glacier ice, some are green or jade and contain organic material which reflects green light due to the sea water frozen on the underside of ice shelves as marine ice. In the same way, stratification or layers of compacted snow, white bubbly ice or clear blue ice from ice shelves and glacier tongues are exposed to form striped icebergs. Their variety of shapes and colours make them photographers' delight in the Southern Ocean.

Once calved, many icebergs remain grounded on the ocean floor and stay close to the continent, lasting many years. Once they are detached, they move northwards, breaking and melting as they move. Due to circumpolar currents, they drift at a speed of 20 km per day. Occasionally, they cross the Antarctic Convergence and melt in warmer waters. In 1894, an iceberg was spotted as north as 26°S. They often take years to melt but are riven by internal fractures as they age and break into smaller icebergs.

Regarded by caution and fear by earlier sailors, icebergs also inspire awe. They can pose serious danger to ships. Icebergs when they are between two and five meters are called bergybits, and decayed fragments, smaller than two meters, are known as brash ice.

The many true or imaginary stories about icebergs hold a fascination for those visiting Antarctica. One such story tells of how an iceberg the size of one cubic kilometre towed to Saudi Arabia arrived in a grand size of one cubic centimeter in the form of an ice cube in a refrigerator. Later, while serving in that country for few years with an automobile company, my colleagues used to ask me to repeat this story of the towing, arrival and final cost of the ice cube to solve the drinking water problem there.

3.13.11 *Ice caves*

When moving ice pushes against an obstacle, a space may be created in the lee of the obstacle. Beneath glaciers, water tunneling through the ice creates sculptured and smooth cavities. In icebergs, wave action can cause similar cavities. These cavities are called ice caves. They present spectacular views in shades of blue, azure and violet colours.

3.13.12 *Sea Ice*

By the end of March each year, as temperatures start to fall after the brief summer season, the sea begins to freeze around all the Antarctic coast to form pack ice. The area of sea-ice is a minimum of around 4 million sq km in March and a maximum of about 20 million sq km in September. Unlike the Arctic where the sea-ice grows to about 3 m thick, under-formed Antarctic sea-ice usually reaches about 1 m in thickness because of disturbances from wave action and ocean currents and because it rarely lasts for more than one year. On an average, the advancing edge of the sea ice moves 4.2 km a day and the total ice cover increases by about 100,000 sq km a day and the final extent reaches as far north as 61°S. Distinctly different than glacier ice, sea-ice is saline and formed by crystal growth rather than from compressed snow.

Sea-ice acts as a barrier to the energy exchange between the atmosphere and the ocean. While in winter it insulates the ocean, it reflects eighty per cent of solar radiation in summer, delaying the warming effect of the sun. When sea-ice melts, it cools both the atmosphere and the ocean. This, in turn, limits the supply of water vapour to the atmosphere, and is one reason why so little moisture reaches the continent.

3.13.13 *Packice and Subtle Variations*

The transformation of the ocean's surface water into sea ice surrounding the continent, to reach the maximum limit of 20 million sq km, undergoes some interesting stages before we witness seemingly uniform frozen sea, generically called pack ice.

When the sea, being saline, begins to freeze at around minus 1.8°C, the first crystals to form are tiny spheres of pure ice. These crystals rapidly grow into thin discs, which turn into hexagonal stars on the surface of the super-cooled brine. In stable waters, the crystals become stacked in long needles of congelation ice. In turbulent conditions, the arms break off the crystals and are mixed in the ocean. Here they coat the surface with an oily sheen, which is known as grease ice. Below the surface, particularly in open or turbulent waters, where the ice crystals are constantly jostled, an unstructured slush is created. This is known as frazil ice. Most areas of sea ice begin as a composite mixture of grease ice and frazil ice to form a soupy layer.

As freezing continues, grease ice forms a thin crust, which is usually broken up by wind and waves into separate masses to form platelets or pancake ice. These platelets collide with each other and are curled up at the edges. They then coalesce, acquiring additional layers of frazil ice from below and a covering of snow on top. This thickening layer continues to be broken up by waves into ice floe. Flat sheets of sea ice together comprise pack ice. Some floes may be tens of kilometres across. When floes are rafted together or forced edge to edge, they form pressure ridges, which are very hazardous to ships. Fast ice is sea ice that forms and remains fast, or attached to the shore. It tends to be the thickest form of Antarctic sea ice, reaching more than 3 m if it survives for several years.

3.14 Polynyas

Areas of open water among pack ice are known as polynyas. They tend to recur in the same locations year after year and are areas of enhanced biological activities. Coastal polynyas are caused by katabatic winds and storms driving sea ice offshore and causing warmer waters to swell up. Open sea polynyas are large and last longer. Caused by deep upswelling of warmer water, some remain open for years at a time.

Earlier explorers would anticipate breaking into open waters when they observed grey skies caused by the low reflectivity of open water as compared with that of the surrounding pack ice.

3.15 Research from Indian Antarctic Stations

Ever since 1981, Indian quests and endeavours in Antarctica have continued uninterrupted over about thirty yearly expeditions. In addition, five special expeditions to the Southern Ocean and an Indian expedition from Maitri to the South Pole were also undertaken.

Indian geologists have systematically mapped about 20,000 sq km of the area in Central Dronning Maud Land (CDML). Three geological maps of Schirmacher Oasis, Wolthat and Orvin mountains (GSI, 1991, 1999, 2006) and a geomorphological map of Schirmacher Oasis (GSI, 2006) have been released so far. Monitoring of a glacier snout and continental ice margin, on a regular basis, has revealed a significant retreat of the ice margin in the last two decades.

A permanent Global Positioning System Station (GPS) at Maitri contributes to the understanding of the inter-plate motion and crustal deformation between India and Antarctica (Ravikumar and Malaimani, 2000). The station has been tied up geodetically with other International Terrestrial Reference Frame stations (ITRF). The seismological observatory at Maitri is a part of the Global Seismological Network and contributes the data to the United States Geological Survey (USGS) and the International Seismological Centre (UK).

Ice-core studies have proceeded well. Cores of 65m, 75m and 55m have been retrieved from the continental and ice-shelf areas (*NCAOR Through the Years*, 2010). They have been analyzed to evaluate the temporal changes in environmental characteristics during the past few centuries. Results show that temperatures at the beginning of this period were 2°C cooler than today (Nizampurkar, 2002). The ice-core studies also indicate that sulfate aerosol deposition is related to volcanic eruption (Thamban, 2006).

Several activities are planned from the new Indian research station Bharati in the Prydz Bay area of East Antarctica which formed a contagious landmass with the Eastern Ghat Mobile belt of India in the supercontinent of Gondwana and is considered a key to an India-Antarctica link. Apart from rocks correlation and an integrated geological-geophysical study of the area, research on the fastest-moving Lambart Glacier in East Antarctica, movement of continental ice margins along the coast, glaciations in the Vestfold Hills and Larsmann Hills area, sediment records available in the lakes, among others, is planned.

3.16 In Conclusion

The evolution, statistical facts and a general study of Antarctic geology and geography make for fascinating reading. With just about two per cent exposed area free of ice, Antarctica's geological record is mainly hidden beneath its ice cap and poses difficulties and problems in research. However, with modern techniques such as deep drilling to take core samples from the seabed and ice streams, remote-sensing techniques, including seismic and radar soundings, have contributed to this research. Considerable study has been carried out ever since the continent was physically discovered and evidence in the form of fossils has been gathered to conclusively prove the existence of Gondwana. Research got a huge boost only after IGY and all exposed rocks are mapped now. Early geographical research focused on the coastline and the physical geography of the continent. Introduction of motor vehicles and aircrafts enabled charting of the coastline and inner areas. Core samples from ice sheets and sediments revealed the details of atmospheric conditions of the past, such as temperature, pressure, precipitation, and other aspects. Ice-core samples at Vostok could enable the mapping of climate changes over 420,000 years. Today, most geographers are glaciologists and climatologists studying physical processes.

In its thirty years of a continuous Antarctic presence since 1983, India has undertaken research activities, first from DG till its abandonment due to snow burial (first six years) and then from Maitri. So far, the activities are concentrated around Maitri. Now, with two fully operational research stations, Maitri and Bharati, in different areas of East Antarctica, about 3100 km apart on the Antarctic coast, India is in a position to enhance its research activities to a considerable extent, both in quantum and range in the field of geology and geography, in areas like climate, biology, environment and other aspects. India can surely build on the research already done and enhance its contribution to global studies in the near future. Working with and in the vicinity of research stations from countries like Russia, China, Romania and Australia may certainly enhance international cooperation, range and quantum of scientific research. Also, with the enhancement in the level of awareness about polar areas in the country, the participation of universities in the geological and geographical research projects may go up considerably.

Chapter 4

The Antarctic Climate: A Test of Endurance

> "Men wanted for hazardous journey. Low wages, bitter cold, long months of complete darkness, constant danger, and safe return doubtful. Honour and recognition in case of success."
>
> —***Sir Earnest Shackleton***
>
> *(British explorer, in an advertisement, almost certainly apocryphal, preceding an expedition)*

4.1 Introduction

Like all other environmental aspects, the climate of Antarctica, too, is unique and the most extreme of any on earth. It is the result of high altitude, constant snow cover, an ice plateau at a great height above sea level, high-wind velocity, extreme low temperatures, and the vast extent of the surrounding Southern Ocean geographically isolating the continent. Given the volume of ice it contains, Antarctica is the largest desert as also the highest, coldest and windiest continent on earth. The mean temperature is so low that fuel oil turns to jelly and boiling water thrown in air spontaneously explodes into ice crystals before reaching the ground. The wind can be so strong that the great Australian polar explorer, Sir Douglas Mawson, during his expeditions in 1911–14, observed that the weather at Cape Denison was little else than one continuous blizzard all year round; a hurricane of wind roaring for weeks together, pausing for breath only at odd hours and starting all over again. He aptly called the continent 'a home of blizzards' (Mawson, 1996). Inland, almost no snow falls. In fact, in the polar plateau the water equivalent of the annual rate of snowfall is less than 5 cm, very similar to the Sahara Desert.

As a result of its unique climate, Antarctica is home to a series of very interesting phenomena, such as the Aurora Australis, ozone hole, pointed summer and coreless winter, mock suns and sundogs, wind and wind-chill factor, blizzards, temperature inversions (all described in the course of the chapter). About the only meteorological happening one is safe from is being struck by lightning, since thunderstorms are unknown (*Reader's Digest*, 1985).

As a result of its extreme climate, no permanent habitation or plant and animal life is possible in Antarctica, except the microscopic one that exists at places. These conditions have evolved and stabilised over millions of years.

4.2 Arctic and Antarctic Contrasts

The Arctic and the Antarctic both serve the same basic thermodynamic function in acting as heat sinks of the earth. However, the similarity ends here as their environments are in stark contrast to each other, the most basic being the distribution of land and sea as shown in Figure 4.1 (Walton and Dudney, 1987).

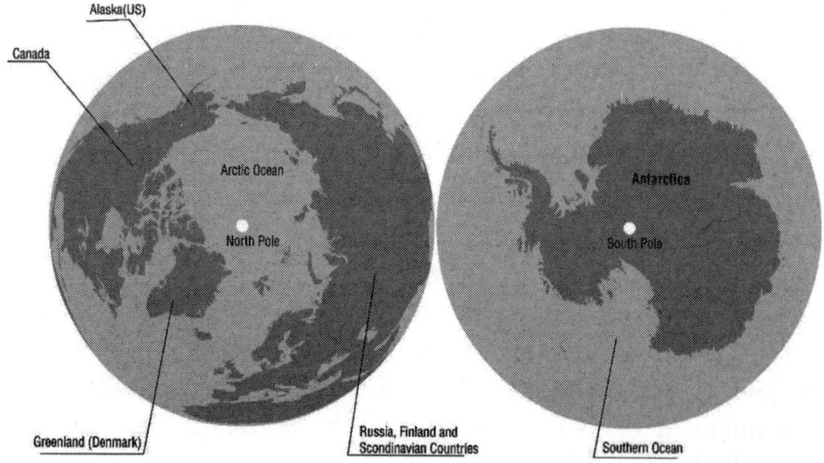

Fig. 4.1: Contrast between the Arctic *(L)* and the Antarctic *(R)*

Whereas the Arctic consists of an ocean basin surrounded by continents, the Antarctic contains a pole-centered isolated continent surrounded by ocean. In the Antarctic, as very little land, except for a few islands, exists between 40°S and 60°S, there is a vigorous westerly circulation known as the West Wind Drift of the Southern Ocean. In contrast, the surrounding continents moderate the wind flow in the Arctic. Due to this, the southern latitudes are known as the roaring forties, furious fifties and screaming sixties, as experienced by early explorers (*Reader's Digest*, 1985).

The earth is closest to the sun during Antarctic summer and most distant during Arctic summer. As a result, about seven per cent more solar energy enters the atmosphere in the Antarctic than in the Arctic during their respective midsummers. Also, while the Arctic is surrounded by the major industrial areas of the world, there is no industrial activity in the Antarctic. The total absence of pollutants and dust to reflect, absorb or scatter incoming radiation results in more incoming solar energy in the Antarctic. This, along with a high mean elevation (about 2300 m from sea level), causes about thirty per cent less atmospheric mass to be available over the Antarctic to attenuate the incoming radiation. The net result is that about sixteen per cent more solar energy is available to heat the surface at the South Pole than at the North Pole in summer.

Thus, other things being equal, temperatures should be higher in the Antarctic than in the Arctic. But it is a paradox that the situation is quite the opposite. The mean temperature in the Southern Hemisphere is about 1.6°C lower than in the Northern. In the Southern Hemisphere, glaciers exist down to the sea level for latitudes 10° closer to the equator and the quantum of area where the climate is too severe to support tree growth (where the temperatures during even warm months are less than minus 10°C) is three times more than in the Northern. The monthly mean temperatures on the south polar plateau range from minus 25°C to minus 70°C, and even on the coastal margins are as low as minus 55°C. In the Arctic region, this temperature variance is from 0°C to minus 35°C.

Several interacting factors, arising out of isolation, strong winds and a massive ice cap, enable this apparent abovementioned paradox. For starters, the Gulf Stream feeds warm ocean currents into the Arctic basin around Greenland and Norway, whilst no warm ocean currents penetrate the West Wind Drift in the Southern Ocean. In addition, the strong westerly winds in the south effectively inhibit the transfer of heat from the equator by atmospheric circulation. Thus, in all, less heat is transferred to the Antarctic than to the Arctic. The Arctic Ocean, beneath its thin crust of ice (about 3m thick), remains relatively warm, at about 2°C. It, therefore, acts as a vast heat reservoir throughout winter, so that as much as five times more heat flows in the atmosphere from the surface at the North Pole than in midwinter at the South Pole. Heat, provided by solar radiation on the Southern Ocean, is rapidly convected to lower levels, maintaining a fairly cool ocean surface within the Antarctic convergence. The sub-Antarctic atmosphere, therefore, remains relatively cool. As the margins of the Arctic are continental land masses, heat is not convected to great depths here but is, instead, quickly transferred back into the atmosphere.

Both polar regions have low atmospheric vapour content because of low temperatures. However, being warmer, the Arctic atmosphere has more water vapour and is thus cloudier. Due to this, there is relatively more melt in the Arctic, thus, reducing the albedo (proportion of the incident-light or radiation that is reflected by a surface). As a result, the typical summer value of albedo becomes sixty-five per cent for Arctic ice as compared to ninety per cent for Antarctic ice. The northern ice is, therefore, more efficient in absorbing solar radiation and, consequently, the temperature difference is further accentuated.

4.3 History of Antarctic Climate

As discussed in the earlier chapter, 500 Ma Antarctica was situated near the equator, as part of the supercontinent Gondwana, with a temperate climate and rich vegetation. In the long period of geological and climatic evolution, Gondwana drifted through the whole range of climatic zones, from the equatorial to the polar. Around 280 Ma, Antarctica first approached the South Pole. Subsequently, the continental drift carried it equator-wards and then towards the South Pole again and it finally arrived in the vicinity of the South Pole around 100 Ma as the central part of Gondwana. Then occurred the gradual separation and drifting away of Africa, India, New Zealand

and Australia, as explained in the earlier chapter. Finally, South America, too, drifted away about 23 Ma, opening the Drake Passage and creating the Southern Ocean around the continent. Antarctica was now totally isolated from the rest of the world as shown in Figure 4.2 (Walton and Dudney, 1987).

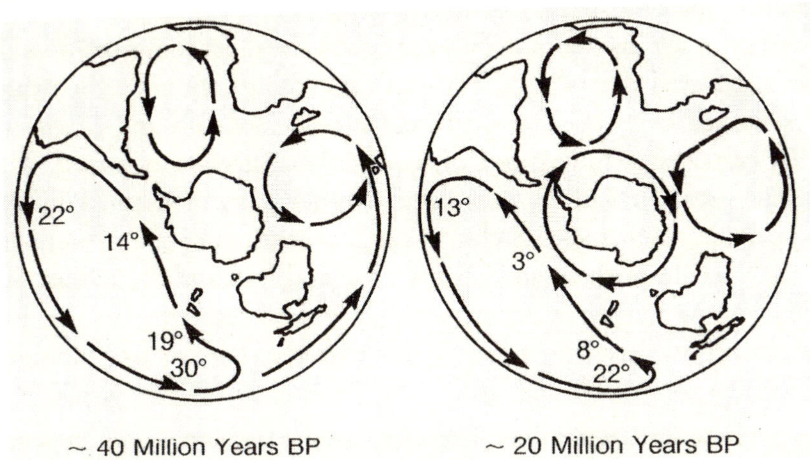

Fig. 4.2: Opening of Southern Ocean and Formation of West Wind Drift

It is from this development that the present climate of the continent started to evolve. Due to the circumpolar ocean currents, the continent was isolated from warm ocean currents which had for so long transported heat from tropical regions. With the changes in the oceanic circulation and the consequent lowering of surface temperature, polar temperatures decreased and the temperature gradient from equator to South Pole increased. This further strengthened the westerly circulation and enhanced the cooling process. All this paved way for the quaternary (relating to the period of geological time from about 2.6 Ma) ice ages, affecting the globe in the last million years.

4.4 Ice Cores and Past Climates

The Antarctic ice sheet contains a wealth of information about the earth's climatic past because climate changes such as temperature, precipitation and atmospheric conditions are preserved in its identifiable yearly layers. To study the layers, long tubes or ice cores are drilled. Analysis of the layers reveals much about the earth's atmosphere at the time the ice was formed. Each water molecule contains one atom of oxygen which weighs between sixteen to eighteen times as much as a hydrogen atom, depending on the temperature of the air at the time it fell as a snowflake. It is not only temperature but other details, too, that can be ascertained. By examining deposits, it was discovered that 18,000-year-old ice contains far more dust than present-day snow. As this has to come from the soils and deserts of other continents, indications of winds at that age are obtained. The measurement of carbon dioxide and other gases contained in the trapped air bubbles gives a clue to the composition of the atmosphere at the time. Impurities, such as methane and lead and the radioactive fallout from nuclear-weapon testing, are also layered into the ice. Levels of both lead and mercury, for example, have at least doubled in concentration from that about 10,000 years ago (Trewby, 2002).

To date the ice obtained from ice cores, various methods are used such as counting of the annual deposition in case of young, shallow ice, decay of elements like carbon and chlorine in case of deeper ice and the flow characteristics of ice sheets in case of ice at greater depths. The National Ice Core Laboratory in Colorado (US) holds over 13,000 tubes of ice, each just over 1 m long (McGonigal, 2008).

Some of the prominent efforts to drill ice cores have been at Byrd Station: 2164 m, Dome C Station: 900 m and Vostok Station: 2083 m (*Books, LLC: Outposts of Antarctica,* 2010). The Russians have gone much deeper at this location (about 4 km) recently, to reach the waters of the sub-glacial lake, as pointed out earlier.

4.5 Present Antarctic Climate

Temperature and wind are the two most important elements in determining polar weather. Considered together, the different combinations of the two characterise the regional climate of Antarctica.

Three basic climatic regions can be distinguished in Antarctica, namely the Interior, the Coast and the Peninsula. The interior is a polar desert and is characterised by extreme cold and light snowfall. The coastal areas have milder temperatures because they are at sea level and are influenced by weather systems that form over the Southern Ocean, which bring warmer and moister air to the coast, higher precipitation and more cloud formations. The peninsula also has a relatively warm and moist coastal climate, but is more variable. When cold air flows north from the continent, temperatures can plummet.

Temperatures decrease from summer to winter with increasing latitude and height above sea level. Over most of the continent, the mean temperature of the coldest month is around minus 15°C to minus 30°C on the coast and minus 40°C to minus 70°C in the interior. The lowest temperature recorded in Antarctica was minus 89.6°C in July 1983 at the Russian station, Vostok *(Books LLC, Outposts of Antarctica,* 2010). In summer, the mean temperature of the warmest month is around 0°C at the coast and minus 15°C to minus 35°C in the interior. Mean surface temperatures during summer and winter in Antarctica and the Southern Ocean are as shown in Figure 4.3 (Walton and Dudney).

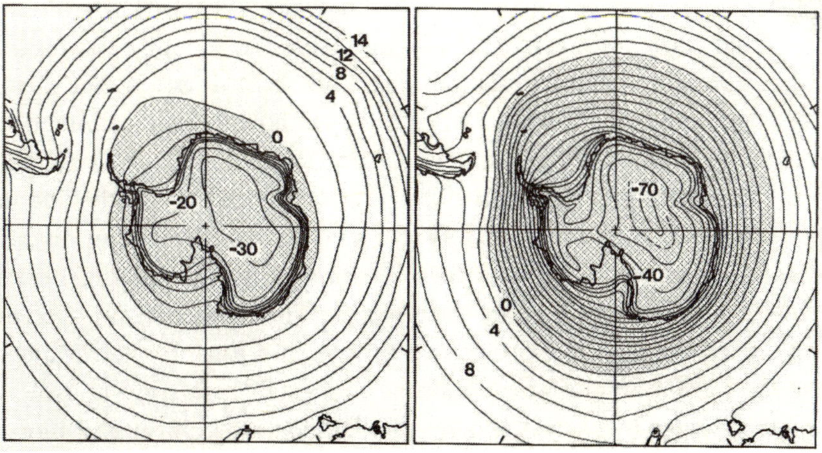

Fig. 4.3: Mean Surface Temperatures for Summer and Winter Periods (Shaded Area is Below 0°C)

The remarkable thickness of the ice sheet combined with wind conditions result in an unusual distribution of summer-winter months, known as pointed summer and coreless winter and temperature inversions as shown in Figure 4.4 (Walton and Dudney).

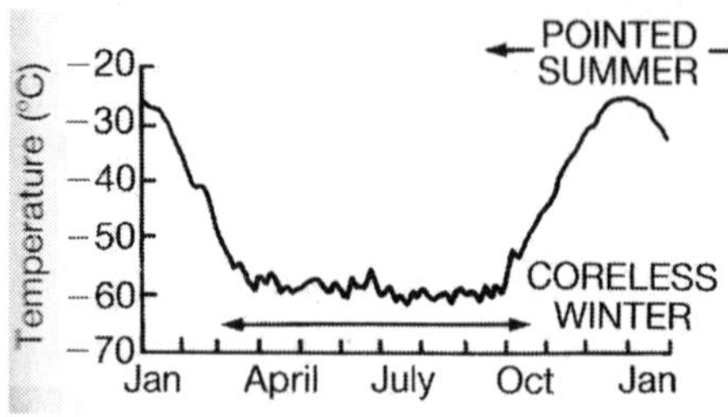

Fig. 4.4: Pointed (Short) Summer and Coreless (Long) Winter

4.5.1 *Pointed or Short Summer and Coreless or Long Winter*

Surface temperatures at the South Pole Station, as shown above, illustrate the occurrence of an unusual distribution of the two seasons. Summer is very short. Only for the period from mid-December to mid-January does the temperature exceeds minus 30°C. The summer warming arises from a combination of increased solar radiation and small temporary reduction of surface albedo, resulting from the surface snow as it ages during spring. The first light snowfall or influx of drifted snow restores the albedo and, with the waning solar input, the temperature begins to fall rapidly. This short duration and shape of temperature variation has earned the nickname Pointed Summer. It was the rapid decline of temperatures after midsummer which caught Captain Scott's party in their ill-fated return from the South Pole in January 1912 (Trewby, 2002). Scott had clearly mentioned in his last days' diary entries as to how their seemingly adequate arrangements made for the return journey were proving inadequate in the face of rapidly declining temperatures and other conditions. By end-March, the sun has set for the winter months and a stable situation known as coreless winter is reached, where, for the next six months and more, temperature varies by only a few degrees.

4.5.2 *Temperature Inversions and Winds*

A remarkable feature of the Antarctic troposphere is the presence of a very pronounced surface-temperature inversion. Variations of

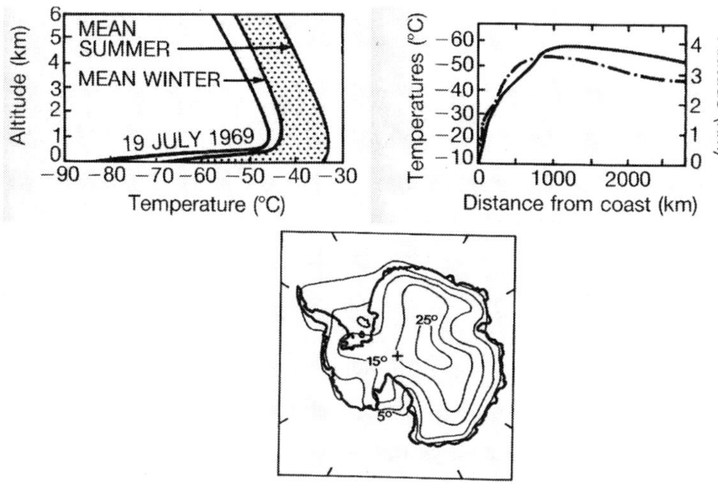

Fig. 4.5: Temperature Inversion at the Polar Plateau

temperature inversions as observed at the polar plateau are as shown in Figure 4.5 (Walton and Dudney). It occurs when the air temperature initially increases upwards from the ground before decreasing, rather than just decreasing upwards, as it does elsewhere. Temperature inversions are normally about 500 to 1000 m thick and the air within them is stable. This effect is most exaggerated over the south polar plateau because the sky is clear and often cloud-free, allowing solar radiation to bounce off the surface. Cooled by the cold snow surface, the lowest layer of air can be 30°C colder than the air temperature at 1000 m. A detailed study about this phenomenon is presented based on the data collected in 2004–2005 (Hudson and Brandt, 2008). However, when this layer is disturbed by strong winds, it moves with the air above. For this reason, surface temperatures in Antarctica are very erratic and can rise and fall quickly over a few hours.

The behaviour of the surface winds over Antarctica is extraordinary and simple rules relating to surface-pressure variations to the resultant winds do not apply. Instead, the key factors influencing the wind-speed and direction are slope of the terrain and strength of the temperature inversion, resulting in an inversion wind. This blows persistently in a direction set by the fall line of the slope, with a speed related to the magnitude of the slope. In simple terms, cold surface air moves down the slope under the action of gravity. Near the base of the coastal escarpments and of the mountain ranges, katabatic and barrier winds occur.

Katabatic winds are associated with a rise or decrease in temperature and are severe. They are further divided into two classes, ordinary and extraordinary. For ordinary katabatic winds, the wind direction remains constant but the speed is highly variable, gusts alternating irregularly with week winds or periods of calm. Extraordinary katabatic winds are those which blow uninterruptedly for days or even weeks with almost unvarying ferocity. They are limited in distribution, particularly to a small sector of East Antarctica in the vicinity of Cape Denison and Port Martin (*Books LLC, Outposts of Antarctica,* 2010). They result from a combination of the nature of upstream topography which produces a convergent flow, the presence of a large drainage area and persistent intense upstream inversion. Mawson has eloquently described his experiences about the extraordinary katabatic winds in his famous book *The Home of the Blizzards: A True Story of Antarctic Survival* in 1915. In the rest of Greater Antarctica, ordinary katabatic winds prevail.

Barrier winds are a manifestation of a strong surface inversion, associated with extended mountain barriers and are generally very fast, in an excess of 70 kmph. Two such barriers exist in the Transantarctic Mountains and the mountains of the Antarctic peninsula. The barrier winds on the eastern side of the peninsula contribute to a very harsh climate as compared to the western side.

Drifting or blowing snow generally accompanies the wind, as loose snow on the surface becomes entrained in the airflow. The concentration of snow particles and the altitude to which they are carried depends upon the amount of loose snow available, the wind-speed and the turbulence in the flow. Drifting snow is a layer of fine snow less than 2 m thick and is likely to be sustained by a wind-speed of about 5 m/second. Blowing snow is much more unpleasant with a thickness varying from 2 m to 100 m and a wind-speed in excess of 20 m/second. Here it is called a blizzard which is a combination of strong winds and falling or blowing snow.

4.5.3 *Wind-Chill Index and its Effects*

A wind-chill index links wind and temperature to show how heat is lost in various conditions. In still air, at minus 20°C, the index is around 550 or very cool; at the same temperature, in a one knot (1.9 kmph) wind, the index is around 900 or very cold; at the same temperature in a wind of 7.5 knots (14 kmph), the index is 1400 or at which exposed human flesh starts to freeze. The effect of wind on the temperature is as shown in Figure 4.6 (Reader's Digest, 1985).

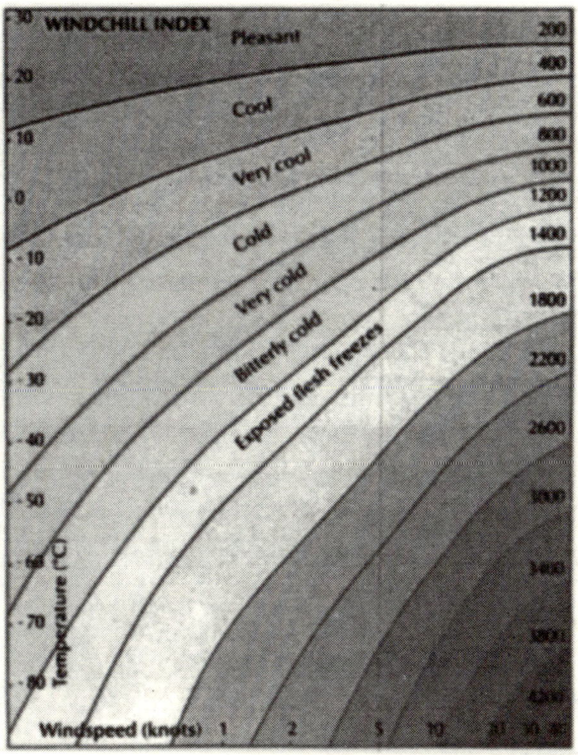

Fig. 4.6: Wind Chill Effect and Index

4.6 Cloud Formation

Clouds only form regularly in areas where air masses absorb sufficient water vapour from the ocean surface. Heavy, low stratocumulus (a low-altitude, often patchy in horizontal layers with rounded tops) clouds form over the Southern Ocean, particularly during summer when cyclonic weather systems and associated fronts bring warm, humid air southwards. Strong winds that blow onshore, from the ocean to the coast, can pile these clouds over coastal regions.

In summer, billowing (swelling outwards) cumulus clouds may form over areas of rock that absorb heat, such as mountain ranges, nunataks and oasis. Because cumulus clouds need to be heated from below to form, they are rarely seen over ice and snow surface. Hogsback clouds also form over mountains when there is strong high-level airflow and often are the first sign that bad weather is on its way.

Cloud formation is minimal or limited over inland Antarctica, which is known for its arid climate. The air in this zone is very clear and the clouds are thin and composed of ice crystals. They have low moisture content and do not reach very high into the atmosphere.

4.7 The Antarctic Ozone Hole

Atmospheric ozone provides protection from harmful ultraviolet (UV) rays of the sun. The reduction or depletion of the amount of ozone in the atmosphere due to the destruction of stratospheric ozone by chlorofluorocarbons (CFCs) is popularly known as an ozone hole. It is noticed prominently over Antarctica and it engages keen attention of the world community due to the obvious danger it signifies.

4.7.1 *Structure of the Atmosphere and the Importance of Ozone*

The atmosphere is a mixture of gases in layers, enveloping the globe. It acts as earth's shield, protecting it from the sun. The layers have different temperatures and physical properties. From the earth's surface, the atmospheric layers are known as troposphere (up to 10 km), stratosphere (10 to 50 km), mesosphere (50 to 90 km), and thermosphere (90 to 500 km). Beyond this lies exosphere or magnetosphere which extends to a height of several thousand kilometres.

Ozone (a form of oxygen) is an extremely sparse constituent of the earth's atmosphere, with only about three molecules of it in every ten million molecules of air. Its concentration is less than all other trace or rare gases except Radon (one part in a million). Whereas ordinary oxygen molecules are made up of two atoms (O_2), ozone molecules are of three atoms (O_3). More than ninety per cent of ozone is found in the lower stratosphere.

Despite its very small quantity, ozone is extremely important to the atmosphere and for the past six hundred million years or so, it has been the atmospheric sentinel, protecting life on the planet from the biological damage caused by the sun's ultraviolet (UV) radiation. The temperature of the stratosphere is also maintained by a balance between the absorption of solar UV rays by ozone and the emission of atmospheric infrared radiation by ozone.

Though ozone is also one of the greenhouse gases, its effect on global warming is minimal (only about three per cent) and hence should not be confused with its major area of impact discussed here.

4.7.2 *Production and Function of Ozone*

In the upper reaches of the atmosphere, air molecules absorb solar UV rays of wavelengths less than 0.24 micron and break up into atomic oxygen. The newly formed atomic oxygen then combines with oxygen (O_2) molecules and become molecules of ozone (O_3). The ozone molecules absorb solar UV rays of wavelengths up to 0.29 micron and dissociate into molecular oxygen. Although mechanisms like auroras, cosmic rays, thunderstorms electric discharges can also produce atmospheric ozone, their contribution is negligible as compared to that produced by UV rays from the sun. Most of the ozone destruction takes place at the surface of the earth when they come into contact with organic matter, mostly vegetation.

4.7 Phenomenon of the Ozone Hole

During the early 1970s, a scientist at the British Antarctic Station, Halley, first became aware that possible damage to the ozone layer might be occurring as a result of human activities (*Books LLC, Outposts of Antarctica,* 2010). Thereafter, observations from a US research programme in 1987 through a high-altitude NASA aircraft and other experiments produced irrefutable evidence that the ozone hole was mainly caused by the chemical destruction of ozone by atmospheric chlorine. It has now been established that pollution from synthetic chemicals, finally, causes the depletion. Pollutants like nitrogen oxide (NOx) from jet aircrafts, fertilisers, carbon monoxide (CO) and carbon dioxide (CO_2) from burning of fossil fuels and methane (CH_4) are responsible to a small extent, but the prime offenders are CFCs containing chlorine atoms. They are produced from the cooling agents in refrigerators, air conditioners and propellants in aerosols. They are inert in normal conditions and are not destroyed at lower atmosphere by rain, oxidation or sunlight. Natural atmospheric circulation transports these chemicals high up into the stratosphere. Here, the CFCs are split and the chlorine molecules are liberated. The process of ozone destruction takes place in the chain chemical reactions as shown below:

Cl (Chlorine) + Ozone (O_3) = ClO (Chlorine Monoxide) + O_2 (Oxygen) (One ozone molecule destroyed)

$3ClO + O_3 = 3Cl + 3O_2$ (One more ozone molecule is destroyed and three Chlorine molecules are generated for chain reaction)

$3Cl + 3O_3 = 3ClO + 3O_2$ (Three more ozone molecule are destroyed)

$3ClO + O_3 = 3Cl + 3O_2$ (One more ozone molecule destroyed)

This catalytic reaction continually repeats itself so that for every chlorine atom released, an estimated 100,000 molecules of ozone are removed or destroyed from the atmosphere before the chlorine atom becomes inactive. In normal conditions, decomposition of CFCs is not possible. They are estimated to remain in inert form for a lifetime of about 75 to 110 years. So, even with a complete reduction of the use of CFCs, it would take over a hundred years for the repair of the rupture already inflicted to the ozone layer (McGonigal, 2008).

Ozone is depleted over Antarctica at a much higher rate than in the rest of the world. This is mainly due to the existence of polar stratospheric clouds (PSCs) which form in very cold temperatures inside the polar vortex, a giant whirlpool of wind that seals the Antarctic stratosphere in winter. In spring, when the combination of light and low stratospheric temperatures occurs, the PSCs provide a surface for the chain of chemical reactions explained above. As the vortex disintegrates in spring, the slow downward circulation drives pools of ozone-poor air over the Southern Hemisphere and can be dangerous to countries therein. This process reaches its maximum in September, eroding more than half of the ozone over Antarctica, resulting in the phenomenon called the ozone hole.

By November, the stratosphere begins to warm sufficiently to evaporate the PSCs. Ozone-rich air seeps back from higher latitudes, the chlorine is dormant for another year, and ozone destruction ceases.

4.7.4 *Extent of the Ozone Hole*

Ozone depletion in the stratosphere depends on latitude and season. For example, the total ozone depletion year round was about five per cent between latitudes 30°S and 60°S during the 1970s to 1990s. In the years leading up to 2000, the ozone loss was observed up to about seventy per cent. The extent of the ozone hole as observed from 1979 to 2006 is as shown in figure (Illustration No. 32) (McGonigal, 2008).

In late September 2006, the ozone hole over Antarctica had reached a record size of around 29.2 million sq km.

In the Northern Hemisphere, ozone losses over the Arctic region during spring have reached up to thirty per cent and are considerably lower than the losses that occur over Antarctica. It is so because the Arctic stratosphere is generally warmer (about 10°C), particularly in March, when sufficient sunlight is available to cause ozone depletion.

Indian stations Dakshin Gangotri (DG, 1984–89) and Maitri (from 1990 onwards) are actively involved in total ozone measurements first through the Dobson spectrometer and then through the Brewer Spectrometer from 1999. Representative data for years 1997, 2002, 2003 and 2004 has been duly analyzed to show variations of column ozone and the extent of ozone holes. They, however, show no firm pattern for depths and recovery cycles (Jain, 2011).

Continuous research on the ozone hole brings out new observations from time to time. A recent report from the NASA/Goddard Flight Centre points out that during the years 2013 and 2014, the size of the ozone hole was 24.1 and 24 million sq km (roughly the size of North America), much less than the maximum size mentioned above. Simultaneous reduction of about nine percent in the concentration of ozone was also observed in the stratosphere.

4.7.5 *International Efforts for Problem Solving*

Evidence that synthetic chemicals are responsible for ozone hole depletion since mid-1970s is overwhelming. In 1985, the Vienna Convention for Protection of the ozone layer was established and signed by twenty countries. A protocol to this treaty, known as the Montreal Protocol, on substances that deplete the ozone layer was established in 1987 to identify and to implement specific procedure to regulate the use and production of ozone-depleting substances worldwide. This has now been signed by 191 countries, the maximum for any international accord (McGonigal, 2008).

The combined stratospheric abundance of ozone-depleting chemicals has shown a downward trend since their peak values in the 1990s. However, as mentioned earlier, even if no CFCs are produced in the future, it may take about a hundred years to eliminate the ozone-depletion effect of existing CFCs.

4.8 Changing Weather Patterns

Antarctica plays an integral role in the world's climate. It is a heat sink, it powers currents in the world's vast oceans and it influences the weather in the mid-latitudes of 30°S to 50°S in the Southern Hemisphere. The continent is surrounded and isolated by the most productive marine environment, the Southern Ocean, which also absorbs heat and carbon dioxide. Through the actions of the wind, evaporation and winter sea-ice formation, salty water accumulates at the ice sheet-sea margin and increases the density of surface water below sea ice. These waters, which are now heavy, descend to the ocean floor and form deep nutrient-rich currents that flow around the world. This is called the vertical mixing of oceans and it prevents stagnation of water and sustains benthic life.

4.9 Changes in Air Pressure, Winds, Ocean Currents and Temperatures

Antarctica is continually buffeted by strong winds in various forms and of different origins, effects and names. These create changing patterns in air pressure, wind speed, ocean currents and temperatures in the continent as also in the surrounding ocean. Observations have revealed a clear trend since IGY which demonstrate how a drop in sea-level pressures has resulted in changes in wind formations, moving of the polar front or ice shelf southwards by 1.5 to 2 degrees, and creating deep low-pressure systems off Amundsen Sea, making the polar trough more cyclone prone.

4.9.1 *Katabatic Winds*

Persistent and powerful gravity winds emanate from the polar plateau which can attain a speed of around 300 kmph by the time they travel, 1600 km down the slope, to the coast. The major effects of katabatic winds are blizzards, sastrugi, scattering of pack ice and creation of heavy low-lying clouds as described in detail earlier.

4.9.2 *Circumpolar Vortex*

Due to limited energy coming from the sun during the darker months of winter, air mass at a high altitude in the Antarctic stratosphere naturally cools and moves in a westerly, circular motion in accordance with the earth's rotation. On an annual basis, this air mass develops into an enormous cyclone called the Circumpolar Vortex, which circles around a near-stationary column of air above the South Pole. Thirty years ago, this vortex tended to break in November but now it breaks in December. This cooling is linked to the ozone hole (escape of heat in the absence of a protective shield) and other greenhouse gases.

4.9.3 *East Wind Drift Current (coastal current) and West Wind Drift Current (circumpolar current)*

Strong airflow off the Antarctic continent creates polar easterly winds or the East Wind Drift that drive from east to west the flowing surface or coastal currents. East Wind Drift is shallower and slower moving than the West Wind Drift, and the two are separated by the Antarctic Divergence.

One of the six major ocean circulation systems, the West Wind Drift is a surface ocean current that flows 24,000 km around the continent from west to east. It is driven by fierce westerly winds over the Southern Ocean, driving the waters east.

4.9.4 *Antarctic Convergence (Polar Front) and Divergence*

The Convergence is roughly a circular belt of water about 40 km wide between latitudes 40°S and 60°S in the Southern Ocean. It marks a definite change in ocean temperatures and chemical composition. It is formed where cold north-flowing Antarctic water meets warmer water flowing south from the Atlantic, Indian and Pacific Oceans. Here a sharp temperature drop of up to 2°C takes place and a frost smoke rises from the ocean surface (water vapour produced from the cooling of warm air).

The Divergence is a zone around the continent at approximately 65°S latitude, where opposing winds and water masses collide, separating the East and West Wind Drifts. Above the Divergence, the circumpolar trough make similar atmospheric boundary between easterly and westerly air flows.

4.9.5 *Temperature Changes in the Southern Ocean*

Evidence shows that the Southern Ocean has warmed substantially since the 1950s. The warming has occurred at all depths and it seems to be caused by the poleward shift of the Antarctic Circumpolar Current.

4.9.6 *Temperature Changes in Sub-Antarctic Islands*

As the ring of sub-Antarctic islands lies in the path of the Antarctic Circumpolar Current, they have experienced warming in some form over the last fifty years. On Marion Islands (one of the most northerly) the mean air temperature has risen by 1.5°C and in the southerly Macquarie Island, an increase in precipitation is observed.

4.10 Greenhouse Effect and Global Warming

Global warming refers to the rising average temperature of the earth's atmosphere and oceans and its projected continuation in the foreseeable future. In the last hundred years, the earth's average surface temperature has increased by about 0.8°C with about two-thirds of it occurring over just the last three decades. Climate models are summarised in the fourth assessment report (AR4) by the IPCC. They indicate that during the twenty-first century the global surface temperature is likely to rise a further 1.1 to 2.9°C for their lowest emission scenario and 2.4 to 6.4°C for their highest emission scenario. Warming is expected to be the most in the Arctic and is associated with the continuing retreat of glaciers, permafrost and sea ice (*IPCC*, 2007). The study of weather patterns in Antarctica, therefore, should be viewed in this light. The Scientific Committee on Antarctic Research (SCAR) has also published a comprehensive report on Antarctic environment and climate change with exhaustive analysis, recommendations and corrective actions as contribution to the International Polar Year 2007–08 (*SCAR Report*, 2008). The fifth assessment report by IPCC published in 2013–2014 cements earlier observations. It puts the temperature-rise figures without any mitigation measures in the range of 2.7°C to 4.8°C. This report and its authentic conclusions formed the basis for the Climate Summit held in Paris from 30 November 2015 to 11 December 2015 (COP 21) under the aegis of the United Nations Framework Convention on Climate Change (UNFCCC). Details about this are given in later chapters.

The greenhouse effect is the process by which the absorption and emission of infrared radiation by gases in the atmosphere of the warm planet's lower atmosphere and surface take place. It was first proposed by Joseph Fourier in 1824. The warming effects of natural greenhouse gases as a percentage of the total effect are water vapour (36–70%), carbon dioxide or CO_2 (9–26%), methane or CH_4 (4–9%) and ozone, O_3 (3–7%) (Wikipedia, Global Warming, 2011).

Warming of the atmosphere caused by greenhouse gases (greenhouse effect) acts in the same way as a glass roof traps heat in a greenhouse. The gases let short-wave solar radiation pass through to the surface, but prevent long-wave radiation emitted from the earth's surface from escaping. This is an essential natural process for supporting life beneath the roof, that is, the atmosphere enveloping the earth. However, the study of air bubbles in ice cores has shown that greenhouse gas concentrations have been increasing steadily over the past two centuries and are now at higher levels than at any time in the last 1, 60,000 years (Trewby, 2002). For example, concentrations of carbon dioxide and methane are thirtyper cent and forty per cent higher respectively, than they were two hundred years ago. A typical car releases tons of carbon dioxide each year. Cows burp as much as 200 kg of methane per year. New greenhouse gases such as CFCs have been added to the air. This extra greenhouse effect is warming the lower atmosphere and changing the earth's climate. The aspect about the tipping point, when the temperature rise crosses the threshold limit of 2°C and its effects are as explained earlier.

4.11 Changes in Continental Ice

It is established that the East Antarctica ice sheet has been gaining ice slightly through small increase in precipitation. On the other hand, the smaller West Antarctica ice sheet is losing its mass mainly due to its ice draining into the Amundsen and Bellingshausen Seas. It is attributed to warmer sea temperatures where the glaciers meet the ocean, causing increased glacial flow rates and melting of ice-shelf bases.

The Larsen Ice Shelf has disintegrated into parts as mentioned earlier. Many other glaciers are retreating mainly by the calving of small icebergs from their ocean margins and behind. The collapse of floating ice shelves has no direct impact on sea level and is akin to an icecube floating in a glass of water. But ice shelves buttress the flow of grounded glacier ice from the land behind them and the recent increase in the speed of glaciers has been observed in places where the peninsula ice shelves have disintegrated.

4.12 Changes in Sea Ice

In the absence of scientific recording mechanisms earlier, the shipboard data available from whalers from the 1950s to 1970s gives a general idea that the summer sea-ice extent has decreased by about twenty-five per cent which is yet to be verified, while satellite imagery started in 1973 has indicated a sea-ice decrease in Weddell Sea and the Antarctic Peninsula. Some evidence has also suggested that the sea-ice season has also been reduced by about a week in the last thirty years.

4.13 Ice Conditions in the Sub-Antarctic Islands

Ice caps occur on a handful of sub-Antarctic islands, which are the Marion, Grande Terre in Iles Kerguelen, South Georgia and Heard Island. Here, in varying degrees, mean temperatures have increased and ice caps have rapidly retreated. Many of the glaciers that previously flowed into the sea are now grounded on land, resulting in major changes to the coastline. In places, glacial lakes are created between the front of the glaciers and the thin glacial coastal strip.

4.14 Melting of Ice and Sea Level

The Intergovernmental Panel on Climate Change (IPCC) was established in 1988 by the World Meteorological Organisation (WMO) and the United Nations Environment Programme (UNEP). Its role is to provide a balanced view on climate change. It has estimated that the sea level rose from 1993 to 2003 at the rate of 3 mm per year and just about ten per cent of this was estimated to be from the Antarctic ice sheets. With present estimates, the decay of East Antarctica ice might result in about a 57 m rise in the sea level over a period of 10,000 years (McGonigal, 2008 and the IPCC fourth assessment report, 2007). Melting ice is projected to raise ocean levels by as much as 20 ft within the next two hundred years, uprooting eighty per cent of the world's population (Copeland, 2007). Many books and articles have dwelt on this subject with data from various scientific sources. The first person to have walked to both the North and South Poles, Robert Swan, has raised dire warnings and consequences as to the likely conditions like annual greenhouse-gas emissions rising to seven hundred billion tons, amounting to 5°C atmospheric temperature rise, considerable drop in oil production, seventy-five per cent shrinking of ice in the Himalayas, affecting water supply to China and India, massive species extinction and other likely effects by 2041 when the Antarctic Treaty comes up for review (Swan 2009). The next few years are sure to see more research on all these aspects, leading to definitive conclusions.

Rapid sea melting in Antarctica has seriously threatened the lives of emperor penguins. This largest species of penguins breed and raise their young exclusively on sea ice. Researchers from Woods Hole Oceanographic Institute (WHOI) have revealed that, on the Dion Islets penguin colony from 1948 to 1970, over 150 breeding pairs were recorded. The number was reduced to twenty in 1999 and the colony was found vanished in 2009 (S Jenouvrier, 2012). Similar observations have been made elsewhere, indicating the same trend.

4.15 Aurora Australis

Aurora is the most beautiful and spectacular of all optical phenomena seen in both the polar regions. Named after the Roman goddess of dawn, it is called Aurora Australis or Southern Lights in Antarctica. Its Northern Hemisphere counterpart in the Arctic is called Aurora Borealis or Northern Lights.

Captain James Cook, venturing below the Antarctic Circle on his last voyage in 1774, was the first European to report the phenomenon in the south (*Reader's Digest*, 1987). The Maoris of New Zealand had long known of it as *Tahu-Nui-A-Rangi* or the great burning of the sky (Peat, 2007). For long, the phenomenon has truly mesmerised explorers.

Aurora is the manifestation of complex processes that occur in the upper atmosphere. No artist's brush or camera can hope to capture the elusive nature of its constantly changing form, the purity of its colours or the complexity of its spatial dimensions. One of Robert Scott's colleagues, Edward Wilson, describes what he saw in July 1911 thus, 'We had a magnificent display of auroral curtains, during which four-fifths of the eastern half of the sky was covered by waving curtains right up to the zenith, where they were all swinging round from left to right foreshortened, swaying curtains forming a rapidly moving swirl, constantly altering its formation. They were very brilliant and showed bands of orange and green and again orange fading in to lemon yellow upwards.' The cause of the aurora was something of a mystery in Scott's time. Complex scientific explanations came much later with better understanding of the structure of the earth's magnetic field and upper atmosphere and events that occur beyond the earth, on the surface of the sun and solar wind.

All the planets are, in reality, frozen grains of matter floating in a sea of ionised gases called plasma. Plasma makes up 99.9% of the universe. Surrounding the earth, at the edge of the atmosphere, this plasma is held and channeled by the earth's magnetic field in a region called magnetosphere.

The aurora is initiated by the sun. Because of its very high temperature, the atmosphere of the sun, the solar corona, is forever expanding outwards, at a velocity of 400 km/sec, far beyond the orbit of the earth. This solar wind is very hot and is made up of charged sub-atomic particles, mostly hydrogen atoms that have each lost their single electron. While travelling into space, the solar wind carries with it a sample of the sun's weak magnetic field.

The earth too has a magnetic field and the electric properties of the solar wind prevent it from crossing this field. The blast or bombardment of the solar wind at a great speed distorts the earth's field. A cavity is created, surrounding the earth and its atmosphere with a comet-like tail extending perhaps for a million kilometers out on the night-side of the earth. This is shown in Figure 4.7 (Reader's Digest, 1985).

The interactions between the solar wind and the earth's magnetic field vary in intensity according to the direction of the field carried by the solar wind. If the field is northwards, opposite to the direction of the earth, then the interaction is weak. But if the field is southwards, then the interaction is very strong. The particles in the solar wind cannot cross the earth's magnetic field but they can follow them and

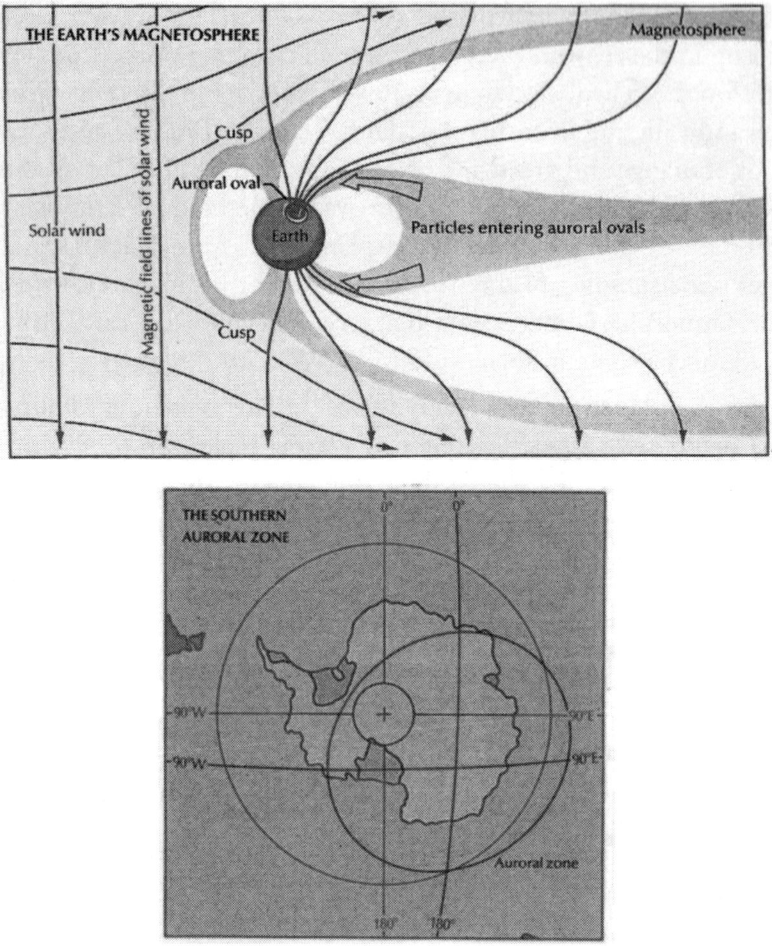

Fig. 4.7: Earth's Magnetosphere, Impact of Powerful Solar Wind and Aurora Oval

enter the magnetosphere at cusps centered near the noon meridian in each hemisphere. These particles give rise to many complex reactions in the magnetosphere. Some particles travel down magnetic field lines into the lower atmosphere and from there into the auroral ovals. There they give rise to vast systems of electrical currents and also excite the polar aurora.

The frequency and the intensity of the aurora are closely linked to the orientation of the magnetic field in the solar wind. Whereas with weak interaction, the aurora almost vanishes, strong interaction leads to a greater number of particles entering the earth's magnetosphere, triggering a series of auroral substorms.

The colours of the aurora are dictated by the atoms and the molecules that make up the thermosphere, like violet for nitrogen and red and green for atomic oxygen.

The aurora occurs almost every night beneath the auroral zone. The Indian stations Maitri and Bharati are located in the sub-auroral and auroral zone respectively. Simultaneous observations from the two stations would add a new dimension to research in this respect (NCAOR, CEE, 2010).

4.16 Other Optical Phenomena

The cold, dry climate and dust-free air of Antarctica mean that there is a complete lack of haze. In such conditions, distant objects seem close and mirages are common. When the air over the polar ice cap is filled with ice crystals, either composed as clouds or simply suspended in air, the light of the sun and the moon is reflected, refracted, diffracted and scattered by the crystals. This produces a variety of optical effects that dazzle visitors to the continent.

In the Antarctic skies, halos form around the sun or moon in the shape of circles, arcs or spots, known as mock or dog suns or moons. In daylight, vertical streaks of light known as sun pillars can occur. Fog bows are formed by the diffraction of light through super-cooled water droplets, while perihelia are created by sunlight passing through ice crystals. Halos, sun pillars and fog bows can occur simultaneously.

4.17 Bases for Monitoring Antarctic Climate

Permanent scientific bases were established in Antarctica mostly as part of and around the 1957–58 International Geophysical Year (IGY). The weather stations of eighteen of these stations have yielded continuous climate records for air temperature, pressure and wind speed for over five decades. Eleven stations have shown slight warming trends while seven have demonstrated slight cooling trends. The data points out regional variation. It shows slight cooling over interior Antarctica. The strong westerly winds that surround the continent generally restrict warm, maritime air from penetrating into too much of the interior of the continent. Based on this, one can safely say that most of East Antarctica is neither warming nor cooling (McGonigal 2008).

In the Dry Valley region, the Long Term Ecological Research Centre (LTER) of US had observed local summer-cooling in the early 1990s. There were reduced stream flows, lowering of lake levels and increased lake-ice thickness. However, in 2001–02, when a relative heat wave struck the region and melted local glaciers, the lake levels were rapidly restored to the levels observed before 1990. The heat wave also resulted in a record high summer air temperature at McMurdo Station.

The most substantial and remarkable climate change signal, however, is coming from West Antarctica. The Vernadsky weather station has recorded significant warming. Over the last fifty years, the mean and winter air temperatures have risen by 3° and 5°C respectively. Most of the glaciers on the peninsula's west coast have retreated in the last fifty years as have ice shelves around the peninsula, including, and dramatically so, the Larsen Ice Shelf in the Weddell Sea. A major recession has also occurred in the winter sea ice in the Bellingshausen and Amundsen Seas. The Larsen Ice Shelf consisted of four embayed segments A, B, C and D. Up to 2000 sq km of ice shelf from A and 3250 sq km from B disintegrated in 1995 and 2002 respectively in just a few days. Warm winds had been crossing the peninsula, melting the surface ice and, in late January 2002, pools of water began to appear on the ice shelf's surface. The weight of water wedged open small crevasses, forcing them through the ice thickness and breaking the shelf apart. Remaining sections of the ice shelf C and D stretching farther south along the coast of Palmer Land are fairly stable at present but are showing signs of thinning. It is predicted that these sections will also recede in the coming decade if the warming trend continues.

4.18 Research from Indian Stations

Continuous climatic records were recorded at Dakshin Gangotri from 1984 to 1989 and from then on at Maitri. Now this is done at both Maitri and Bharati. Though located in East Antarctica, all three stations have different environmental conditions. Whereas DG, built on the shelf, was buried in snow and had to be abandoned, the other two stations, built on oasis areas, face no such danger. Maitri has to be logistically maintained with tractor or sledge convoys from the old DG location on the ice shelf, serving as a supply base. Being on a coast promontory itself, the maintenance of Bharati may be comparatively easier due to the very short distance from the berthing place of the ship.

Recorded data on a few climatic parameters at DG and Maitri during the last winter (1989) at the former and the second winter (1990) at the latter are shown at Table 4.1 (Khadilkar, 1990 and NCAOR Technical Publication No. 6).
Salient points from the above data and other reports are as under (Koppar, 2004):

- At DG, the highest maximum temperature reaches up to 4° to 6°C in December-January and the lowest minimum temperature goes up to minus 50°C in August – September. The mean temperature of the year is in the range of minus 15° to minus 20°C.
- At Maitri, the corresponding figures are 8° to 9°C in December-January, minus 30°–35°C in August. The yearly mean temperature is minus 8.5° to 10°C. The highest maximum temperature recorded so far was 12.2°C in 1996.
- The temperature regime at Maitri is distinctly warmer than at DG. Data on blizzards show that they are more severe, frequent and long lasting at DG. Fog is a rare phenomenon at Maitri, whereas at DG, thick fog is experienced in December-January.

Micro-meteorological data is recorded hourly and transmitted on a real-time basis. Maitri is included in the WMO network of synoptic stations with the index number 89514.

Ozone monitoring is being done since 1987 and the profile by Laser Heterodyne and Radio Spectrometer has shown 30% to 68% depletion in concentration in the height range of 13 km to 40 km (Jain, 2004). Sun photometer, UV related studies and optical aeronomical experiments have been conducted for monitoring auroral

Table 4.1: Climatic Data Recorded at DG and Maitri

Sr. No	Month	*Air Temperature (°C)*						*Wind Pressure (Knots/Hour)*				*Blizzards (No of days)*					
		Highest		*Lowest*		*Mean*		*Highest*		*Mean*		*No of Blizzards*		*No of Days*		*Longest Duration (Hrs)*	
		DG	*M*	*DG*	*M*	*DG*	*M*	*DG*	*M*	*DG*	*M*	*DG*	*M*	*DG*	*M*	*DG*	*M*
1.	January	1.0	7.8	–12.6	–7.6	–3.9	0.9	35	40	15	9.4	3	–	9	–	84	–
2.	February	–0.5	3.7	–20.5	–9.2	–7.4	–2.9	50	62	20	15.9	8	–	16	–	102	–
3.	March	–1.8	–0.2	–27.9	–18.5	–13.9	–7.1	48	72	16	19.3	6	–	12	–	90	–
4.	April	–4.5	–1.0	–31.5	–21.6	–19.0	–10.1	58	60	15	18.6	4	–	9	–	86	–
5.	May	–17.5	–2.3	–42.0	–30.8	–29.6	–12.3	50	61	10	19.7	3	–	7	–	32	–
6.	June	–10.0	–5.3	–44.0	–29.5	–25.2	–16.3	52	62	12	17.8	5	–	6	–	71	–
7.	July	–14.0	–0.6	–46.0	–21.2	–28.1	–10.1	45	70	11	26.4	8	–	13	–	44	–
8.	August	–14.0	–8.8	–43.0	–32.0	–27.6	–20.5	33	61	11	11.5	4	–	4	–	17	–
9.	September	–11.8	–9.2	–46.0	–31.5	–27.6	–19.2	35	50	9	13.3	2	–	4	–	25	–
10.	October	–5.7	1.0	–35.5	–22.8	–18.8	–9.2	42	85	12	19.9	7	–	12	–	47	–
11.	November	–1.3	6.2	–24.2	–16.4	–9.7	–4.1	50	50	17	16.5	7	–	13	–	91	–
12.	December	–3.7	9.0	–14.5	–5.1	–4.6	1.0	25	41	8	11.8	NIL	–	NIL	–	NIL	–

* In last three columns for blizzards, the data for Maitri is not available so not given. However, blizzards are found to be much more frequent and severe at DG than at Maitri due to the proximity of the former to coast (Koppar, 2004).

emission. Regular monitoring of greenhouse gases such as CO_2, CH_4 and CO is also done. The Indian contribution in this respect has been brought out by certain scientists (Rajendran, 1993).

The interaction of the sun's atmosphere with earth's magnetic field and magnetic storms are studied. Instruments like Riometer/Flux magnetometer have been installed to observe aerospace interaction.

From Bharati, too, research on all the above aspects is planned and data will be obtained when the station is fully functional. As of now, recorded data from the nearby Chinese station Zongshan shows the area is marked by strong katabatic winds. Daytime temperature from December to February frequently exceeds 4°C with a mean monthly temperature above 0°C. Pack ice is extensive and snow cover is thick. Severe weather with storms is experienced. Extreme minimum temperature recorded so far in the region is minus 40°C (NCAOR, CEE, 2010).

4.19 In Conclusion

Antarctic climate can be termed as the most detested and feared aspect of its environment for all explorers and visitors. Explorers, since the early ages, described it and found it uninspiring and gloomy. Mawson famously called it the home of the blizzard well over a century ago. Due to the dangers posed by blizzards, periodical record-keeping and research of the Antarctic climate was started by sealers and whalers in the surrounding ocean in early explorations, and on the continent proper in the beginning of the twentieth century. However, climatic research got a real boost in and after the IGY and, today, all Antarctic stations have weather-recording stations in addition to various Automatic Weather Stations (AWS) at remote places. Research about optical phenomena, delving deep into climatic variations through history, weather forecasting and other aspects have been constant features of the meteorological research. The ozone hole once detected added an entirely new dimension in the field of atmospheric research in Antarctica since the late 1970s and remains one of the key areas of research today. The importance and recognition of this aspect can be seen by various conventions and agreements concluded as part of the

ATS (Antarctic Treaty System) to reduce ozone depletion by reducing and eliminating man-made reasons for it. Wider acceptance of the gravity of this issue can be seen by the fact that the UN Convention for the Protection of the Ozone Layer (1985) and the protocol on substances that deplete the ozone layer (Montreal Protocol, 1987) are signed by 191 countries. Catastrophic effects of climate change and global warming give conclusive warning as to why the pristine environment of Antarctica must be maintained. Climate research in Antarctica can play a very significant role in this respect.

Attention to the future assessment reports of the IPCC and the sincere execution of the Paris Agreement by the world community would go a long way to address the dangers of the climate change.

With two permanently maintained research stations and elaborate meteorological observatories in Antarctica, India has fitted itself well into the World Meteorological Organisation's (WMO) network of synoptic stations, with Maitri assigned the index number 89514. India has been contributing to the research on the ozone hole since 1987. As Maitri and Bharati are located in the sub-auroral and auroral zones respectively, simultaneous complimentary research in this respect from both the stations would help in understanding the development and movement of the longitudinal propagation of auroral current systems. These and other aspects of scientific research mentioned elsewhere would enhance the Indian contribution to global Antarctic research (NCAOR CEE, 2010).

CHAPTER 5

The Life that Clings on Ice, Abounds in the Ocean

"Take it all in all, I do not believe that anybody on earth has a worse time than an Emperor Penguin."

—Apsley Cherry-Garrard (1886–1959)
(Member of Scott's last expedition in 1910–13
in the introduction of his famous book
The Worst Journey in the World)

5.1 Introduction

The harsh combination of a permanent and massive ice cap over ninety-eight per cent of area, rapidly changing temperatures, strong drying winds, irregular water and nutrient supply, frequent snowfalls, frost and soil movement due to freezing and thawing, is hardly a recipe for any plants or animal life to exist. Only a few plants and invertebrates can survive on the ice-free land of Antarctica which consists of oases on the continent proper, maritime Antarctica and the sub-Antarctic islands. Plant and animal life grow in accordance with the conditions to create and sustain them.

On the other hand, the marine environment in the Southern Ocean is far richer and stable. Conditions have remained virtually unaltered over thousands of years and ocean currents bring in a constant flow of new species. A giant self-sustaining food web, consisting of interacting and interdependent food chains, has been built up over thousands of years. Due to this environmental interdependence, the adverse effects of climate change and the human exploitation of Antarctic life may severely affect the world environmental and ecological balance with catastrophic results.

5.1.1 *Research over the Years*

Considerable research activities in the continent and the surrounding ocean have taken place but, in the early years, they were individual efforts as parts of expeditions, chiefly for physical exploration. Joseph Hooker seems to be the first to make botanical observations in 1839–45. The first lichen and the first of the only two flowering plants were found in 1895 and 1899 respectively. Early zoological research mainly described and classified the various species encountered. Invertebrates (a small fly, mites and springtails) were first found in 1900 by Emile Racovitza. While early explorers studied seals killed for dog food, Edward Wilson made an observation about birds in Scott's ill-fated expedition of 1912. It was only after the International Geophysical Year 1957–58 (IGY) that a coordinated and structured biological research on the adaptation techniques of flora and fauna and their anti-freeze systems in a cold-desert ecosystem gathered momentum. Branding of birds and fitting of transmitters started only in the 1970s. The study of the impact of effluents from research stations on undersea ecology and of tourism on penguin behaviour and their colonies, too, began. The considerable progress made in the research on various aspects of Antarctic life is described in this chapter.

5.2 Plants and Animals (Invertebrates) in Various Antarctic Zones

In terms of plant and animal life, all land in the Antarctic region can be divided into three distinct zones, as described below. Area and islands referred in the text are as shown in Figures 3.7 and 3.8 in Chapter 3.

5.2.1 *Life in the Continental Antarctic Zone*

The continental Antarctic zone is the circle of East and West Antarctica with the exception of the west coast of the peninsula. Near the coast of continental Antarctica, mosses, lichens, liverworts and algae can grow but are generally sparse. Apart from snow algae, plant life is limited to ice-free areas where they can grow in sand, soil, rock, weathered bones and feathers of dead animals or other plants.

A population of over one million per square meter has been found in the case of mites, springtails, nematodes and others. The same has been observed in the case of algae growing on feathers of dead penguins and in mosses, and of algae in areas enriched by the debris of penguin rookeries.

Areas capable of supporting life are the Dry Valleys of Victoria Land – cold deserts with saline lakes, small areas of the continent where the winter snow melts in summer, providing water for a variety of organisms and the exposed rock-faces on isolated mountains. On the high East Antarctic ice sheet at the South Pole itself, there are probably no plants at all. There are no vertebrates in continental Antarctica. Though penguins, seals and petrels must come ashore to breed, they live and feed at sea. Some microscopic invertebrates live where moisture is available on land and in lakes. Certain bacteria, algae, fungi and arthropods have also been found in the most barren environments close to the South Pole (Walton and Everson 1987).

5.2.2 *Life in the Maritime Antarctic Zone*

This zone consists of the west coast of the peninsula and its offshore islands, together with the islands south of the maximum extent of sea ice – South Sandwich, South Orkney, South Shetlands and Bouvetoya. In the maritime zone, mosses and liverworts (bryophytes) are reasonably abundant and widespread. However, only two species of non-woody flowering plants (soft-stemmed angiosperms) grow and then only sparsely. Microscopic invertebrates and larger invertebrates such as springtails and free-living mites live in the soil and among rocks and vegetation.

5.2.3 *Life in the Sub-Antarctic Islands Zone*

This zone is made up of six island groups near the polar front or Antarctic Convergence – Marion and Prince Edward Islands, Iles Crozet, Iles Kergulen, Macquarie Island, McDonald and Heard Islands and South Georgia. These islands support abundant but treeless vegetation, verdant in summertime but brown when foliage dies in winter. The islands to the south and southeast of New Zealand

(Snares, Auckland, Campbell, Antipodes and Bounty islands) support similar vegetation with the addition of upright shrubs and trees. All the major plant groups, except cone-bearing trees (gymnosperms), occur in this zone. Microscopic invertebrates and larger invertebrates such as snails, earthworms, flies, beetles and spiders are plentiful on these islands.

5.2.4 *Survival Techniques of Antarctic Plants and Animals*

Organisms of the Antarctic have developed astonishing survival capabilities in the face of the harsh environment. Plants survive by maintaining the vital processes of photosynthesis and respiration at temperatures even below minus 10°C. Most of the potentially suitable sites for vegetation are occupied by penguin rookeries, whose trampling and excrement kill many plants. Sheets of green algae can survive such conditions, however, and brightly pigmented algae cause green, yellow and red snow on glaciers and ice caps. Microscopic fungi occur in the soil and, with bacteria, are responsible for breaking down dead plants to form peat and releasing nutrients into the ecosystem.

Ice damages cells and causes death. In order to survive the freezing temperatures of Antarctica, invertebrates must prevent or control the formation of ice in their bodies. They avoid freezing by a physical process known as 'super cooling', whereby their body fluids are maintained in a liquid state in temperatures below their normal freezing point. They have a constant struggle to maintain this unstable condition. The presence of food material in the gut provides some particles around which ice will form. So, in order to survive, they must strike a balance between freezing and starving. Their ability to synthesise glycerol, an anti-freeze agent, enables them to survive temperatures of minus 35°C. Some, such as tardigrades, accomplish this by losing water and shrivelling up when the temperature drops and reabsorbing it when conditions are more favourable. Others, like rotifers, lay drought-resistant eggs that remain dormant for long periods, hatching only when conditions improve.

5.2.5 *Continental Bio-Diversity*

Currently, the estimated numbers of species found in the three Antarctic regions are as given in Table 5.1 (McGonigal, 2008). The data is uptil 2006 and due to continuous research the figures keep changing frequently.

Table 5.1: Estimate of Antarctic Species (Plants and Invertebrates)

Plants	*Sub-Antarctic Islands*	*Maritime Antarctica*	*Continental Antarctica*
Flowering plants	56	2	0
Ferns and relatives	16	0	0
Mosses and relatives	~400	~100	~30
Lichens	~300	~150	~125

Invertebrates	*Sub-Antarctic Islands*	*Maritime Antarctica*	*Continental Antarctica*
Millipedes (Myriapoda)	3	0	0
Mites (Acari)	52	24	21
Flies (Diptera)	44	2	0
Springtails (Collembola)	92+	?	24
Crustaceans (Crustacea)	41	10	14
Snails (Mollusca)	3	0	0
Annelid worms (Oligochaeta)	4	0	0
Nematode worms (Nematoda)	22	40	10
Tardigrades (Tardigrades)	40+	17	6
Rotifers (Rotifers)	102	46	41

5.3 Antarctic Plants

Mosses, algae, fungi, lichens and flowering plants comprise the few species of Antarctic plants.

Moss is a class of bryophyte. It is a small plant with simply constructed leaves and no woody material. The southernmost moss species collected is *ceratodon purpureus,* found at 84°30′ S at Mount Kyffin in Southern Victoria Land. Most of the Antarctic moss species are found north of 70°S in the peninsula. Here, due to piercing sun rays on a clear summer day, temperatures on rock-faces, even under the dense coat of rapidly growing mosses, can rise up to 40°C even though the air temperature may be close to zero. Many varieties of mosses have tightly packed stems and shoots which prevent water loss. However, mosses are not as hardy as lichens and are thus less widespread. For most of the year, mosses survive the cold, drought and nutrient deprivation in a state of suspended animation. Moss banks may build to several meters in depth with all but the top 2.5 cm permanently frozen and turned to peat. Such peat banks may be thousands of years old. A footprint in a moss bank will remain imprinted for centuries (*Reader's Digest*, 1985).

Algae are a great group of thallophytic sea weeds. Numerous species of unicellular algae are found in Antarctica and cryoplankton, the unicellular algae, lives and reproduces in snow banks and glaciers. Algae have chlorophyll for photosynthesis but lack true stems, roots and leaves. Patches of cryoplankton are found in summer on snowfields near the Antarctic coast, particularly on the peninsula. The so-called red snow is coloured by algae that contain red and green pigments. Commonly, this occurs close to penguin colonies which supply enough nutrients for these algal populations to thrive, forming large mats of green algae such as the nostoc commune form. Algae are also found in the melt-water pools that form briefly in summer and in the water columns and bottom communities of fresh-water lakes.

Antarctic flora includes some species of macro-fungi (toadstools), but filamentous macro-fungi occur in soils. Few species are endemic. Macro-fungi are found only on the northern end of the peninsula, which is generally warmer and moister than the rest of the continent. However, they have been found to survive even in the extreme conditions of dry valleys where temperatures may drop to minus 80°C in winter. Here, in 1978, biologists had found widespread microbial activity, including fungi, algae and lichens, within light-coloured semi-translucent rocks. The microorganisms live in a micro climate inside cracks and pores in the rocks, and some are believed to have survived here for 200,000 years (Trewby, 2002).

Lichens are plants formed by the symbiotic association of fungi and algae or cyan bacteria (blue-green algae). They can survive even in the cracks and pores inside rocks under conditions of extreme cold and drought in Dry Valleys. Lichen distribution extends as far south as 86°30′ S. They are able to photosynthesize while frozen at temperatures as low as minus 20°C and can absorb water when covered by snow. During drought, they survive in a dry and inactive state. Lichen growth rates are extremely slow. In the northern part of the peninsula and nearby islands, they grow up to 1 cm per century. In the harshest Dry Valley region, they may take a thousand years to grow this much. Some lichens may survive for at least two thousand years. Lichens are sensitive to atmospheric pollution. Some have been found to contain traces of dichloro diphenyl trichloroethane (DDT) and oregano chlorine (Trewby, 2002).

There are only two flowering plants that grow below 60°S. They are the Antarctic hair-grass (*Deschampsia Antarctica*) and Antarctica pearlwort (*Colobanthos Quitensis*). Both are found on the western side of the peninsula, growing in small clumps near the shore. Antarctic hair-grass is more common and needs the soil created by lichens and mosses in which to flourish. Pearlwort grows in compact cushions about 25 cm across. Together with moisture and soil, both need the nutrients that they can obtain near bird-breeding grounds.

5.4 Antarctic Animals (Invertebrates)

Almost all of Antarctica's permanent animal inhabitants can only be viewed through a magnifying glass or microscope. The largest is a wingless midget, *Belgica Antarctica*, which grows to a maximum length of 12 mm. Most soil mites are barely visible, being only half to one-third millimeter long. Rotifers are around one-quarter of a millimeter long and are visible to the naked eye only when they are present in large numbers, sometimes straining pools red. Tardigrades are about half as long as rotifers. Seventy-five per cent of rotifers are endemic.

Most of the invertebrate parasites found in Antarctica – forty species of chewing lice, four species of sucking lice, nine species of feather mites, two species of nasal mites and two species of ticks – are all temporary residents which live on their seal or bird hosts. They escape many of the difficulties associated with living in Antarctica that must be faced by any free-living species. Only one of the parasites, a flee, remains all-year round in the nest of its bird hosts. Even entirely aquatic creatures, like whales, are hosts to a great variety of external parasites, such as lice, barnacles and water fleas.

5.5 Life in the Southern Ocean

As explained earlier, the terrestrial and fresh-water ecosystem in Antarctica tend to be relatively simple. Wide fluctuations in climatic conditions throughout the year militate against the successful establishment of a wide diversity of plant and animal forms.

The marine environment, on the other hand, is far richer and more stable. Conditions have remained virtually unaltered for thousands of years and ocean currents have brought in a constant flow of new species, gradually building up a diverse web of interacting plants and animals. The nutrients, plants and animals in an ecological system form an interdependent network. This is rightly called the giant food web of the Southern Ocean, linking all life in the Antarctic. Primary production by phytoplankton supports bacteria, protozoans and zooplanktons. Other key groups in the food web are copepods, crustaceans, squid and fish. Together with krill, they form the food base for birds, seals and whales.

5.6 The Giant Food Web of the Southern Ocean

All life in the Antarctic is linked by a giant food web. Primary production by phytoplankton and micro-plankton supports bacteria, protozoan and zooplankton. Other key groups in the food web are copepods, carnivorous zooplankton, some crustaceans, squid and fish. Together with krill, these organisms directly or indirectly form the food base for the more conspicuous birds, seals and whales.

Most organisms have more than one food source and are eaten by more than one predator, so that a food web is composed of a number of food chains or microbial loops interacting together. A food chain always begins with a primary producer, a plant or microorganism that is able to fix light energy into food. The primary producers are then eaten by primary consumers in the food chain: the herbivores, a group that includes zooplankton, comprising krill. Secondary consumers may be carnivores and include fish, seals, whales and birds. In the Southern Ocean, birds and mammals replace fish as the main consumers. For example, phytoplankton-zooplankton-krill-whales is one of the prime food chains. Protozoa-crustacean-demersal fish/octopods-ross and fur seals is another example of a food chain.

The giant food web of the Southern Ocean, encompassing all non-living (like nutrients) and living (plants and animals from microscopic to giant size) beings form an ecosystem based on interdependence and self-sustenance. It truly exemplifies the ancient Sanskrit saying '*Jeevo Jeevasya Jeevanam*', meaning how lives of living beings depend on one another. It helps to maintain the harmony and balance between all major parts of the global environment, namely land, ocean, cryosphere, climate and biosphere. Human presence in the Antarctic for just about two hundred years has threatened this enduring system which existed peacefully for millions of years.

The giant food web and the complex interdependence of various components is shown in Figure 5.1 (Walton and Everson).

5.7 The Growth of Phytoplankton as the Primary Production in the Southern Ocean

Phytoplankton are floating, microscopic, single-celled plants that make up the base of the oceanic food web. Phytoplankton grow abundantly during the extended daylight of summer and feed huge populations of zooplankton. Diatoms are unicellular algae. There are about a hundred species of diatoms and they form ninety-nine per cent of phytoplankton. Their bloom gives sea water a greenish tinge and strains ice floes a red-brown. Bacteria feeding on organic material dissolve mucus on the sticky surface of diatoms. The glassy silica walls accumulate as thick deposits at the bottom of the sea.

Around the continent of Antarctica, the maximum and minimum seasonal extent of sea ice is four and twenty million square kilometre in summer and winter respectively. The sea ice which forms and melts every year is called seasonal ice. Depending on this and the availability

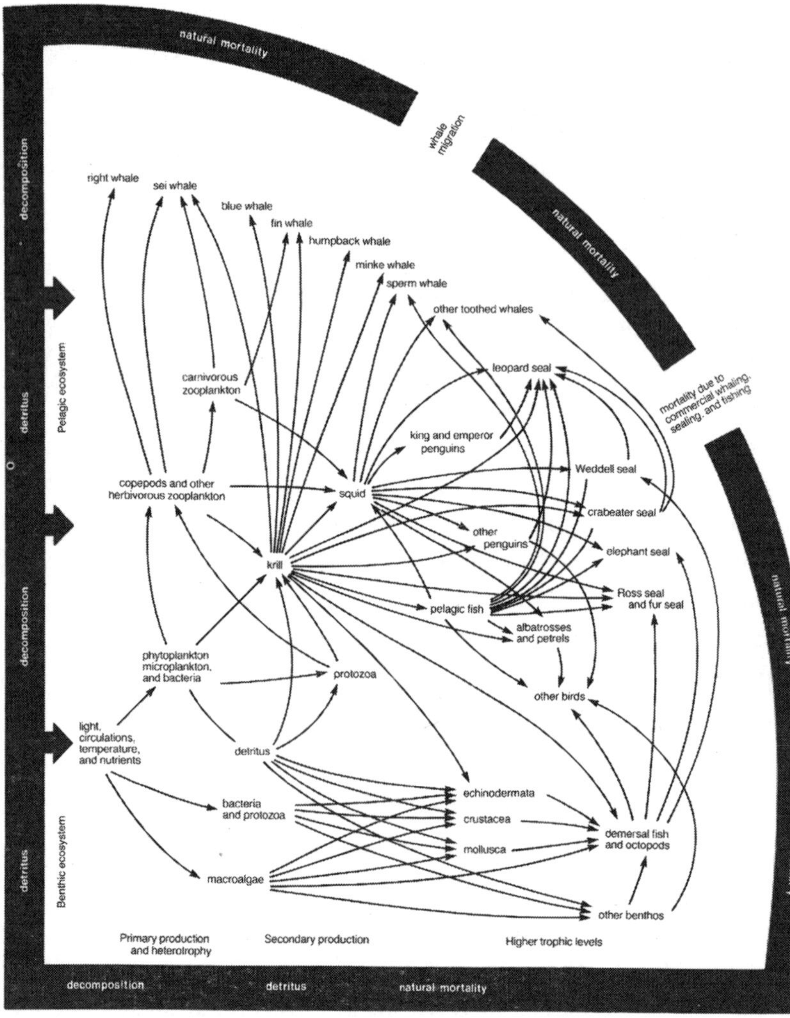

Fig. 5.1: The Giant Food Web of the Southern Ocean

of nutrients and light for photosynthesis (converting carbon dioxide and water into sugars or oxygen), the growth of phytoplankton varies in the ocean in various biological zones.

5.7.1 *Nutrients*

Photosynthesis and growth require both sunlight and nutrient-salts to form enzymes and structural molecules. There is an abundance of such nutrients in the Southern Ocean. South of the Antarctic Convergence, nutrient levels in the surface waters are very high, making it one of the richest oceans in the world. Upwelling of the deep occanic water at the Antarctic Divergence, the East Wind Drift and the stormy turbulence of the Southern Ocean, all help to bring nutrient-rich water to the surface. Here, it is available for photosynthesizing phytoplankton and thus enters the marine food web.

In maritime and sub-Antarctic habitats, minerals are derived from three major sources. Frost action accompanied by chemical and biological weathering causes minerals to be released from rocks and soils. Detritus is a mass of substance gradually worn off in the form of loosened fragments of rocks. A second source is aerial deposition in rain, snow and marine-derived aerosols. The third source of nutrients (especially nitrogen and phosphorous) is from the faeces of birds and seals who come to land to breed and moult. All these eventually reach the ocean in the form of melt-water run-off to add to the nutrient content of the ocean.

As a result of oceanic circulation, upswelling of water takes place in the Southern Ocean, too, bringing large amount of nutrients from the bottom to the surface. Here, along with sunlight, it creates excellent conditions for photosynthesizing, leading to an abundant growth of the primary producers of the food web or phytoplankton. This is then consumed by secondary producers or zooplanktons, completing the process of nutrients entering the food web. Further in the process, decomposing organisms such as bacteria and fungi enable the nutrients to be recycled within the ecosystem.

5.7.2 *The Process and Cycle of Phytoplankton Bloom*

Phytoplankton, which draw carbon dioxide from the atmosphere and produce oxygen, form the ocean's pastures on which all other living things depend.

As soon as the temperature of sea water falls to freezing point, a layer of ice known as pancake ice forms on the surface, insulating the water beneath. The pancakes gradually grow larger and soon collect a covering of snow which forces the ice down deeper into the water.

When the area between the ice and the snow reaches water level, sea water seeps between the layers, carrying nutrients and small algae which start to grow by photosynthesis. As the season progresses from autumn to winter, with days getting shorter and nights longer, the time and quantum of sunlight gets progressively reduced. Eventually, a turning point is reached when more energy is lost by organisms at night than is gained by them in photosynthesis during the day, interrupting their growth during winter months.

In the following spring, when sunlight returns, the algae in the ice grow very rapidly. The area of winter pack ice is so large that the crop of ice algae is nearly as great as that of the phytoplankton in the water below. As the pack ice starts to break up in summer, the entire annual crop of ice algae becomes available to animals. These algae, now living freely as phytoplankton, find all that they need to grow – open water, lots of nutrients (compounds like plant-fertiliser stirred up from deeper water by the wind and ocean circulation, and intense sunshine. What happens next is a seasonal bloom or a population explosion of phytoplankton in the water.

This goes to show how algae exploit an unlikely environment in the continent for their growth.

5.7.3 *Zones for the Growth of Phytoplankton*

The Southern Ocean south of the Polar Front or Antarctic Convergence can be divided into three zones that differ significantly in the extent of phytoplankton bloom or oceanic pasture. The three biological zones are shown in Figure 5.2 (McGonigal, 2008).

The three zones are:

1. ***Permanently Open Ocean Zone (POOZ):*** POOZ is nutrient-rich, but has relatively low primary production. Nowhere in the world are shore beds of kelp as luxuriant and prolific as they are in the surf zones around the sub-Antarctic islands. The phytoplankton here are generally tiny and called nanoplankton. They are grazed by several groups of animals but usually not by the Antarctic krill.
2. ***Seasonal Ice Zone (SIZ):*** South of the POOZ, SIZ is the most productive zone for phytoplankton. It is ice-covered in winter

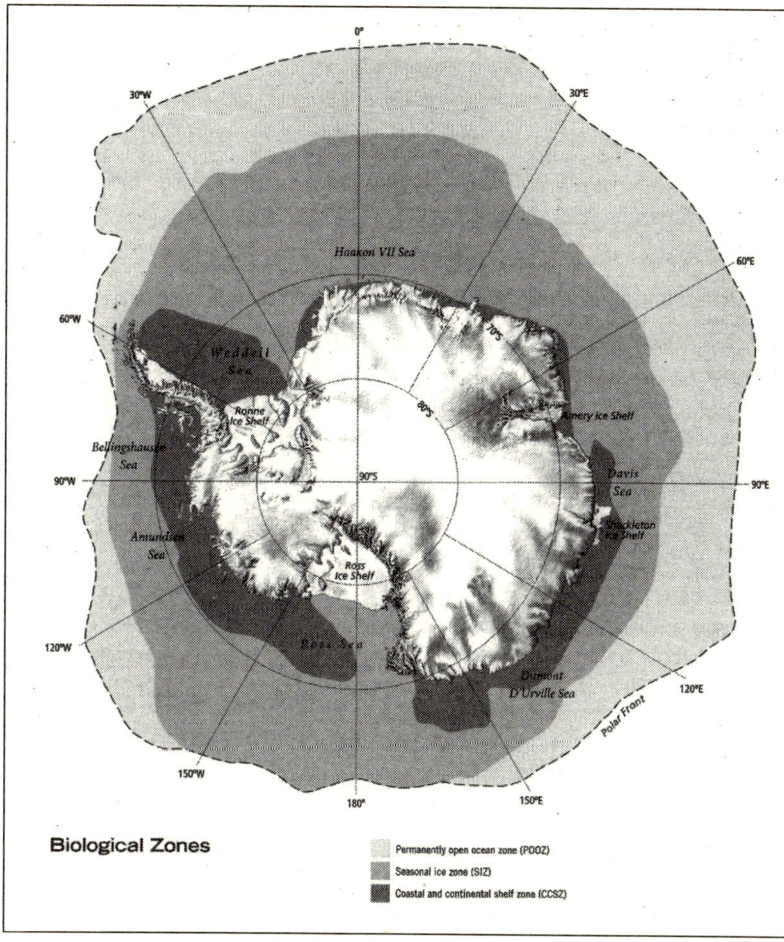

Fig. 5.2: Biological Zones of Southern Ocean for Growth of Phytoplankton

(twenty million square kilometres), but most of it, due to sea ice melting becomes open water in summer months (sixteen million square kilometres). Phytoplankton blooms are common in the shallow, less saline water produced by the retreating ice edge in spring and early summer. They are grazed by dominant and abundant planktonic animals, mainly krill. Baleen whales, crabeater seals, penguins and other birds feed on massive krill stocks.

3. ***Coastal and Continental Shelf Zone (CCSZ):*** The most southerly ocean region with an area of four million square kilometres is also

known as the permanent pack ice zone, although the region is not entirely covered by pack ice at the time of its maximum retreat. Phytoplankton blooms may be intense here but they are short-lived. Krill are uncommon and replaced by their smaller relative, *euphasia crystallorophias*. Because of the absence of krill, birds and mammals are less abundant here than in the SIZ.

5.7.4 *Impact of the Phytoplankton Cycle and Ocean Circulation on Global Environment*

As the sea water freezes, cold, highly saline water is excluded from forming ice (icehas a lower salinity than the sea water from which it is produced). This cold, saline water sinks to the sea floor and flows to the Northern Hemisphere as Antarctic bottom water. There is a concern that global warming may reduce the amount of sea ice around Antarctica and, therefore, diminish the production of phytoplankton and Antarctic bottom water. If the vertical mixing of the ocean lessens, it could lead to stagnation of the bottom water, depleting it of oxygen and making it inhabitable for most living things. This can have a catastrophic effect. Reduction in phytoplankton may adversely affect the entire food-web functioning.

5.8 Primary Consumers and Secondary Production of the Food Web

Though very small or microscopic and performing the same role as consumers of primary production and a prey to other secondary production and higher animals, primary consumers and secondary production of the food web are very different in their living and escape methods.

5.8.1 *Copepods*

These are crustacean-like krill that feed on phytoplankton and, in turn, are consumed by zooplankton, fish, birds and whales. They are the largest biomass on earth, competing with krill. They are generally microscopic and only a few of the seventy species are more than one millimeter long. Some of them graze on protists (common name for single-celled plants and animals), others are carnivores and some are omnivores. Estimates indicate that copepods consume more than eight times as many protists as krill do. Most copepods are frcc living and found from the surface to great depths. Some copepods are parasitic.

5.8.2 *Protozoa*

These are single-celled animals that feed on phytoplankton, bacteria and detritus in the water. They are highly diverse in shape, size and lifestyle. Some types even retain the photosynthetic system from the phytoplankton they eat.

5.8.3 *Echinodermata, Crustacea and Mollusca*

Also known as spiny skins, *Echinodermata* are starfishes that usually move by tube feet. The *Echinodermata* phylum contains five distinct classes of organisms in the Antarctic benthos (sea floor): sea stars, feather stars, brittle stars, sea urchins and sea cucumbers. There are about forty-five species of sea stars also known as starfish. Females brood their young rather than spawning eggs in water. They prey on sponges, mollusks, other echinoderms and various zoophytes. *Echinodermata* have a unique method of digestion. Most have no teeth or feeding appendages; instead they push their whole stomach out of their body cavity and over their meal, the tissues of which are then dissolved by the sea star's digestive juices. When it has finished eating, the sea star simply withdraws its sac-like stomach, now full of particles, back into its body.

The Antarctic zooplankton is primarily composed of species of the crustacean class of organisms, particularly krill and copepods, making crustaceans a critically important component of the Antarctic food web. They have two pairs of antennae, a jointed external skeleton, gills and a body divided into a head, thorax and abdomen, of which the head and the thorax are often fused. They usually go through several larval stages. The Antarctic benthos community includes a number of crustacean species like the giant isopod which is a predator of other benthic animals. Another forty isopods species along with a hundred species of amphipods are present in the food web. Small shrimp-like crustaceans are also found, with nineteen out of thirty-seven species endemic to the Antarctic region. Mysids can form large swarms and are an important part of the diet of many fish, birds and crabeater seals.

About nine hundred species of molluscs form part of the giant food web. Mollusc is a family of soft-bodied invertebrates, the bodies of which are usually protected by a shell. A few species of molluscs are without a shell. Antarctic molluscs are usually small with a few exceptions like *yoldia eightsi* (shell-length of 3.5 cm at an estimated age of sixty-five years, based on growth-ring calculations), *yoldia* with a shell-length of 4.5 cm at about 150 years and the Antarctic scallop with a shell-length of about 10 cm.

5.8.4 *Demersal Fish*

The coastal bottom-dwelling or demersal fish of Antarctica are a varied and endemic group of about fifteen families dominated by members of the sub-order *notithenioidel* which include Antarctic cod and plunder fish. Demersal fish feed mainly on invertebrates commonly found in the benthos, as well as fish larvae and young fish. Some young demersal are pelagic, or live in the surface waters of the Southern Ocean and are thus able to take advantage of the summer phytoplankton abundance. Once they mature, they inhabit the benthos.

5.8.5 *Antarctic Fish*

Worldwide, there are more than 20,000 species of fish. Of these, only about 120 are found in the Southern Ocean (McGonigal, 2008). In marked contrast to other parts of the global ocean, the Southern Ocean contains a few pelagic (open-sea) fish like the Antarctic herring or lanternfish. The sea-ice zone is the habitat of the adults of many of these, but they spend their first years of life in deep water. During the day, they can be found at depths of 200 m, but migrate near the surface at night. Up to 30 cm length, they feed on copepods, euphausiids, and juvenile fish and are, in turn, eaten by marine mammals and birds. Some of them have luminous organs on their head and body. Few can grow to more than 50 kg.

Unlike the Antarctic invertebrates, the body fluids of fish are less saline than sea water, with a freezing point of minus 1°C. They have evolved anti-freeze compounds called glycoprotein, which depress the freezing point of water in their body fluids.

They generally grow slowly, take three to eight years to become sexually mature, have long life spans and low metabolic rates. They produce only large, yolky eggs, so their reproduction rate is generally low. These characteristics put the Antarctic fish at serious risk of overfishing. Marbled rock cod, mackerel icefish and the Patagonian tooth fish all have suffered predictable patterns of exploitation – initially high catches that rapidly decline to an uneconomic point, whereupon the fishing is transferred to a new target species, and the pattern is thus repeated.

5.8.6 *Squid*

Squid are one of the most important components of the ecosystems of the Southern Ocean. Around seventy species of squid are found in the Antarctic waters, about half of them endemic. Their size ranges from 15 cm to 10 m, from the small *brachioteuthis* to the large *mesonychoteuthis hamiltoni.* There also is a giant squid, *architeuthis dux*, which reaches 18 m. Squid have ten sucker-studded arms and large eyes that allow them to see well in low light. They use large hooks as well as suckers to capture prey. There is a strong relationship between a squid's body and its beak size. They feed on small fish and crustaceans like krill. They produce egg masses in water deeper than one kilometre. Young squid grow rapidly, reach sexual maturity within a year, spawn, and then die. Some species may live for several years. They eat voraciously, consuming up to thirty per cent of their body weight per day.

Squid form a major part of the diet of seals, toothed whales except orcas, and birds, particularly albatross. They number nearly a hundred million ton and about forty per cent of this number is eaten by their predators. Up to 18,000 squid beaks have been recovered from inside a single sperm whale (Trewby, 2002). Their population can fluctuate due to their short lifecycle and the fact that they spawn only once and die soon after.

Squid are especially susceptible to overfishing. Presently there is no commercial fishing in the Antarctic waters.

5.8.7 *Krill*

Krill is a Norwegian whaling term meaning small fry. It is used to describe about eighty species of open-ocean crustaceans. However, the dominant species is a transparent shrimp-like creature called *euphausia superba*, the Antarctic krill. From the survival and numbers' angle, krill is one of the most abundant and successful animal species on earth and an estimated five hundred million tons of Antarctic krill are in the Southern Ocean. Without this most important component, the Antarctic ecosystem and giant food web would collapse.

Adult Antarctic krill are about 6 cm long. They have a shell (exoskeleton). They grow by moulting the old shell, expanding the new one as it is forming and then growing into it. This is done repeatedly until they reach maturity. For much of the year, krill go without food as their principle food, phytoplankton, is scarce due to the absence of sunlight for photosynthesis in winter. In laboratory studies, krill have coped with two hundred days of starvation. Rather than rely on accumulated fat reserves, they use up muscle protein. They live for five to ten years and sexually mature at three years of age. They continue to moult about once a month, but instead of growing, as they should when food is abundant, they shrink. Also, at the end of summer months, mature krill lose their sexual characteristics, which they regain in spring for spawning in summer. So a five-year-old mature krill, when it has lost its sexual characteristics and has shrunk, cannot be distinguished from a sexually immature two-year-old krill.

The primary source of food for krill is microscopic phytoplankton that blooms in the surface waters during spring and summer. Krill trap these abundant drifting plants with the fine hairs on their thoracic appendages.

Spawning takes place near the surface from late December to March. The females lay up to 10,000 eggs, depending on food availability. The eggs sink to a depth of 1000 m, where they hatch. The embryos go through a series of non-feeding larval stages before rising to the surface as larvae needing to feed. This process of descent and ascent takes between three to five months. In the second summer, they develop into juveniles and can form swarms.

Krill move in swarms which can contain around 30,000 individuals per cubic meter and are easy targets for squid, whales, seals, penguins and fishing trawlers. Often, krill swarms consist of all males, or all females, or all juveniles. A single swarm of krill encountered near Prydz Bay in East Antarctica, calculated to weigh about 57,000 tons, was composed entirely of large and sexually immature males. A super swarm seen in 1981 off Elephant Island covered about 450 sq km and was estimated to contain more than two million tons of krill (McGonigal, 2008).

Concerns about the commercial fishing of Antarctic krill, which began in 1970, led to the signing of the Convention for the Conservation of Antarctic Marine Living Resources (CCAMLR) in 1981, which restricts but does not ban krill fishing with the aim to conserve stocks and maintain ecological balance.

5.9 Penguins

Flightless, aquatic birds of the southern hemisphere, penguins are among the most specialised and well adapted of all creatures. They are true ocean dwellers and extremely accomplished at swimming and diving. Their distinctive black and white coats, streamlined torpedo shaped bodies and upright swagger have made them one of world's most favourite and photographed birds.

Of the eighteen living penguin species, seven can be classified as Antarctic or sub-Antarctic. Whereas two species, the Emperor and Adelie, breed on the Antarctic continent below the Antarctic Circle, three species, the Chinstrap, Gentoo and Macaroni, breed on the Antarctic Peninsula, and two species, the King and Rock Hopper, breed on the sub-Antarctic zone. Genera-wise they can be classified as *pygoscelis* or brush-tailed (Adelie, Gentoo and Chinstrap), *eudyptes* or crested (Macaroni and Rock Hopper) and *aptenodytes* or the largest/ heaviest (Emperor and King).

One specie, the Royal, is confined to Macquarie Island and is similar to the Macaroni Penguin. The remaining penguin species live in warmer zones as far north as the equator (Galapagos Islands) and are not found in the Northern Hemisphere.

Fossil evidence shows that there were once many more species of penguins than are in existence today. They appear to have evolved from gull-like or petrel-like flying birds of the Eocene Era (40 to 50 Ma) and passed through a stage when they could both fly and swim. In 10 to 15 Ma, there were penguins even larger than Emperor Penguins, with the largest, from Seymour Island, standing at 1.7 m.

Although the indigenous inhabitants of the northern coastal lands, such as New Zealand and Tasmania, Australia, would have known them, the first recorded sighting of penguins was in 1499 by Vasco de Gama's voyage (*Reader's Digest,* 1985). Penguin colonies became convenient food sources for sealing and whaling stations in the nineteenth century for both meat and eggs. Commercial exploitation was also tried. A penguin oil factory was operated for about twenty years on Macquarie Island before closure in 1919 (K Scott, 1993).

Like all birds, penguins are warm blooded with a body temperature that is close to humans. They are equipped with dense plumage, each waterproof feather overlapping its neighbour and trapping a layer of air next to the skin. The feathers come complete with the underfelt of woolly down and there is a thick layer of fat between the skin and underlying muscles. The fat also serves as an energy store. Well-insulated penguins can, sometimes, become overheated, and ruffling their feathers allows the insulating air layer to escape. Bare patches on the face, flippers and feet also get rid of surplus heat.

Penguins are superb swimmers and divers. Their triangular tail and webbed feet form a rudder, while their flippers, powered by huge wing muscles, provide the motive power to push them through water at great speed. They porpoise in and out of the ocean, coating themselves with air bubbles to reduce friction, and to snatch breaths of air. With retinas sensitive to violet, blue and green light, penguins' eyes are focused underwater. They have no teeth but spiny tongues and powerful mandibles to grasp their slippery prey. They absorb a large quantity of sea water, but have special glands that extract salt from the blood and excrete through the nose.

Crustaceans, squid and fish are their main diet. Penguins themselves are preyed upon by marine mammals, including leopard seals, sea lions and whales. Some species of birds scavenge their nesting sites.

5.9.1 *Brush-Tailed Penguins*

They are mainly black and white and derive their name from the prominent tail of stiff feathers which sweeps the ground. Similar in shape, size and behaviour, they are:

1. *Adelie Penguins.* [Height: 71 cm. Weight: 5 kg. Population: 2.6 million pairs and 10 million immature pairs (McGonigal, 2008). A small patch of dark-orange on the base of their bill distinguishes them from other penguin species.]

   The most abundant and widely distributed specie, Adelie live and breed on the shores of continental Antarctica or the Antarctic islands. They are rarely found north of 60°S. Adelie rookeries will form anywhere they can reach, be it hillside, knolls, beaches, peninsulas, as long as it is free of ice-ridges. Colonies of 20,000 to 30,000 birds are common. They winter on pack ice where the temperature is higher than on land and, in October, begin moving south, often travelling 80 km or more, to their breeding sites. The males arrive first to occupy the nest spaces a few days before the females. The pairs rebuild their nests at exactly the same location year after year. In November, after the eggs are laid, males incubate the eggs as the females go to the ocean to feed on a diet of krill and larval fish. They return after ten days or so for their shift. The partners alternate the duties for the rest of the thirty-five-day-incubation period. After the chicks hatch, they are warmed by their parents for two to three weeks before forming crèches. In February, young birds leave the colony, followed a month later by adults, after they have moulted. Adelie have remarkable homing abilities and they navigate by the sun.

2. ***Chinstrap Penguins.*** [Height: 72 cm. Weight: 3.8 kg. Population: 6.5 to 7.5 million pairs, with some populations increasing rapidly (McGonigal, 2008). They have a ring mark around their neck which distinguishes them from Adelie and Gentoo.]

   Chinstrap Penguins mainly live along the western shores of the peninsula and nearby islands. They nest on rocky, sloping ground. Noisy and aggressive, they often infiltrate Adelie colonies and take over their nesting sites. Females take the first incubation watch and parental duties are shared once the chicks hatch.

3. ***Gentoo Penguins.*** [Height: 75 cm. Weight 5.5 kg. Population: Over 300,000 pairs. Lower risk: near-threatened (International Union for Conservation of Nature, IUCN). They have a triangular white flash above each eye and a bright orange bill and feet to distinguish them from Adelie.]

   These penguin species are spread over a wide geographical range on the Antarctic and sub-Antarctic islands but move no further than 65°S. Two-thirds live on the sub-Antarctic island of South Georgia and the rest in the far south. Two different Gentoo species have been identified – the southern group or *P Papua Ellsworthii*, and the northern or *P Papua Papua*. The *P Papua Ellsworthii* is fully migratory and follows a clear path each year. Its feet, flippers and bill are smaller than that of the *P Papua Papua* which does not migrate. Gentoos establish themselves on mounds amongst tussock grass and collect stones and scraps of vegetation to make their nests. They follow a similar incubation cycle like that of the Adelie and the Chinstrap. They are the fastest swimmers of all the birds and are capable of bursts of speed of up to 27 kmph. They feed mainly on krill and other fish and can dive up to 100 m.

5.9.2 *Crested Penguins*

They are sometimes called crested, a reference to their jaunty, yellow eyebrow plumes. Their life-pattern is very similar within the species and the relationship between the species is also very close, resulting in occasional interbreeding and successful reproduction.

1. ***Macaroni Penguins.*** [Height: 71 cm. Weight: 5 to 6 kg. Population: About 12 million pairs, some in decline. Vulnerable (IUCN) (McGonigal, 2008)]

   Confined to the inner and outer Antarctic islands, Macaroni Penguins are found in millions on Heard and South Georgia Islands. After spending winter at sea, they return to their colonies between September and November to engage in a lively courtship before building their sparse nests.

   Like the other *eudyptes* species, Macaroni Penguins lay a very small first egg which is usually lost after the second egg is laid six days later. Both parents take shifts in the incubation period of a month and hatching takes place between November and January. The chicks fledge after sixty-five days and both parents go to sea to feed for a few weeks, returning to the beaches in March to moult.

2. ***Royal Penguins.*** [Height: 70 cm. Weight: 5 to 6 kg. Population: About 850,000pairs. Vulnerable (IUCN), probably due to their confined existence in the large colonies on Macquarie Island and nearby islets (McGonigal, 2008)]

   Many scientists believe that the Royal Penguin is a subspecies of Macaroni, the only difference between them being the colouration of face, chin and throat.

   Whereas most penguins start breeding by three to six years of age, Royals have prolonged adolescence, and it may be as much as eleven years before they breed. They build their nests by excavation from sand and stones.

3. ***Rock Hopper Penguins.*** [Height: 45–58 cm. Weight: 2.3 to 3.4 kg. Population: About 1.1 million pairs, some in decline. Vulnerable (IUCN)]

   They are the smallest of the Antarctic penguins and breed on the cliffs and scree slopes of islands close to and north of the Antarctic Convergence. Very similar to Macaronis, they are found most often in their company.

5.9.3 *Largest Penguins*

This wholly Antarctic genus comprises the two largest species of penguin, namely Emperor and King. Their bills are long and slender and curve gently downwards. The two species are closely related, but they live in vastly different conditions. Neither species builds a nest, even though King Penguins live in a region where nest-building materials are readily available. Their hatching strategies are totally different.

1. ***Emperor Penguins.*** [Height: 100–130 cm. Weight: 38 kg. Population: About 500,000, thought to be stable (McGonigal, 2008)]

   They have blue-grey backs shading to a black tail and their under parts are white flushed with yellow which strengthens to deeper yellow curving ear patches. The chicks have black heads with a white eye mask. They live in about forty known colonies situated on winter fast ice and never set foot on land. The Emperor's body is designed to withstand temperatures of minus 10°C before it must use body energy to keep warm.

   After pairing for six weeks, the female lays a single egg and transfers it to the male. The male holds it on his feet and covers it with a brood patch to keep it warm. The female heads to the sea to fish. For the next nine weeks the males incubate the egg through the winter in constant darkness, surviving by huddling together when the temperature often drops to about minus 60°C with winds blowing at up to 180 kmph. The fathers give the first feed before the females come back and persuade transfer of the chicks. Having lost half of their body weight, the males now walk about 100 km to reach the sea for much-needed food. If the female is late in returning, the male simply abandons the egg or the chick and makes for the sea to feed before it is too late for him to survive. After the chicks begin to hatch, they need constant feeding, and the parents take turns for seven weeks. They take out the food from their stomach and feed the chick. After a while, crèches are formed and both the parents go off to the sea to feed.

Emperor Penguins dive to an average depth of 300 m but can go up to 400 m. Similarly, their aerobic dive limits are about four minutes, but two single dives have been recorded of about twenty-two minutes, both while breaking long fasts.

3. ***King Penguins.*** [Height: 95 cm. Weight: 13 kg. Population: About 2 million, most populations increasing (McGonigal, 2008)]

   They are the second-largest living penguin species. Unlike most penguins, Kings have big splashes of colour in their feathering. They have a beautiful spoon-shaped patch of bright-orange on either side of the neck. There are two sub-species, one lives on South Georgia and Falklands and the other lives on Iles Crozet and Kerguelen, Prince Edward, Heard, and Macquarie Islands.

   Unlike the Emperors, after a very elaborate courtship, the pairs are formed. King Penguins gather in large numbers to reproduce in a staggering three-year cycle. The egg hatches in late January after it is incubated for two to three weeks by the parents in turns. The chick does not fledge till until next November at the earliest, but first it must survive the long winter. The parents take turns in fishing and guard duty till the chicks are so big that they need two parents to hunt and bring back enough food. By the end of summer, a healthy chick reaches a weight of around 11 to 12 kg. As winter approaches, food becomes scarce, parents can feed only sporadically and by the time spring arrives without plentiful food, the chicks have lost about one-third of their weight. So the parents are not able produce another egg until much later in the following breeding cycle. The hatching takes place earliest by March. So this later-born chick weighs much less by the start of winter than the chick born in January, and is less likely to survive.

King Penguins generally dive to depths of about 50 m, but dives up to 300 m have been recorded.

5.9.4 *Other Penguins*

Three species of crested and one of yellow-eyed penguins occupy New Zealand waters. A specie of little penguins is found in Australia and

New Zealand and four other species are found in the equatorial Pacific regions of South Africa and South America. They are least Antarctic.

5.10 Antarctic Seals

Seals form an important component of the giant food web. The scientific name for seal is *pinniped,* meaning fin-footed, and this is the characteristic that makes them ideally designed for an amphibious existence.

Antarctic seals are divided into two groups, five species of *Phicidae* or true or earless seals (Southern Elephant, Leopard, Ross, Weddell and Crabeater) and *Otariidae* or eared seals (Fur). Both have excellent hearing and call one another while underwater. Weddell and Ross Seals sing complex songs with whistles, moans and clicks. They have specially adapted eyes for underwater vision. On land, fur seals are more agile but in water, both the groups are supreme movers. They can dive to more than 600 m, stay under water for well over thirty minutes and feed on fish, krill, squid, crustaceans and penguins. Four of these species are ice-habitat specialists, breeding on the sea ice in spring. Leopard and Ross Seals tend to be solitary, whereas Weddell and Crabeater Seals form groups or colonies. The other two, Fur and Elephant Seals are found north of the pack ice zones and breed in dense colonies on beaches, with dominant males (bulls) maintaining harems of females (cows). They are well adapted to the cold polar environment with thick blubber layers that act both as food reserves and insulation. Antarctic seals have no terrestrial predators and show little fear of humans. Studies of seals are made with the help of radio transmitters but this is effective for one year for one seal as they moult each year (Seals: AAD, 2012). Males and females vary greatly in size and weight in some types.

5.10.1 *Elephant Seals*

Males weigh up to 8000 lbs and are 18 to 20 ft long. Females average about 900 lbs and measure up to 10 to 12 ft. (Landis 2001). Their population is about 750,000. Males have large proboscises that they inflate to threaten other males.

5.10.2 *Leopard Seals*

They weigh about 700 to 800 lbs and are 10 to 12 ft long. Their population is between 220,000 and 440,000. The females are slightly larger than the males. These seals have a spotted chest and belly fur.

5.10.3 *Ross Seals*

They weigh about 500 lbs and are about 8 ft long. They have a population of about 220,000. With large heads and slender bodies, these seals are very rarely seen. Fewer than one hundred sightings have been recorded since James Ross discovered them in 1841 (Landis 2001).

5.10.4 *Weddell Seals*

They weigh about 900 to 1100 lbs and are 10 to 12 ft long. Their population is about 800,000, and their head is small compared to their body.

5.10.5 *Crabeater Seals*

They weigh about 500 lbs and are 8 to 9 ft long. A population of about 15 to 40 million, these seals are characterised by large brown eyes and features reminiscent of dogs with cropped eyes.

5.10.6 *Fur Seals*

The females weigh about 90 lbs, males 420 lbs. Both are about 4 to 6 ft long and their population stands at more than four million. They have luxuriant silvery-brown ruffs. The females are about one-third the size of males.

5.11 Antarctic Whales (McGonigal, 2008; Landis, 2001)

They are the largest and most powerful components of the giant Southern Ocean food web. The main species found in the Antarctic waters are in two groups as described below.

5.11.1 *Baleen Whales*

The name is derived from the horny substance called baleen contained in plates on either side of the whale's mouth. The plates, fringed with bristles, serve as a giant sieve for straining krill and other sea creatures. Baleen whales breed in tropical waters in the north, migrating to the Antarctic waters to feed on a rich diet of krill during summer months. After migrating back to temperate waters, females give birth to a single calf, a year after mating. The food reserves, stored in the form of blubber, sustain them for rest of the year. The calves live on their mother's milk for six months, accompany them on the next migration and start feeding independently. In case of Humpback and Southern Right Whales, the calves depend on their mothers for up to a year. They feed in two ways. In 'swallow feeding', they first engulf a mouthful of food and water and then squeeze the water out by contracting the grooves in their throat. Right and Sei Whales swim through a swarm of krill while keeping their mouths partly open and skim food from the ocean. Humpbacks swim in a circle and trap the krill inside an air-bubble net. They then swim up through its centre, engulfing the food. There are six species in this group of Antarctic whales, varying greatly in weight, size and other body parts, like seals.

1. ***Blue Whales:*** With a length of up to 30 m and weight of about 180 tons, these are the largest living animal on earth. They are a shy band and rarely seen. The enormity of their size is illustrated by the fact that a blue whale calf drinks about 600 litres of milk every day. By the 1980s, they were near extinction, with only about a thousand remaining from the pre-whaling population of 200,000 (May, 1988). Their strength has somewhat increased as of now.

2. ***Fin Whales:*** The most common specie, the Fin Whale grows up to 25 m with an average weight of 40 to 50 tons. It is the fastest swimmer among all the baleen whales.

3. ***Humpback Whales:*** These can grow up to 17.5 m with a maximum weight of 48 tons. Their skin is covered with warts and bumps, earning their name. With extraordinarily long flippers, they produce the longest and the most varied songs in the animal world.

4. ***Minke Whales:*** With an average length of 8m and a weight of 6 to 8 tons, they live for about fifty years.
5. ***Right Whales:*** Fat and stocky with a finless back, this whale can grow up to 18 m with a weight of 96 tons. It has a large head which accounts for quarter of its length. It hosts a number of parasites, like barnacles, worms and lice.
6. ***Sei Whales:*** With a length of 18.5 m and a weight of 29 tons, they have a life span of seventy years.

5.11.2 *Antarctic Toothed Whales*

They differ from other mammals in having only one nostril, though two nasal passages are present. Three species are found in Antarctic waters.

1. ***Orca Whales:*** They have a length of 9 m and weigh 7 to 8 tons. They are voracious pack hunters, feeding on squid, fish, sharks, sea birds and other whales. They are very strong and particularly fond of the tongues of large baleen whales (May, 1988).
2. ***Southern Bottlenose Whales:*** They have a cylindrical body, tapering towards the tail. The melon lump on their heads contains oil. A rarely seen specie, it grows up to 7 m with a weight of 3 to 4 tons. They feed on squid and cuttlefish.
3. ***Sperm Whales:*** The largest of the toothed whales, its males grow up to 18 m and weigh 70 tons, while its females grow up to 11.5 m and weigh 17 tons. The blunt head of this whale contains a wax-filled case on the skull. It can change the density of the spermaceti wax while diving to alter the buoyancy of the body. It feeds on squid. At a depth of 1000 m, these whales use sonar to locate their luminous prey.

5.12 Antarctic Sea and Shore Birds

There is a much greater proportion of sea to land between latitudes 50° and 70°S. As a result, only nineteen flying species in Antarctica depend upon the twenty-four island groups as breeding grounds. Others breed along the west coast of the peninsula and isolated nunataks on the ice sheet. The sea birds show a variety of feeding methods at the surface layer, which extends to 3 m, like shallow dipping, shallow diving, paddling along the surface with immersed bills, etc. Except penguins, 55% of the sea-bird population eats plankton, 20% (mainly albatrosses) eat squid, 20% have a mixed diet of plankton and squid and 5% eat fish. It is estimated that birds consume at least a hundred million tons of krill annually of which 90% is eaten by penguins (May, 1988). All sea birds can be grouped into two groups, namely *Procellaroformes* or *Tubinares* (sea) and *Charadriformes* (coastal) as described below.

5.12.1 *Albatrosses and Petrels*

The *Procellaroformes* or *Tubinares* group consists of ocean-ranging albatrosses and petrels. They have obvious tubular nostrils (with glands to excrete excess salt) on top of their beaks or on either side of it. This group contains six species of albatross, twenty-three species of petrel, prion and shearwater, four species of storm petrel and two species of diving petrel. The species range in size from the wandering albatross with a wingspan in excess of 3.3 m and weight between 7.7 and 9.5 kg, to the storm petrel which weighs a mere 30 to 60 gm. Very vague estimates show the population of albatrosses to be 750,000 pairs and all petrels together at 150 million (May, 1989).

5.12.2 *Coastal Birds*

The Charadriformes group is basically coastal birds rather than deep-water sea birds. There are many species within this group: one of gulls, two each of cormorants, sheathbills and skuas and three of terns. They dive for fish, squid, benthic worms, mollusks and crustaceans. The skuas are very aggressive and notorious predators and nest near penguin colonies and raid for eggs and young birds, returning to the same breeding grounds with the same partner. They lay two eggs in a rough nest of bones, pebbles and moss, but hatch only the first. Terns breed on the peninsula and islands in small colonies, their eggs and chicks are camouflaged by the surroundings. The two above species are supplemented by the Arctic tern, which migrate more than

40,000 km from the Arctic to feed here before returning. Arctic terns spend much of their lives in daylight and transit.

5.13 The Spectre of Exploitation

Due to the physical barriers, Antarctica was the last continent discovered. So the wildlife in the surrounding Southern Ocean, unharmed and secure, flourished and developed over millions of years through the medium of the oceanic giant food web. However, human greed gave way to the cruel exploitation of its unsuspecting biological resources, the history of which goes back to well over two centuries. This exploitation can be broadly termed into categories like sealing, whaling and fishing. Early exploitation was so easy and indiscriminate that some species came to near extinction before it was halted. Fortunately, the Antarctic Treaty (AT) and its instruments came into being to remedy the situation.

5.13.1 *Sealing*

British sealers first crossed the Antarctic Convergence in 1778 and the Amcricans in about 1792. Initially, the profits were enormous. In 1797, an American ship worth $3000 gathered 45,000 fur seal skins at Falklands. These were sold in Canton for $90,000. The crew then bought Chinese goods and sold them in New York for $260,000 (NSF, *Antarctica, Past and Present*, 1997). In 1820–21, at least fifty-five vessels were hunting seals in the South Shetlands. As further discoveries around the peninsula were made, fur seals (for fur) and elephant seals (for blubber and oil) were slaughtered to near extinction by the mid nineteenth century. As a result, sealing became uneconomical and was stopped for a while and seal stocks improved to some extent in a few years. In early last century, the British re-established the hunting of elephant seals with a maximum quota of six thousand per year on South Georgia, which continued for a while. With protection provided by the ATS, the situation improved and no seal hunting has taken place in the Antarctic since 1964. The Convention for the Conservation of Antarctic Seals (CCAS) provides total protection for Fur, Elephant and Ross seals, whereas for the other three species, an annual hunting quota is fixed. However being uneconomical, no seal hunting is done at present.

5.13.2 *Whaling*

Whaling began in Antarctic waters in the nineteenth century. With steamships, harpoon guns and shore-processing stations in the sub-Antarctic islands, the industry was enlarged greatly. For example, in 1912–13, the number of whales caught was 10, 760. It kept increasing and, in 1931, the number rose to 40, 199. As the stocks kept dwindling, the International Whaling Commission (IWC) was established in 1946 but many nations flouted the IWC regulations and whaling continued unabated. As one by one species faced extinction, IWC provided full protection to the Blue and Humpbacks in the 1960s, extended the same to Fin and Sei in the 1970s and, finally, established a total whaling moratorium in 1986. By this time, however, nearly 1.5 million whales were harvested (McGonigal, 2008). However, as per the exception made by the IWC, whales may be captured for scientific investigation purposes. Three IWC members still hunt whales in significant numbers. More than 25,000 whales have been killed since 1986. Whereas Japan conducts what it terms scientific whaling, Norway and Iceland openly defy the ban (*Fact box about IWC*, 2010). These whales end up in restaurants and one whale can have a market value of $1,000,000 at a restaurant table (*Cool Antarctica*, 2012). Japan had hunted up to a thousand whales in Antarctica annually as per the IWC report. As pointed out earlier, due to the ruling of the International Court of Justice (ICJ) in 2014, in a case taken up by Australia, this practice has now stopped at least in the Southern Ocean.

5.13.3 *Fishing and Krill Harvesting*

Commercial fishing in Antarctic waters began in 1905 around South Georgia. Full-scale trawling in the 1960s saw a decline in the population of rock, cod and ice fish due to overfishing. It has since given way to more profitable long-line fishing for the Patagonian tooth fish. But the real interest was generated by the fish which forms the most important component of the food web, the krill. Estimates in the 1980s suggest a standing stock of 500 million tons, out of which whales eat about 43 million tons, fish 60 million, birds 40 million, seals 80 million and squids about 100 million tons per year. Without disturbing the balance of the food web, it is thought that humans could take 80 to 150 million.

With this in mind and based on the Biological Investigation of Marine Antarctic Systems and Stocks (BIOMASS) from the Scientific Committee on Antarctic Research (SCAR), all signatories to the AT signed a Convention for Conservation of Antarctic Marine Living *Resources (CCAMLR).* This is in force since 1982 and covers all species except seals and whales for whom separate conventions, like CCAS and IWC exist.

Though exploratory fishing for krill started in the 1960s, it has not picked up as thought, with the maximum catch still standing at 527,000 tons in 1981–82 (May, 1987). Currently, Japan, Poland, Korea, Argentina and Ukraine together harvest about 110,000 tons krill per year. Recently, some countries like India, UK, Norway, US, Canada and Australia have shown some interest. The low volumes of fishing may be due to taste, high fluoride levels, etc, and presently the catch is mainly used as animal feed. They have, however, been used as cheese spread (Russia), krill sticks (Chile) and soup products (Norway) [McGonigal, 2008]. As compared to the world fishery with an annual catch of about eighty million tons, Antarctic fishery is just a fraction as of now (*NSF, Antarctica Past and Present*, 1997).

We may hear more on the subject of krill harvesting in the coming years.

5.13.4 *Accidental Birds Killing*

Though birds are not commercially important, in the initial stages of Antarctic exploration, penguins were killed for oil. A factory at Macquarie Island of Australia had processed 150,000 King Penguins from 1895 to 1919 before the practice was abandoned (McGonigal, 2008). Incidental killing of birds has also happened on a very large scale. Long-line techniques for catching fish have killed between

20,000 to 70,000 albatrosses, 5000 to 11,000 giant petrels, and 80,000 to 180,000 white-chinned petrels in four years between 1996 and 1999. Despite efforts from CCAMLR, the problem still exists, though due to the number of illegal ocean trawlers, the exact data is not available (Trewby, 2002).

5.13.5 *Continuing Research*

New studies and surveys about global warming and its effects on the extent of sea ice, the giant food web, icebergs and their effect on the Antarctic wildlife make frequent appearances in world news. A recent study in East Antarctica points out that the population of Adelie penguins which nest along the coastline, has increased by 135 times in the last 19,000 years. This is as a result of deglaciation which reduces sea ice, making foraging for food for their chicks easier. But this could be balanced by the sea-ice reduction – less growth of phytoplankton – reduction in krill numbers cycle (Younger, 2015). In a probable freak incident in the Commonwealth Bay area of East Antarctica in February 2010, a massive iceberg 100 sq km in size got grounded. As a result, Adelie penguins had to walk 60 km more to find food, killing most of the penguins in the process and reducing their population of about 1, 60,000 to a mere 10,000 by December 2013 (University of NSW, Australia and West Coast Penguin Trust, New Zealand, 2016). The effect of sea ice shrinking maybe devastating to the entire population of polar bears in the Arctic. As per current projections, global warming may completely free parts of the Arctic for five months in a year by the middle of the century. The longer duration the sea ice disappears for, the longer the bears will have to go without food. As a consequence, the current population of 26,000 may be reduced by as much as thirty percent (Bolten, 2015).

5.14 Research Activities of the Indian Antarctic Program (IAP)

With only one year-round research station, Maitri, operational since 1989 about 90 km from the coast, Indian activities were limited and Indian scientists focused on sea-ice micro-organisms, living benthic communities and bio-diversity in the Schirmacher Oasis area. The recent establishment of Bharati on the coastal promontory about 3000 km away from Maitri, would certainly add to the quantum and range of activities of the IAP in this aspect. Salient points of their attention and research are as under.

5.14.1 *Research Activities at Maitri*

- Psychrophilic bacteria (organisms that can live and grow in temperatures that are about minus 10°C) play an important role in sub zero activity in Antarctica. So far, around 125 new species have been discovered from Antarctica, of which India has contributed around twenty new species (Shivaji, 1989: Reddy, 2002a, b; 2003 a, b, c, d). Two new species, *Arthobacter Gangotriensis* and *Plannococcus Maitriensis,* are named after the first two Indian stations, DG and Maitri (Alam, 2003; Gupta, 2004).
- Inventory of the extent of biodiversity for areas around Schirmacher Oasis.
- Systematic mapping of the above area for species diversity distribution in different micro-climatic conditions.

5.14.2 *Planned Activities from Bharati (NCAOR CEE, 2010)*

In addition to the aspects of the biological research activities at Maitri, the following additional areas will get an impetus from Bharati:

- Physical, chemical and biological oceanographic studies in the Indian Ocean sector of the Southern Ocean. A long-term and systematic approach regarding the time-series observations of hydrographic parameters at different locations in the Prydz Bay area will be carried out to understand quantity, causes and year-to-year variability of fresh water input in the area.
- Biological studies at the new site have the potential to explore new vistas in:
    1. A biodiversity study of sea-microbial community.

2. The conservation of terrestrial and aquatic biota (fungi, lichens, bryophytes, micro fauna like protozoans, nematodes, tardigrades), and aerobic bacteria.
3. Monitoring of the wild life population and behavioural studies.

5.15 In Conclusion

As can be seen from the details mentioned earlier, biological research in Antarctica has come a long way and will go further in the coming years. Exhaustive details about plants and animal life in the continent proper and the surrounding ocean have been gathered and studied. The interdependence of all these, the need to maintain a balance between the entire ecosystem and the effects of environmental changes, are all well understood with the available information. The breakup of ice shelves and the spread of plants in the Antarctic Peninsula, for example, suggest regional warming. Global warming with all its implications will likely affect various aspects of Antarctic life – the extent of sea ice, growth of phytoplankton, vertical mixing of oceans, ozone hole and their individual and collective impacts on elements of the giant food web. Anthropogenic aspects, as discussed due to the spectre of exploitation, will add immensely to the seriousness of the impact, necessitating urgent global attention. Due to the obvious importance of the aforementioned aspects, intense research on them will attract the attention of scientists in the near future,

With two year-round research stations, India will be engaged in various enhanced research activities, as explained earlier. The close proximity of Bharati to the Australian, Chinese, Romanian and Russian stations will also go a long way in fostering significant scientific collaboration and cooperation.

CHAPTER 6

International Presence and World Governance in Antarctica

> "It is in the interest of all mankind that Antarctica shall continue forever to be used exclusively for peaceful purposes and shall not become the scene or object of international discord."
>
> *(Essence of the Antarctic Treaty reflected in the opening sentence of its Preamble)*

6.1 Introduction

The theory of a sphere-shaped earth with equally distributed land masses and climatic zones was developed by Greek philosophers in the fourth century BC. Physical exploration of the Southern Hemisphere, however, commenced much later. Ferdinand Magellan undertook the first circumnavigation of the world in 1519–22. In 1520, he sailed through the strait that bears his name today. As shipbuilding and navigation techniques of European countries gradually developed, the next three hundred years or so saw a growth in the geographical knowledge of the Southern Ocean. Many sub-Antarctic islands were discovered during sealing and whaling voyages. The great navigator and explorer, Captain James Cook, in his voyages during 1768–75, charted New Zealand completely, discovered the east coast of Australia and claimed the islands of South Georgia and South Shetlands for England. In the process, he became, unknowingly, the first person to cross the Antarctic Circle. Many expeditions and discoveries later, the Russian Admiral Thaddeus von Bellingshausen sighted the new continent on 27 January 1820. Three days later, the leader of another expedition, two thousand kilometres away on the coast, the Briton Edward Brans field, too, sighted the continent. For the next eighty years, sealing and whaling expeditions continued in search of new pastures. Through all this, the Antarctic coastline took shape but no exploration was made on the continent proper.

Thus, even till the beginning of the twentieth century, maps of Antarctica were blank except for a few wavy lines representing partial coastlines, ice shelves and landfalls. While the British forged inland towards the South Pole through the expeditions of Robert Scott and Ernest Shackleton, a flurry of other expeditions added new coasts and inlands to the Antarctic map. German Erich Dagobert von Drygalski, on 21 February 1902, named a hitherto unexplored stretch of coast on Antarctica's eastern side, Wilhelm II Land. The same day, his ship, *Gauss*, was beset by ice and remained ice bound for the next thirteen months. The Norwegian ship, *Antarctic*, was crushed due to pack-ice pressure near the tip of the Antarctic Peninsula. The ship with its crew and the Swedish expedition team on it were forced to endure an unscheduled second year on Snow Hill Island. The Scottish national Antarctic expedition of William S. Bruce had to retreat from Weddell Sea and winter on Laurie Island in South Orkneys. After maintaining a meteorological station there till 1904, they offered it to the British government who rejected it on financial grounds. So they offered it to the Argentines who have since occupied it as Orcadas Station. Frenchman Jean Baptiste Charcot mounted a rescue mission for the ship *Antarctic* and charted over 600 miles in 1904 and returned to survey about 1250 miles of coastline four years later in the Antarctic Peninsula. In the process many new islets, bays and mountains were discovered in the continent (May, 1988, McGonigal, 2008)

6.2 Territorial Claims of Seven Countries

Due to a flurry of these expeditions in the first decade of the twentieth century, Antarctic exploration suddenly acquired an international, nationalistic flavour as opposed to efforts which were only individual till then. As charts started giving shape to the elusive continent on paper, one country took steps to protect its Antarctic interests. On 21 July 1908, Great Britain, the front-line colonial nation in the world, proclaimed its sovereignty over certain areas of the continent. By 1942, six more countries had laid formal territorial claims in Antarctica in different circumstances.

It will be seen that areas contained in all the claims are pie-shaped, from 60°S latitude to the South Pole, except the one from Norway as shown in Figure 6.1 (*Reader's Digest*, 1985).

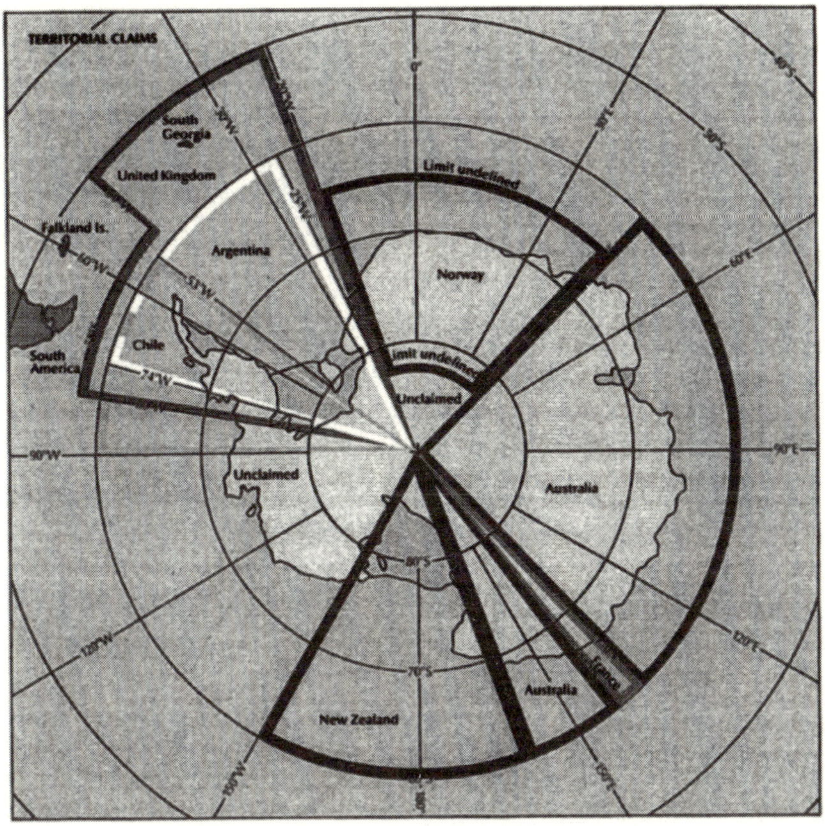

Fig. 6.1: Territorial Claims Made by Seven Countries

6.2.1 *The United Kingdom (1908)*

With the pioneering voyages of Cook (1772–75), Bransfield and Smith (1819–20), Weddell (1820–23), John Biscoe (1830–32), Ross (1839–43), the Challenger Expeditions (1872–76), the heroic expeditions of Borchgrevink (1898–1900), Scott (1902–04 and

1911–13), Bruce (1902–4), Shackleton (1907–09, 1914–16 and 1921–22) and Rymill (1934–37), the British Imperial Expedition (1920–22) and the Discovery Voyages (1924–51), UK had all along been in the forefront of Antarctic exploration.

With this background, UK registered the first claim to the Antarctic territory in 1908. Since it inadvertently contained parts of Argentina and Patagonia (sparsely populated region at the end of South America) as well, it made a revised claim in 1917. The British claim contains South Orkneys, South Shetlands, South Georgia, South Sandwich Islands (all Scotia Arc Islands, forming the north, east and south border of the Scotia Sea) and the Antarctic Peninsula. The claim, which overlaps with others' includes all Antarctic lands between 20°W and 80°W to the Pole. UK calls this area the British Antarctic Territory (BAT).

6.2.2 *New Zealand (1923)*

The New Zealand claim, called the 'Ross Dependency', dates from a British Order in Council of 30 July 1923. However, apart from an involvement in the British-Australian-New Zealand Antarctic Research Expedition (BANZARE) of 1929–31, New Zealand's early activities were nominal and the claim is mainly a colonial heritage.

New Zealand claims 50°S sector (150°W to 160°E), an area below its national boundaries that embraces the whole of the Ross Sea.

6.2.3 *France (1924)*

French interests in Antarctica date back to the sixteenth century, through the explorations of Palmuyer de Gonneville, Jean Baptiste's discovery of the sub-Antarctic island of Bouvet (1739), Kerguelen (1769), Dumont d'Urville (1840) and Charcot's expeditions (1903 and 1908–10).

After correspondence and clarifications with and from UK, France formally claimed an arc of six degrees, from 142°E to 136°E, calling it Terre Adelie, on 27 March 1924. It is the smallest of the seven claims made so far.

6.2.4 *Australia (1933)*

The early history of Australia is closely linked to the opening up of the southern polar region. This search received further impetus from the sealing and whaling operations of the nineteenth century. Its special Antarctic interests were formulated in the 1880s after the Challenger expeditions in 1872–76, Douglas Mawson's own expedition of 1910–13 and his organisation of the BANZARE expeditions of 1929 and 1933.

In 1933, Australia inherited from UK the sovereign claim over what they call the Australian Antarctic Territory (AAT), south of 60°S latitude and between longitudes 45°E and 160°E, excluding the French claim of Terre Adelie. The Australian claim, therefore, incorporates about thirty per cent area claimed in Antarctica and is the largest of all the seven claims which have been made so far.

6.2.5 *Norway (1939)*

Norwegian whalers were very active in the Southern Ocean since the end of the nineteenth century. By mid-1930s, Norway shared with Britain over ninety per cent of the catch. Over-exploitation saw the region's annual whale catch dwindle from 30,000 in the 1950s to less than 2000 in the 1970s. Norway ceased whaling in 1968 only after it became uneconomical. One of the whalers, Carsten Borchgrevink, became the first to set foot on the continent from the ship *Antarctic* on 24 January 1895. In addition to contributions of other expeditions, Norwegian Roald Amundsen became the first human to reach the South Pole on 14 December 1911.

When Nazi Germany announced the territorial motives of the Schwabenland Expedition, Norway quickly pre-empted a German claim to sovereignty by making one herself. It laid claim, in 1939, for lands between 20°W and 45°E with undefined northern limits as shown in the figure. The claim was basically made in an attempt to protect Norwegian whaling interests along the coastline of Dronning (meaning Queen in English) Maud Land.

6.2.6 *Chile (1940)*

Chile's Antarctic territorial claim is based on a mixture of historical inheritance, geographical contiguity, ecological interdependence and supportive acts of regional administration. One of the earliest manifestations of a 'Chilean Antarctica' concerns Charles V of Spain who, in 1539, allegedly granted the adventurer Pedros Sanchez de la Hoz the territory from the Strait of Magellan to the South Pole, both inclusive. In 1831, the first President of the Republic, General Bernado O'Higgins, submitted a letter to the British government confirming that the Chilean territory extended at least to the 65°S latitude and, perhaps by implication, to South Pole.

Chile formally laid claim to Antarctic territory from 53°W to 90°W in 1940.

6.2.7 *Argentina (1946)*

Argentina's claim to territorial sovereignty rests on an amalgam of geographical contiguity, colonial inheritance, priority of discovery and administrative acts dating back to 1904. It claims inheritance of the Spanish title to undiscovered southern polar regions where the original Spanish rights were determined by the Spanish/Portuguese Treaty of Tordesillas, drawn up in 1494. Though Argentina always claimed jurisdiction of the area, it defined the current limits of her Antarctic claim in 1946 in response to the rapidly growing number of British and Chilean bases in the peninsula region.

Argentina claims the Antarctic area between 25°W and 74°W.

6.2.8 *Unclaimed Area, Overlapping of Claims, Occasional Conflicts, Non-Claimants and Non-Recognition*

Despite all these claims, about fifteen per cent of the continent (from 150°W to 90°W) remains unclaimed. Also, while all the other nations' claims are in the form of a slice, from 60°S latitude to the South Pole, Norway does not define the outer limit of its claim.

A look at the territorial claims of the seven countries in the above figure reveals that the claims of UK (60° arc), Chile (37° arc)

1. Sand leads to snow. The author in the Thar Desert.
The experience of leading an expedition here led to the opportunity to go to Antarctica.

2. Walking to the ship, *MS Thuleland,* with some family members before departure.

3. Educating the team on the ship. A month-long voyage is an opportunity to learn many things if one is not afflicted by sea sickness.

4. A tough phase for the ship navigating through thick pack ice.
(Courtesy: NCAOR Report)

5. When the pack ice thins, the ship can berth at the shelf if the wind is friendly. (Courtesy: NCAOR Report)

6. Strong winds frequently force ships to abandon the shelf and slide back to anchor in the ocean. Seen here is *MS Thuleland* and its reflection, with a helicopter in the background. (*Courtesy: Prabhat Kumar Gupta, the author's friend and colleague*)

7. A helicopter lifting the load from a ship.
(Courtesy: NCAOR Report)

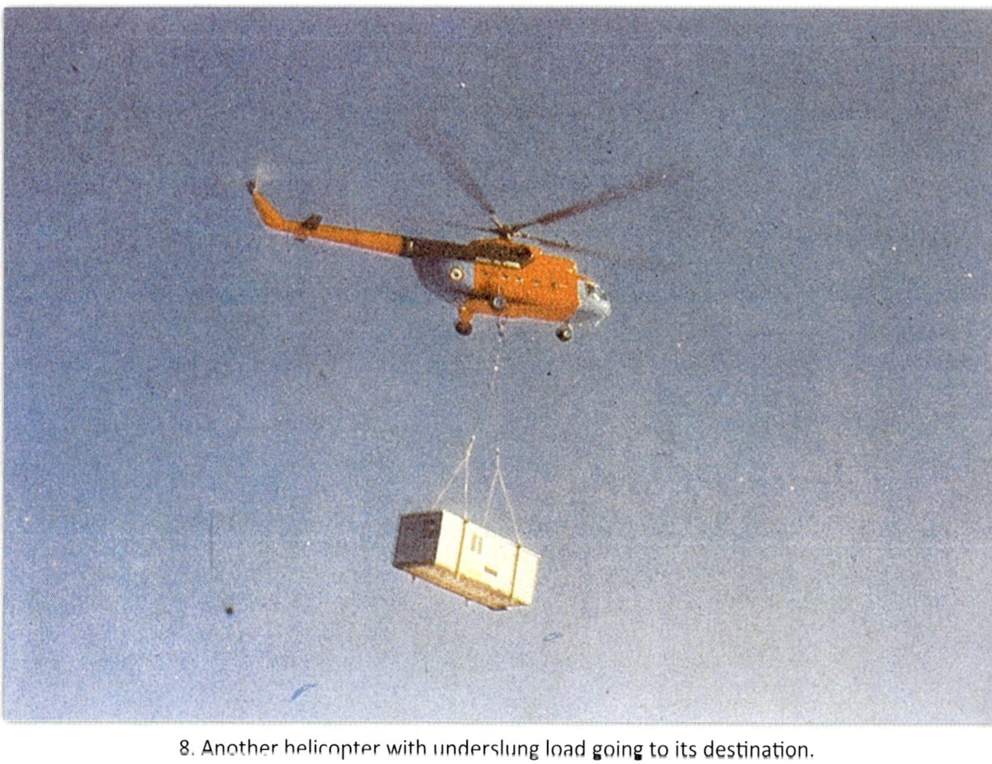

8. Another helicopter with underslung load going to its destination.
Wind is a major factor in Antarctic transhipments. *(Courtesy: NCAOR Report)*

9. A container placed on empty barrels as seen on a blizzardless day at Dakshin Gangotri. It is also used for parking snow scooters.

10. The same container, half buried in snow a few days later, being shifted to a new location to be placed on a fresh set of barrels. The minimum distance for parking containers, vehicles, sledges and others on the shelf is 50 meters apart.

11. Whenever a container is opened, the snow has to be cleared first.
It creates reverse pressure and one has to be very careful while unlocking the doors.

12. Icebergs come in many shapes and colours. They vary greatly in length and breadth, from a few meters to several kilometres. *(Courtesy NCAOR Report)*

13. A grounded iceberg near the shelf. Icebergs remain like this for years before commencing their northern journey in the ocean to disintegrate. A giant iceberg about a few hundred kilometres long, took six years to disintegrate.

14. A congregation of icebergs, most of them dwarfing the ship, seen in the foreground. They can pose grave danger to ships which need to be skilfully navigated to avoid them. Due to an optical illusion, grounded icebergs looked like nearby housing colonies from Dakshin Gangotri.

15. A container placed on the ice shelf to herald India's presence during the first Indian expedition. Later, the full-fledged year-round station Dakshin Gangotri would come up here. *(Courtesy NCAOR Report)*

16. The author with his best friend in Antarctica, Dr Eugeny Uranov, station commander of the Russian Novolazarevskaya Station, after he presented him with a gift and a copy of the Dakshin Gangotri magazine *Himvat* to which Dr Uranov had contributed an article on the Soviet Antarctic expeditions (it was still the Soviet Union as the flag suggests).

17. This is how Dakshin Gangotri appeared with its flashing beacon from a distance. Mutual sighting of beacons used to be a matter of delight at the station as also for convoy returnees.

18. Dakshin Gangotri on an unusually calm day.

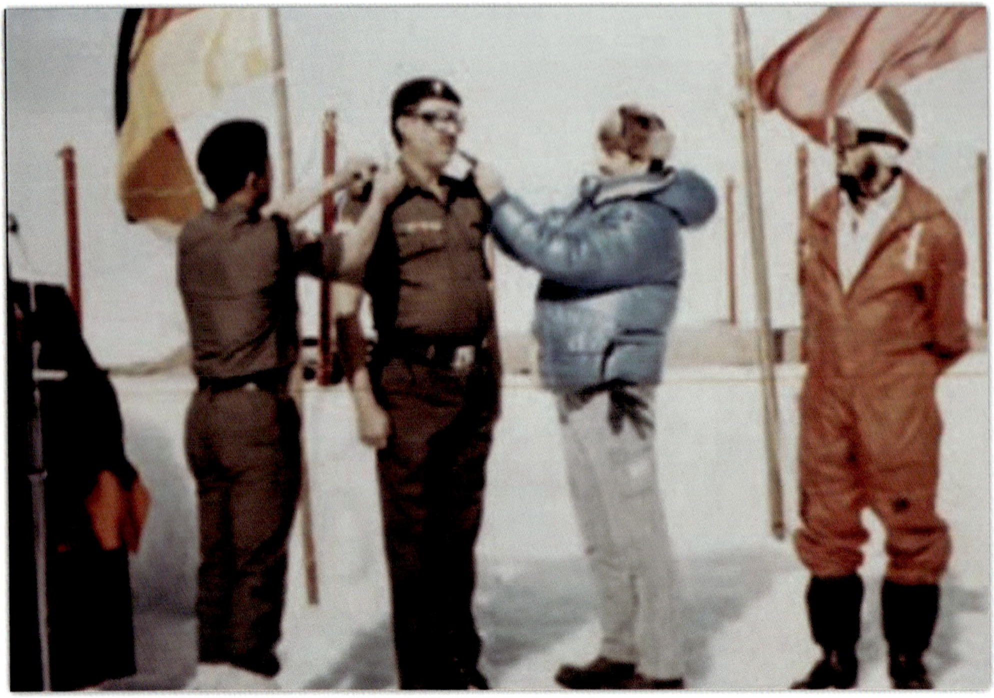

19. On India's Republic Day in 1990, the author *(left)* and the new summer team leader, Dr Rasik Ravindra*(third from left)*, promoting Subedar Premchand Garlyal *(second from left)* to the rank of Subedar Major after the flag hoisting, in front of Indian, German, Russian and Swedish teams.

20. The newly promoted Subedar Major of Dakshin Gangotri, saluting the author, his station commander. Attired in army uniform, both barely survived the extreme cold and were nearly frozen by the time the ceremony was over.

21. The author presenting a gift to Gerald Tereshevsky, leader of the then East German Station, Georg Forster,during the Republic Day ceremony.

22. The floating ice shelf is very unstable due to fissures. These crevasses or bottomless holes have varying widths, from a few centimetres to many meters. There have been several instances where entire vehicles and sledges have disappeared in to them without a trace.

23. The author on a snow scooter, surveying the work of sub-teams. It is great fun to ride a scooter on the ice shelf in summer when the snow is soft, but, in winter, with the ground becoming very hard, glassy and frictionless, it is a very risky proposition and should be avoided.

24. Everyone is delighted when the first penguin is sighted after the winter. When informed, the author dashed outside to welcome it. However, it is very unwise to go out without adequate clothing or to hold a penguin. No one questioned me, but I released the bird immediately and apologised to my team.

25. Emperor Penguins strolling on the ice shelf.

26. The seals on pack ice are totally undisturbed by any provocation.
(Courtesy NCAOR Report)

27. Sixteen members of the Dakshin Gangotri wintering team (the author is standing in the middle of the back row).

28. A game of chess in progress between the Dakshin Gangotri and Novolazarevskaya teams.

29. A statue of Roald Amundsen at the Frammuseum in Oslo.
Amundsen's stature as a polar explorer is even higher than the towering statue.

30. The International Antarctic Centre at Christchurch is a very attractive tourist destination. There is keen interest about Antarctica in New Zealand.

31. These small penguins are not from Antarctica. They reside in Phillip Island near Melbourne and are a major tourist attraction. As photography is not permitted, this is from their website and is permitted for reproduction.

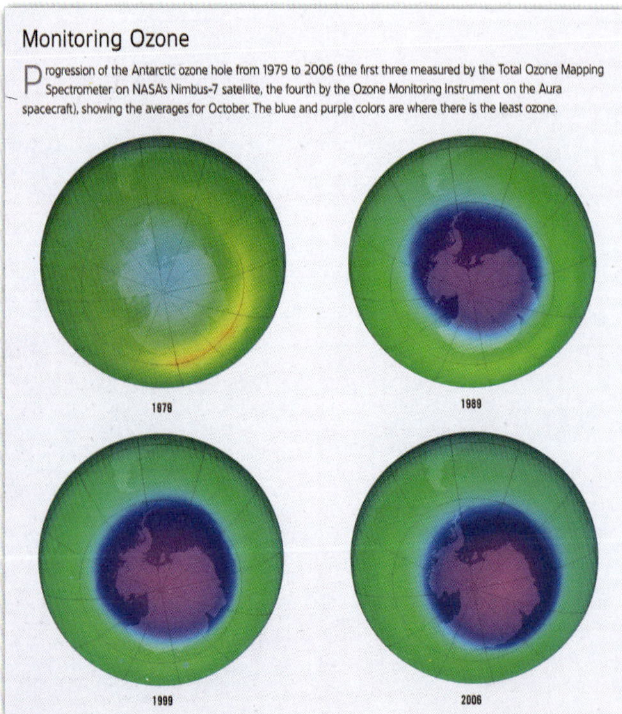

32. Size of the Ozone Hole over Antarctica as Recorded in 1979, 1989, 1999 and 2006 (A photograph from NASA taken from McGonigal, 2008).

and Argentina (49° arc) overlap with one another. During and immediately after World War II, all three countries engaged in a flurry of base-building in their claimed territories. In 1948, both Argentina and UK sent warships to the region and armed conflict seemed likely. In February 1952, Argentinean troops fired over the heads of members of a British party who were rebuilding the Hope Bay station. The 1982 conflict between UK and Argentina over South Georgia and Falklands does not count as an Antarctic conflict as both lie outside the Antarctic Treaty (AT/Treaty) area, that is, to the North of 60°S latitude. However, the three nations whose claims overlap squabbled over expeditions and establishment of stations in the Antarctic Peninsula and nearby sub-Antarctic islands, resembling a three-ring political circus. Apart from countries whose claims overlap, all other claimant nations have recognised one another's claims. The US, Russia, Japan, Germany and Belgium, at different times and levels, have made discoveries and claims in Antarctica but, for various reasons, have not asserted them. Japan, in a 1951 peace treaty after WW II, agreed to forgo any claims it had to the continent (Landis 2001). Speculation about Brazil's claim, the declaration by Uruguay and South Africa about the continent as a zone of national interest, though on paper at present, all may further complicate the situation.

However, in geopolitical terms, two non-claimant nations (US and erstwhile USSR) were just as important as claimants in the development of the unique solution represented by the AT. In 1924, the US adopted the Hughes Doctrine (named after Charles Hughes, Secretary of State from 1921 to 1925), stating that mere discovery is not sufficient for making territorial claim and actual settlement is required. The 1930s saw major US expeditions led by Admiral Richard Byrd and Lincoln Ellsworth when new areas were charted. The US could have based a claim on the explorations of Wilkes, Byrd and Ellsworth but has not done so nor has it recognised any other claims. When Norway claimed Dronning Maud Land in 1939, the then Soviet Union refused to recognise any territorial claims to Antarctica, even though it could have asserted an all-encompassing claim based on Bellingshausen's 1819–21 circumnavigation and first sighting of the continent. Subsequently, Hitler ordered the annexation of the Norwegian claim by Germany and sent two Dornier Walflying boats to drop aluminium darts capped with swastikas in the region.

Post WW II tensions between the US and the USSR ushered in a new era of Cold War. With a view to prepare itself in a possible polar area conflict, the US Navy's Antarctic development project (Operation High Jump) was launched during 1946–48. Though scientific pursuits were given prominent publicity, only 24 men out of over 4700 participants and 55 observers were civilian scientists. The operation included 13 ships, 19 aircrafts and 4 helicopters and was principally a massive logistics exercise to prepare the US armed forces in polar operations (*Antarctica Past and Present,* 2009).

6.3 International Geophysical Year (IGY-1957–58) and the Concept of Scientific Research Stations in Antarctica

After the International Polar Year (IPY) of 1933–34, the next IPY was to be observed after fifty years in 1983–84. Globally, due to WW II and the commencement of the Cold War immediately afterwards, technology in all fields advanced phenomenally. The same factors, however, caused the geopolitical situation to nose-dive all over again.

With this background, several leading scientists of the time met in March 1950 to discuss the future of scientific research in Antarctica. It was decided and suggested to the International Council of Scientific Union (ICSU) that, instead of waiting for the next IPY in 1983–84, there should be an International Geophysical Year (IGY) with a widened research scope, especially considering recent advances in rocketry and radar.

After due consideration, ICSU took the idea further and formalised that IGY and not IPY would be observed from 1 July 1957 to 31 December 1958, a period of eighteen months, with a total emphasis on scientific research, devoid of politics, in Antarctica. This enabled the ICSU to keep focus on polar areas and technological advances both.

Preparations for the IGY started in right earnest with the first Antarctic conference in Paris in 1955 (Landis 2001). Members from twelve nations ushered in a new era of cooperation that extended beyond arbitrary political boundaries. Ironically, the military personnel from the US, the erstwhile USSR, Great Britain, New Zealand, France, Australia, Norway, Chile, Argentina, Japan, Belgium and South Africa established new bases and augmented existing ones. By the time the IGY started on 1 July 1957, over forty research stations on the continent and twenty on the surrounding sub-Antarctic islands were ready with scientists and support personnel.

Scientific investigations on a massive scale during this eighteen-month period produced forty-eight hefty tomes on astrophysics, oceanography, glaciology, meteorology and zoology. One of the most astonishing discoveries was the unsuspected thickness of the ice sheet. Teams from the participating countries trekked hundreds of kilometres into the interior, periodically detonating bundles of TNT along their routes. By measuring the time interval between the explosion and the return of the shockwaves that bounced off the underlying rock, the thickness of the ice could be calculated. These results were doubted by the scientists themselves but numerous experiments by different teams finally established the fact that the ice sheet indeed extended to a depth of two and half miles at places. The data also produced the first sound pictures of submerged mountains and deep troughs. Long-distance flights by aircrafts surveyed areas running into millions of square kilometres. Along with support flights, overland journeys to build bases and numerous geophysical surveys, little of Antarctica was left unseen. As mentioned in earlier chapter, the first overland crossing of Antarctica was made by Vivian Fuchs and his team. Fuchs and Edmund Hillary became the first persons to complete the entire crossing and to reach South Pole overland since Robert Scott (Trewby 2001).

The IGY was, indeed, a remarkable exercise in international scientific cooperation in Antarctica, particularly as it happened within a few years after the end of WW II. Realisation had also dawned that due to the unique and harsh environmental conditions precluding colonisation, Antarctica could serve as an ideal scientific laboratory to the benefit of the entire planet. The concept of permanent scientific research stations on the continent away from the international political turmoil had finally arrived.

6.4 Scientific Committee on Antarctic Research (1961)

As part of the IGY, ICSU established a special committee on Antarctic research in February 1958 to coordinate the research of twelve IGY nations working in Antarctica. This was turned into a permanent group and renamed the Scientific Committee on Antarctic Research (SCAR) in 1961. SCAR is an independent, non-governmental interdisciplinary body of the ICSU. It supplements the Antarctic Treaty System (ATS) by acting as its unofficial scientific body. SCAR can make suggestions and respond to requests for advice and enjoys observer status at ATS meetings. Though not part of any decision-making, as also the fact that many member countries of Antarctic Treaty (AT) are not its members, SCAR plays a pivotal role to coordinate knowledge-gathering on most scientific aspects. The relationship between various SCAR committees, which provide an active and non-political forum for discussion of various Antarctic aspects, has evolved over the years. This can be seen with its latest organisation chart on its website. Important aspects are as shown in Table 6.1.

Table 6.1: A Broad Organisation of SCAR

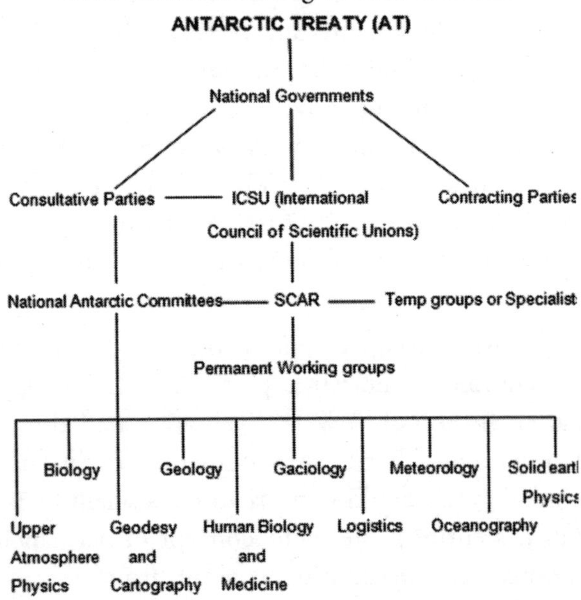

SCAR has been a successful force for international cooperation, especially for research in Antarctica and the Southern Ocean. The committee operates through permanently established working groups in various disciplinary fields such as biology and geology, logistics and oceanography, and temporary groups of specialists charged with solving specific problems. One example of an effective research programme coordinated by SCAR was the Biological Investigations of Marine Antarctic Systems and Stocks (BIOMASS) in the 1980s which played a significant role in official discussions and negotiations.

SCAR has three types of memberships based on the level of research activities in Antarctica for any country, namely full membership (countries having active scientific research programme), ICSU union membership (countries having an interest) and associate membership (countries planning future research programme). Many countries are members of the AT but not of SCAR and vice versa. As of now, SCAR has thirty-one full, nine ICSU and eight associate members. In addition, there are honorary members of SCAR, those individuals who have, over many years, rendered outstanding service to SCAR.

Without being the official organ, SCAR supplements the ATS by acting as its unofficial scientific advisory body and plays a significant role in its successful functioning.

6.5 The Antarctic Treaty System (ATS) (1959/1961)

The AT and related agreements with specific purposes over the years are collectively called the Antarctic Treaty System (ATS). They together regulate international relations with respect to Antarctica, earth's only continent without a native human population. For the purpose of the ATS, Antarctica is defined as all land and ice shelves south of 60°S latitude. (Wikipedia, 2010, Antarctic Treaty System)

6.5.1 *Making of the Antarctic Treaty*

As the IGY 1957–58 progressed, it was increasingly seen as a benchmark for international scientific cooperation. Taking a clue from this, the US President Dwight D Eisenhower stated that

Antarctica should be used only for peaceful purposes and not as a subject of political conflict. He, therefore, invited the other eleven IGY participating countries to Washington to discuss the future of the continent (Triggs, 1987).

The Antarctic Treaty was presented at a formal conference on 14 October 1959 after over sixty secret meetings between representatives of the twelve countries. The Treaty was opened for signatures on 1 December 1959 and went into full force on 23 June 1961 after ratification by all the countries.

The opening sentence of the preamble sets the tone of the document. The twelve signatory governments recognised that, 'it is in the interest of all mankind that Antarctica shall continue forever to be used exclusively for peaceful purposes and shall not become the scene or object of international discord.' The fourteen brief articles of the Treaty emphasise freedom for scientific investigations and encourage exchange of scientists and research data. Nuclear explosions, disposal of nuclear waste and military aggression are banned and onsite inspections of stations by member-nation committees are permitted if reasonable advance notice is given. The Treaty addresses territorial disputes by freezing the status quo. No signatory country had to surrender its past claims; those countries that may have a base for a claim, the US and USSR, for example, did not have to renounce their right to make one. However, no new claims can be made while the Treaty is in effect.

In line with the main objective as mentioned above, the Treaty forbids any measures of military nature, but not the presence of military personnel in scientific and support roles.

The complete text of the AT is shown in *Appendix II* (website of the AT: ats.aq). The essence of all its fourteen articles is summarised as under:

Article 1: The Antarctic Treaty area will be used for peaceful purposes only. Military activity such as weapon testing is prohibited, but military personnel and equipment may be used for scientific research or any other peaceful purpose.

Article 2: The spirit of freedom for scientific investigation and cooperation shall continue.

Article 3: Grants permission to freely exchange information and personnel in cooperation with the UN and other international agencies.

Article 4: The Treaty does not recognise, dispute or establish any current territorial sovereignty claims, at the same time declaring that no new claims can be asserted while the Treaty is in force.

Article 5: Prohibits nuclear explosions or disposal of radioactive waste.

Article 6: The Treaty area includes all land and ice shelves south of 60°S latitude.

Article 7: Observers may have free access, including aerial observation, to any area and may inspect all stations, installations and equipment only after an advance notice for all these activities is given.

Article 8: Allows for jurisdiction over observers and scientists by their own states.

Article 9: Frequent consultative meetings may take place among member nations.

Article 10: The Treaty states will discourage activities by any country in Antarctica that are contrary to the Treaty's objectives.

Article 11: Disputes will be settled peacefully by the parties concerned or, ultimately, by the International Court of Justice (ICJ).

Article 12, 13, 14: Deal with upholding, interpreting and amending the Treaty among involved nations.

6.5.2 *Functioning of the Antarctic Treaty*

The yearly Antarctic Treaty Consultative Meetings (ATCM) form the forum for the administration and management of the Treaty regime.

The Antarctic Treaty Secretariat was established in Buenos Aires, Argentina, in September 2004 by ATCM. It supports the annual ATCM and the Committee for Environmental Protection (CEP) meetings, facilitates the exchange of information between members, provides and disseminates public information and acts as the apex body of ATS (ATS 2012).

Another important organ is the Committee of Managers of National Antarctic Programmes (COMNAP). Formed by the national Antarctic programmes in 1988, it enjoys an observer status at both ATCM and CEP meetings and provides relevant technical advice using the members' pool of expertise. It works through a number of groups on various areas such as ship and air operations, environmental management or training. This includes in particular a Standing Committee on Antarctic Logistics and Operations (SCALOP). Headquartered in Hobart, Tasmania, Australia, COMNAP provides a central point of contact and operates common communication infrastructure and acts as a central information repository and maintains coordination between members (COMNAP 2012).

Though the ATS remains the principal body in Antarctic governance, various agreements conducted under its aegis and forming an integral part of it, as also those from outside, come into play in various situations, making the functioning of matters Antarctic quite complex and calling for a high standard of diplomacy.

Key agreements from both the categories are summarised below.

6.5.3 *Agreements and Conventions Forming a Part of the Antarctic Treaty System (ATS)*

Various agreements to protect the pristine Antarctic environment have evolved over the years, based on recommendations from member countries. They represent how the thinking and execution processes have evolved since the Treaty came into force in 1961 and over the years since its ratification by its members. They are:

- Antarctic Treaty (AT/Treaty) (1959/1961).
- Agreed Measures for the Conservation of Antarctic Fauna and Flora (AMCAFF) (1964/1980).
- Convention for the Conservation of Antarctic Seals (CCAS) (1972).
- Convention for the Conservation of Antarctic Marine Living Resources (CCAMLR) (1980/1982).

- Convention for the Regulation of the Antarctic Mineral Resource Activities (CRAMRA) (1988) (Not in force due to non-ratification and unlikely to be revived).
- The protocol on environmental protection to the Antarctic Treaty (Madrid Protocol) (1991/1998).

6.5.3.1 *Agreed Measures for the Conservation of Antarctic Fauna and Flora (AMCAFF) (1964/1980)*

After the AT came into force in 1961, general rules for conduct of scientific expeditions in Antarctica were drawn by SCAR to minimise human impact on species and the environment. These were incorporated in the AMCAFF which came into effect in 1964. The main points of these measures are (Trewby 2001):

- Prohibit harming, killing, wounding or molesting any wildlife (excluding whales which are covered under International Whaling Commission (IWC) discussed separately), except by permit from SCAR.
- Minimise harmful interference to the environment of mammals and birds.
- Prohibit introducing non-indigenous species, parasites and diseases.
- Create Specially Protected Areas (SPAs), Sites of Special Scientific Interests (SSSIs), and Specially Managed Areas (SMAs) primarily as safeguards against damage due to tourism. No one can visit these areas without a special permit.

The measures were entered into force in 1980.

6.5.3.2 *The Convention for the Conservation of Antarctic Seals (CCAS) (1972/1978)*

This convention was adopted in 1972. It specifies total protection for Fur, Elephant and Ross seals. The catch limits for other seals, with a total estimated population in brackets, are 175,000 (15,000,000) for Crabeater, 12,000 (220,000) for Leopard and 5000 (730,000) for Weddell Seals (*Reader's Digest,* 1985).

6.5.3.3 *The Convention for Conservation of Antarctic Marine Living Resources (CCAMLR) (1980/1982)*

This convention was adopted in 1980 but came into force in 1982. It was negotiated due to concern over the prospect of overexploitation of marine resources in the Southern Ocean. At the time there was a lot of interest in fishing for krill and, therefore, CCAMLR is sometimes also called the 'krill convention'. This convention contains some unique innovations in thinking and execution methodology. It adopts an ecosystem approach to conservation, meaning that, along with aiming to conserve species directly targeted for harvesting, it also attempts to take into consideration the impact on associated and dependent species and the need to restore depleted stocks by establishing an ecosystem-monitoring programme. It defines conservation as 'rational use', a compromise needed to gain an agreement from states interested in fishing as also from those motivated by conservation. The area to which CCAMLR applies is different from the Treaty. Whereas the Treaty extends up to 60°S latitude, the jurisdiction of CCAMLR follows the Antarctic Convergence, which fluctuates between 40°S and 60°S latitude. Because of disputed claims to territory in Antarctica, CCAMLR includes Article IV of the Treaty but with the addition of a phrase allowing coastal states to exercise jurisdiction. This compromise is called the 'bifocal approach' because its ambiguity allows claimants to think that there is no challenge to their claims, and non-claimants to think that it applies only to the sub-Antarctic islands where sovereignty is undisputed (Trewby, 2001).

The convention has created an institutional body known as the Commission for the CCAMLR that operates on a permanent basis from Hobart, Australia. It faced problems in the 1980s due to severely depleted fish stocks from unregulated exploitation in the 1970s and tooth fish fishing in the 1990s. It also faces problems from long-line fishing techniques accidently killing thousands of albatrosses, giant petrels and other birds in the process. However, the expected interest in fishing krill has not materialised due to difficulties in processing the fish for human consumption and the collapse of subsidised fishing by the Eastern Bloc countries after the end of the Cold War.

The Commission bases its decision process on the consensus and advice of a committee of experts.

6.5.3.4 *The Convention for the Regulation of the Antarctic Mineral Resource Activities (CRAMRA)*

The realisation that, at some stage, Antarctic minerals would attract commercial interests, whereupon control would be necessary, consultative parties (CP) to the Treaty discussed the issue of mining in the 1970s and started formal negotiations in 1982. Key issues were the disputed claims to sovereignty in Antarctica, and the need to accommodate the interests of the non-aligned countries and potential mining states. CRAMRA was signed on 2 June 1988. It contains a detailed set of rules and procedures for regulating mineral-resource activity in Antarctica. Guidelines for establishing liability for accidents and environmental damage are included. It maintains the compromise on sovereignty in the Treaty but the decision-making is to be by a three-quarter-majority vote rather than consensus.

Fearing the potential for damage to the Antarctic environment, protests against CRAMRA occurred from various sources, including NGOs like Greenpeace. On 22 May 1989, Australia announced it would not ratify the convention and it was followed by France (Trewby, 2001). With the adoption of the Protocol on Environmental Protection to the Antarctic Treaty (Madrid Protocol) in 1991, it appears unlikely that CRAMRA will ever enter into force.

6.5.3.5 *Protocol on Environmental Protection to the Antarctic Treaty (Madrid Protocol) (1991/1998)*

Following the collapse of CRAMRA, the Protocol on Environmental Protection to the Antarctic Treaty, known as the Madrid Protocol, was negotiated. It was signed on 4 October 1991 and came into force on 14 January 1998, after its ratification by all CPs to the Treaty.

This Protocol is a comprehensive framework for the environmental protection of Antarctica and its ecosystems. It effectively prohibits all activities relating to minerals and oil in Antarctica except for scientific purposes, for a period of fifty years. It includes twenty-seven articles, a schedule on arbitration and five specific annexes on Environmental Impact Assessment (EIA), fauna and flora, waste disposal and management, marine pollution and area protection and management. A sixth annexe on liability arising from environmental emergencies was adopted in 2005 but is yet to enter into force.

A Committee for Environmental Protection (CEP) has been established to provide advice and recommendations to the CPs. The committee meetings take place along with the ATCM every year. Some problems exist with the implementation of the Protocol in the form of non-reaching of consensus on the sixth annexe so far, non-ratification of the agreement and the annexes by some countries and uneven compliance of the annexes. The Protocol will be open for review in 2048 (*Cool Antarctica 2012, Madrid Protocol*).

Details about the annexures and implementation of this Protocol are given later while discussing the Indian Antarctic Programme (IAP).

6.6 Conventions, Laws and Treaties Outside ATS but Relevant to the ATS

Several separate conventions, laws and treaties also address Antarctica and its environment. They are complimentary to the ATS and relate to the ocean and land the environment. However, they are binding only on those states that are members of these agreements. Some of these are the IWC, IPCC and others, as given below (Trewby, 2001).

6.6.1 *International Whaling Commission (IWC) (1946)*

The immense value of whales, the exploitation and near extinction of certain species, leading to the establishment of IWC, the functioning of IWC, its limitations and present position, have already been pointed out in the earlier chapter on Antarctic life. But, a ruling by the ICJ in 2012 to stop the hunting of whales by Japan under the pretext of scientific research, and an enhanced awareness about such issues, have further encouraged the conservation of whales. The IWC maintains a secretariat in Cambridge (UK), and holds an annual meeting to take regulating decisions (*Cool Antarctica, 2012, Human Impacts*).

6.6.2 *Intergovernmental Panel on Climate Change (IPCC) (1988)*

IPCC was formed by the World Meteorological Organisation (WMO) and the United Nations Environmental Protection (UNEP) agency. It does not conduct any research but assesses the latest scientific, technical and socio-economic data on climate change and reports once every four years. In addition to the climate assessment reports, IPCC also publishes special reports on specific topics like 'emission scenarios', renewable energy resources and climate change mitigation and managing the risk of extreme events and disasters to advance climate change adaptation.

So far, five reports have been published by the IPCC in 1990, 1992, 2001, 2007 and 2013–14. As of now, the fourth and the fifth reports are very relevant describing the present-day climate-change crisis the world is facing.

The fourth and the fifth report contain three working group reports (WG I, II and III) and a synthesis report. Whereas WG I details the aspect of physical science, WG II and WG III detail 'impact, adaptation and vulnerability' and 'mitigation' respectively. In the chapter on Antarctic climate, we have already seen the likely effects of global warming and the rise in sea levels, described in the fourth assessment report. Observations of the fifth report are similar. Some of the major conclusions of the fourth and the fifth reports are as under:

- Warming of the climate system is unequivocal and many changes since the 1960s are unprecedented.
- Atmospheric concentration of carbon dioxide (CO_2), methane (CH_4) and nitrous oxide (NO_X) have increased to record levels in the last 800,000 years.
- It is extremely likely that human influence was the dominant cause for global warming between 1950 and 2010.
- Without appropriate mitigation measures, an increase in the global mean temperature may be in the range 2.7 to 4.8°C. (IPCC, Wikipedia 2015).

The latest IPCC assessment report formed one of the major basis for discussion in the recently concluded climate summit in Paris in November/December 2015 (COP 21).

6.6.3 *United Nations Framework Convention on Climate Change (UNFCCC) and Conference of Parties (COP): COP 1 (Rio) to COP 21 (Paris)*

Alarmed by the reports about the possible grave consequences of climate change if left unchecked, member nations crafted UNFCCC in 1992 at Rio. Under its aegis, the first COP 1 was held in 1995 at Berlin. The first climate agreement was made in Kyoto at COP 3 in 1997. Even though the measures to rein in climate change were modest, the execution was very poor. While Russia decided to pass on, the US decided to remain out of the Kyoto Agreement (COP 10 at Buenos Aires 2004). As the limits agreed in Kyoto will expire in 2020, efforts to have a binding treaty were made in subsequent conferences. The conference at Copenhagen in 2009 (COP 15) failed to reach an agreement but the developed nations agreed on greenhouse gas limits. In 2010 at Cancun (COP 16), rich nations agreed to help poor nations. This journey was chequered but in the intervening period, damming reports from many organisations, including the IPCC as mentioned above along with evidence on ground, linked natural calamities with climate change. This also impressed on all the governments that unless something drastic was done, the world may march towards a climate change induced disaster in the foreseeable future.

With this background, COP 21 was held in Paris from 30 November 2015 to 11 December 2015. The participation of 196 countries, active involvement of over 140 heads of states and nearly 40,000 delegates underscored the importance attached by the world community to the issue of climate change.

After a fortnight of tough negotiations, COP 21 concluded with the adoption of a thirty-one-page document termed the Paris Agreement. It basically seeks to limit global warming to well below 2°C and commits a grant of US $100 billion a year from year 2020 by the developed countries to the developing countries. Though the wording in all such international agreements is vague and leaves loopholes in interpretation, the Paris Agreement has made a good beginning. Also, for this agreement to become legally binding, ratification from at least fifty-five countries who produce fifty-five per cent of the world's greenhouse gases is mandatory.

So far the progress on this score is quite impressive. In September 2016 biggest polluters (US and China) and within a month European Parliament pledged full support. This has taken the number of countries well above the requirements of 55 countries and 55 percent of pollution contributors. This would ensure implementation of the agreement on schedule.

Some of the recent developments in world politics like an environmental sceptic like Donald Trump assuming office of US President and declaration by the architect of Paris Agreement and current French President Francois Hollande that he would not seek re-election in 2017, however, should cause anxiety. One can only hope that realities of office, environmental compulsions, global pressure and good sense may finally prevail and Paris Agreement does not suffer in the interest of future generations.

The success or otherwise of COP 21 will solely depend on the generous interpretation and sincere execution of the terms of the agreement by concerned members. Some of these terms are Intended Nationally Determined Contributions (INDCs), Common but Differentiated Responsibilities (CBDRs), transfer of funds and technology for renewable energy, compromises on sustainable lifestyles and energy consumption and all other voluntary actions to cut carbon dioxide emissions much above the INDCs. This may require rising above narrow national interests by many countries for global good. It is sincerely hoped that through proper implementation of COP 21, the world will be benefited in the ecological interests of future generations.

It is famously said that the devil is in the details and it is hoped that the proverbial devil does not affect the implementation of COP 21.

6.6.4 *Other International Conventions Relevant to Functioning of the ATS*

- *Convention on Prevention of Marine Pollution by the Dumping of Wastes and other matters (1972):* It came into force on 30 August 1975 and prohibits noxious wastes being dumped at sea.
- *Convention on the International Regulations for Preventing Collusions at Sea (1972):* It came into force on 15 July 1977 and provides for common standards for lights and sound signals on ships.
- *International Convention for Prevention of Pollution from Ships (1973):* It came into force on 2 October 1983 and finds reference in Annexure 4 of the Madrid Protocol.
- *UN Convention on the Law of the Sea (1982):* It came into force on 16 November 1994 and regulates activities in all the oceans of the world.
- *Convention for the Protection of the Ozone Layer (1985) and Protocol on Substances that Deplete the Ozone Layer (1987):* Entered into force on 22 September 1988 and 1 January 1989 respectively, both these treaties reiterate the concern about depletion of the ozone layer and the ozone hole above Antarctica.
- *UN Framework Convention on Climate Change (1992):* Entered into force on 21 March 1994 with a goal to stabilise greenhouse gases in the atmosphere to limit the effect of climate change.
- *Convention on Biological Diversity (1992):* It came into force on 29 December 1993 and aims for the conservation of and sustainable use of biological diversity.
- *Convention on the Control of Trans-boundary Movements of Hazardous Wastes and their Disposal (1989):* Entered into force on 5 May 1992, parties to this convention have agreed not to send hazardous waste south of 60°S latitude.

6.7 The Status of Various Countries with Respect to ATS

The status of various countries with respect to their membership of ATS and some of the important agreements/instruments like CEP, COMNAP, CCAMLR, SCAR and Hydrographic Committee of Antarctica (HCA) is shown in *Appendix III*.

Currently there are 53 members of the AT of which 29 enjoy CP status and have the right to participate in the ATCM meetings for decision-making. The other 24 members are acceding parties and are allowed to attend. Decision-making CPs include the 12 original signatories plus 17 countries that have demonstrated their interest in Antarctica by carrying out substantial scientific activities there. Out of the member countries, some are not members of SCAR and some are not members of CCAMLR. Due to the loose organisation, compliance becomes voluntary, posing problems at times.

6.8 Research Stations and Bases in Antarctica

There have been bases in Antarctica since 1899 when Carsten Borchgrevink erected two prefabricated wooden huts on a rocky cape in Victoria Land. He and many others used such structures occasionally but they were not intended to be permanent. The first scientific research station was established in 1903 when Briton W S Bruce built a meteorological station in the South Orkneys Island. As his government showed no interest to accept and maintain the station after he finished his expedition, he handed it over to Argentina. It is now the oldest permanently inhabited base in Antarctica, known as the Orcadas Scientific Research Station.

This scenario changed after countries realised that if their territorial claims were to have any weight in future disputes over Antarctic sovereignty, they would need support in the form of regular presence and activity. Starting with bases on the sub-Antarctic islands, Australia became the first country to build a large, permanently maintained base on the continent itself in 1954 by establishing Mawson. IGY started the burst and by 1958 there were sixty permanent and seasonal/summer stations belonging to twelve countries.

Presently, many countries maintain permanent research stations throughout Antarctica. While some stations are year-round, some are occupied only during Antarctic summers. A total of thirty countries, maintain about seventy stations both year-round/permanent or summer only/or seasonal. Permanent stations are manned throughout the year. Obviously, the strength is much more during summer and greatly reduced during winter when only essential services to sustain the stations and for scientific programmes to be carried out is required. There are a few unmanned and irregular (for transhipments only) stations, some are replaced or closed temporarily. In addition to these stations, about thirty field camps are established each austral summer to support specific projects. All these figures keep changing as countries keep closing/decommissioning old stations and rebuilding old ones and establishing new ones.

The population of people supporting science on the continent and nearby islands varies from approximately four thousand during summer season to about a thousand during winter.

Some countries like Argentina, Australia, Chile, China, France, India, Japan, Russia, UK and US maintain multiple stations. Many other countries maintain one station each. All these stations vary greatly from one another in terms of size, strength of summer and winter personnel, availability of airports, logistics network and other parameters.

Details of each station, like country, area, permanent or summer only, availability and year of opening is shown in *Appendix IV*.

Locations of all the stations are shown in Figure 6.2 (Australian Antarctic Data Centre and SCAR Antarctic Information, 2012).

It can be seen from the above data that a heavy concentration of stations (fourteen year-round and eleven seasonal/summer stations) is seen in the Antarctic Peninsula. This is because of its proximity to South America (particularly for countries like Argentina and Chile), a less severe environment and the convenience of logistics support:

- Most of the stations in East Antarctica are located on ice-free oasis areas near the coast and a few are on ice shelves.
- Few stations are located on ice shelves and fewer on the polar ice plateau deep inside the continent due to the risk of snow burial, exacting logistics requirements and need for eventual replacement.
- Nearly thirty stations have aircraft-landing facilities. Out of these, only five runways are for wheels, nineteen for skis and six for aircrafts fitted with either.

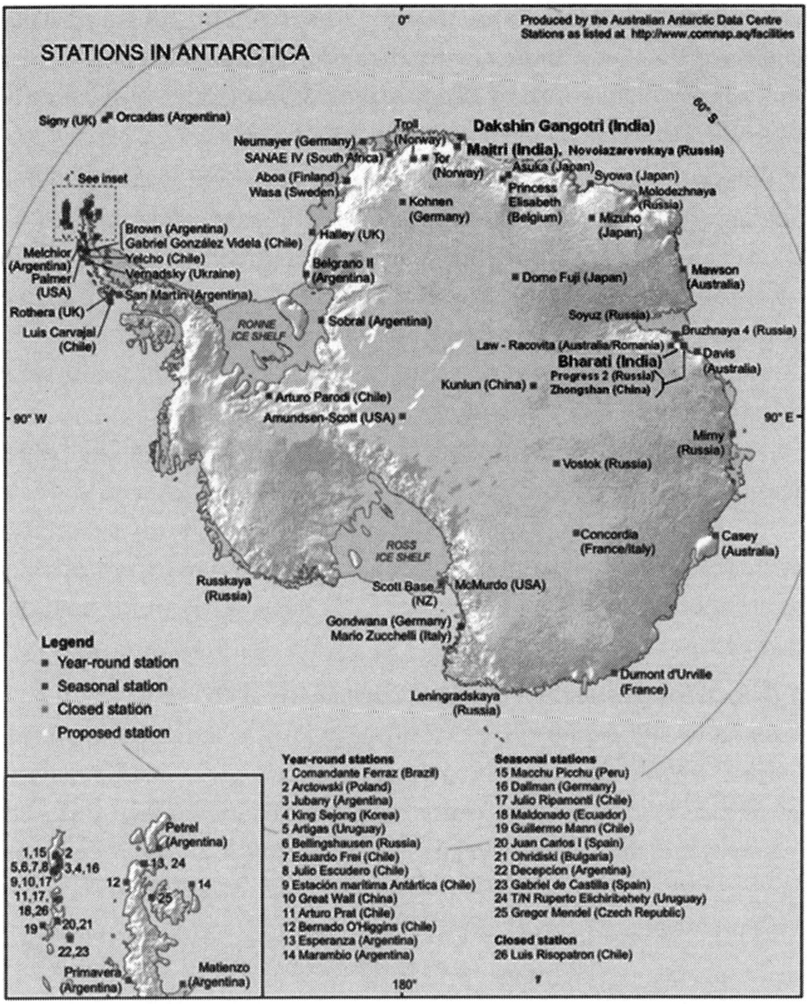

Fig. 6.2: Locations of Research Stations in Antarctica

- The population of all the stations during summer and winter is about 1,000 and 4000 respectively.

6.8.1 *Locations and Structures of Antarctic Bases*

The location of Antarctic bases/stations can be divided into two broad categories. The first one is where there is no danger of snow burial and the stations can exist over a long period with good maintenance and preservation (like Maitri and Bharati). The second category stations are those which are built on ice. They are short-lived and buried under snow in a few years after commissioning, the only question being when (like DG).

The first category locations include the Antarctic Peninsula which enjoys geographical proximity with South American countries like Argentina and Chile and the majority of snow-free oases areas, isolated coastal promontories and a few inland oases areas near the ice shelves. These are good for establishing bases as the environmental conditions here are less severe and there is an absence of permanent ice or snow. Stations established on inland oasis areas, but not near the coast, have to be maintained from supply depots on ice shelves with the help of tractor and sledge convoys. For example, the Indian station Maitri and the nearby Russian station Novolazarevskaya, both situated on Schirmacher Oasis, have to be logistically maintained from a supply depot about 85 km away on the ice shelf near the abandoned Indian station DG.

The second category includes coastal ice shelves all around and glacier ice areas deep inside the continent, known as the polar plateau. Whereas owing to katabatic winds and crevasses, stations on ice shelves face very tough survival conditions, locations deep inside the continent, on the polar plateau, face daunting logistics to maintain them.

Making a base in Antarctica is difficult. Even bases built on areas free of permanent ice face problems of accessibility and isolation. These buildings tend to be conventional and are anchored to a rock platform or built on adjustable stilts. Many modern-day stations consist of buildings on skis. Skis enable the building to change location when required, especially in the event of possible snow accumulation.

However, it is the bases built inland on the polar plateau or on floating coastal ice shelves that pose unique construction problems. All Antarctic ice is unstable and in constant motion, making the construction of a foundation very difficult. A building also gives off heat, causing it to gradually sink into ice. Strong winds blowing large amounts of loose surface snow over the continent pose another challenge. The blown snow accumulates in drifts around any object such as buildings and never melts. The weight of this ice and snow starts to distort the walls and roof and, eventually, as the snow covers the whole building, it causes it to collapse. This is coupled with the yearly increasing snow and ice levels that, ultimately, sink structures built on ice, forcing decommissioning and abandonment after a few years. As a result, very few stations in Antarctica are built on ice. For example, the US built the Amundsen-Scott Station on the South Pole in 1956. The original station, Old Pole, was abandoned in 1975, the second one, Dome, was decommissioned in 2008. The present station is functional since 2008 with a facility for adjustable elevation with stilts. The Russian Vostok (1956), French Italian Concordia (2005), Japanese Dome Fuji (1995) and Chinese Kunlun (2009) are other stations in this category of stations on the polar ice plateau and they require tremendous logistics expertise to maintain them. The British Halley I was originally built on ice shelf in 1956. After five snow burials, the present Halley VI is built on telescopic stilts. In the station complex, three buildings can be raised above the rising snow with stilts. The garage and the accommodation building are mounted on skis to facilitate their movement to another location when faced with burial due to rising snow levels. Stations like the German Neumayer III and the South African Sanae III experienced the same fate – of existing for a few years before forced abandonment. While the Germans have built Neumayer IV on ice again with better technology for adjustable height and longer endurance, the South Africans built Sanae IV in 1997, 200 km away on an ice-free nunatak area, which does not have the danger of snow burial, unlike all its predecessors (*Outposts of Antarctica,* 2010). Detailed information on the Indian experience with regard to its first station DG, which also had to be abandoned due to snow burial, is given later in the chapter on the Indian Antarctic Programme (IAP).

Some radical building designs were developed in the 1960s to increase the lifespan of the bases. The Australian station, Casey, was built in 1969 on a long, raised platform, facing the prevailing winds. Snow blew underneath the platform rather than building up against it. The US Byrd station was built in a large trench cut out of ice and covered by a steel arched roof to shield the buildings from wind and snow. But ice keeps moving and the walls of the trench gradually shift inwards, limiting the usefulness of this technique. In the South African Sanae station, buildings were put in a galvanised steel tunnel for better carriage of ice and reuse of the buildings.

A typical modern base consists of many buildings: living quarters, laboratories, weather station and stores with at least a year's food supply. There might be connecting corridors for easy movements in bad weather. There is often an emergency shelter some distance away in case of fire, a constant danger in the very dry Antarctic air.

Earlier, locations of stations were decided based on the country's political power to assert, convenience and logistics capability or constraints, and not on environmental considerations. For example, when the US built its station, Amundsen-Scott, on the geographical South Pole, USSR built its own station Vostok also on a polar plateau, at the Pole of Inaccessibility – meaning centre of the Antarctic continent (a point that is at the greatest distance from all the coasts, 85° S and 65° E at an altitude of 3718 m). After its abandonment, Vostok was rebuilt at its present location at the south magnetic pole. Other stations on the inland polar plateau, more or less, underscore the same. The situation has now changed due to the cluttering of stations in the peninsula and the Madrid Protocol, which has stabilised the scenario over the years.

6.9 Adverse Environmental Impact of Research Stations

A large number of bases, while carrying out research work, affect the Antarctic environment in a variety of ways. The impact of these activities was properly addressed for the first time in 1985 when SCAR presented a report entitled, 'Man's Impact on the Antarctic Environment'. The report can duly be called one of the major achievements of SCAR over the years which paved way for the success in negotiations and, ultimately, to the signing of the Madrid Protocol in 1991. It acknowledged that the majority of research stations were established in their current locations because they were the most convenient places for either logistics or scientific reasons and there were no environmental considerations whatsoever. It also affirmed that there was often significant damage caused to the locality as a direct result of the existence of the station. The major contributors to pollution and environment damage are construction of bases, construction material, power generation, fuel and toxic spills and waste disposal. Though solid wastes such as discarded machinery have no biological significance, rubber and plastic wastes carry long-term consequences. Small plastic particles have been found in the digestive tracts of sea birds. Whereas domestic and human waste, kitchen refuse and natural wood can all be disposed of by placing in the sea with minimum treatment, all radioactive waste, wastes containing high levels of heavy metals and harmful organic compounds should be removed from the Treaty area. In effect, all such waste should be taken back by countries to be disposed off. The report pointed out that the environmental impact is not limited to the terrestrial ecosystem and living organisms alone. Pollution of the electromagnetic spectrum by excessive radio frequency radiation (which interferes with scientific observations), disturbance or removal of objects of scientific interest (fossils, meteorites, etc), and damage to historic sites, too, affect the environment.

The SCAR report represented the real situation on ground since humans arrived on the continent. The US operated an open solid-waste dump at McMurdo base with open burning, using huge amounts of fuel. Dumps at Wilkes and Casey remained there for years, posing ecological hazards after the former was abandoned in favour of the latter in 1969. There was no separation of toxic material from other non-combustible wastes. Tinned and bottled food, machine parts, buildings, chemicals, including more than two hundred boxes of tinned caustic soda had spilled on snow, metal drums and other scrap. Flares and even explosives were scattered over one square kilometre area. Scavenging birds, skuas, had removed food scraps and were found dead. Australia mounted a major clean up in later years to remedy the situation to some extent, but garbage such as abandoned buildings still remains.

The Antarctic Treaty bans nuclear explosions as well as the disposal of nuclear waste in the Treaty area but it does not forbid the use of nuclear power. For ten years, from 1962, the US operated a 1.8 MW pressurised water plant at McMurdo as an economical source of power before it stopped working. It was an expensive and environmentally damaging story of shut downs, fire damages and radiation leakages. Over the next three years, it was demolished and the reactor, with 101 large drums of radioactive earth, was shipped back to the US for burial at a cost of US $1 million. It took another six years of cleaning up and shipping back of 11,000 cu m of soil before the site was released for unrestricted use (McGonigal, 2008).

All this has obviously changed due to the Madrid Protocol. ATCM does not sanction the establishment of new research stations without first satisfying itself with a comprehensive evaluation of their environmental impact and mitigating measures in line with the Protocol and its annexes in force since 1998.

However, due to the complex and loose nature of the AT, the success of the Protocol, as of the AT itself and its other agreements, depends upon the voluntary compliance from the AT parties. Actions taken by India before establishing its latest permanent base, Bharati, are discussed separately in the chapter on IAP.

6.10 Impact of Tourism (McGonigal, 2008, Trewby, 2001, Kriwoken, 2000)

Tourism in Antarctica began in the late 1950s and is now a significant commercial activity in three main forms: ship cruises, over-flights and extended land expeditions.

Lars-Eric Lindblad pioneered expedition cruising and started Antarctic tourism on a regular basis in 1966 with an Argentinean vessel and launched a liner, *Lindblad Explorer*, specifically built to carry tourists to Antarctica in 1969. Over ninety per cent of tourist activities take place in the Antarctic Peninsula region and on the offshore islands to the west of the peninsula. The season runs from mid-November to mid-March. As for the remainder of the year, pack ice conditions and winter isolates the continent from tourists completely.

The earliest tourist flight to Antarctica was from Chile in 1956, but it was not until 1977 that regular flights were instituted by Qantas and Air New Zealand, using 747s and DC 10s. Over the next two seasons, some 11,000 passengers flew over the continent, an eleven-hour journey of which some ninety minutes were spent over the continent. After the 1979 Air New Zealand DC 10 crash at Mount Erebus that killed 257 people, over-flights ceased until 1994.

Land expeditions operate out of Patriot Hills in the Ellsworth Mountains and include activities like mountaineering, camping and skiing. They are basically private expeditions in small groups and are still very few and far in between due to obvious reasons of risks and costs.

To prevent tourists from leaving a negative impact on the environment and to self-regulate their own activities, Antarctic tour operators established the International Association of Antarctic Tour Operators (IAATO) in 1991. IAATO developed its own guidelines for safe and responsible conduct of tourism and a code of conduct for tourists, which were both later modified by Treaty parties. This provides for a proper conduct of tourism and non-governmental activities in the Treaty area. There are very strict guidelines on personal conduct vis-à-vis the wildlife, putting pressure on tour operators and travellers both to do the right thing and help maintain the environment. IAATO attends ATCM proceedings and presents papers on tourism as observers or invited experts. At the XXXV ATCM meeting, in conjunction with the XV CEP meeting in Hobart (Australia) from 11to 20 June 2012, IAATO presented the following information regarding the number of visitors to Antarctica in last few years as under (iaato.org):

Tourist Season	*Number of Visitors*	*Tourist Season*	*Number of Visitors*
2007–08	46,069	2011–12	26,509
2008–09	37,858	2012–13	34,354
2009–10	36,875	2013–14	37,405
2010–11	33,3824		

The above figures only show that from a few hundred tourists in 1969–70, Antarctic tourism has come a long way. With a member-strength of 111, IAATO represents a large majority and can influence the impact of tourism. It is committed to the provisions of the AT and Madrid Protocol for protection of the pristine environment of Antarctica.

Tourism has grown, as have the concerns about the potential impact on the environment, safety risks to tourists, and possible disruption to scientific research activities in case of the need for rescue operations. The bulk of tourist visits are concentrated in a few locations, and the desire for photographs can put pressure on the fragile environment, crushing minute plant life or disrupting bird breeding.

One argument is that a disaster involving a large number of tourists may divert the transport and rescue crews that support research. The opposite argument is that a visit to Antarctica is often the highlight of people's lives – they come back changed, with an urge to protect the environment, having experienced nature at its most primeval. IAATO advocates that, through a legitimate activity under the Treaty, fortunate travellers return home as ambassadors for the continent's environment and its continued protection for future generations. So a well-managed tourism industry can enhance public awareness of Antarctica, the value of its unique environment and the importance of Antarctic science (IAATO 'Home, News and Antarctic Tourism Fact Sheet', 2012).

6.11 Concepts for the Protection of a Pristine Antarctic Environment

6.11.1 *World Park Concept (1972)*

The idea was first mooted in 1972 by New Zealand to the ATCM in 1972. Except for conditional support from Chile and advocacy from pressure groups like the ASOC and Greenpeace, it got no support and proved to be a non-starter. However, it generated wider support for the ban on mining, which was included in the Madrid Protocol. It has faded in the background since then.

6.11.2 *Antarctic and Southern Ocean Coalition (ASOC) (1977)*

This organisation was founded by Friends of the Earth International (a federation of about sixty environmental groups around the world) in

1977. ASOC monitors and coordinates environmental advocacy about Antarctica and the Southern Ocean. It is a loose coalition of over 230 environmental groups in forty-nine countries to facilitate exchange of information, but does not direct their activities. Based in Washington DC, its Antarctic project was founded in 1982 with the goal of holding governments legally, politically and morally accountable for their actions in Antarctica and the Southern Ocean. It has observers at the meetings of CCAMLR, ATCM and IWC. The objective of ASOC is to protect the last unspoilt wilderness, preserving its value to science and keeping it as the last frontier for human endeavours. It initially campaigned to make Antarctica a world park. Its major activities are a special campaign to ensure proper implementation of the Madrid Protocol, and a campaign against whaling and unregulated and unreported fishing in the Southern Ocean.

6.11.3 *Greenpeace*

It is a pressure group founded in 1971 and represented in over thirty countries. It aims to protect biodiversity, prevent pollution and end nuclear threats. It established a world park base at Cape Evans from 1987–92 to monitor scientific bases and investigate the effects of human activities on the Antarctic ecosystem. Greenpeace spearheaded a campaign in the 1980s to prevent CRAMRA from coming into force, in its stead, the organisation promoted a ban on all mining, which was enacted for fifty years with the coming into force of the Madrid Protocol in 1991/98. It campaigned for Antarctica as a world park and works with the ASOC on environmental issues in Antarctica.

6.12 Situation in the Arctic and its Likely Effects on the ATS

The smooth functioning of the ATS, the effective implementation of the Madrid Protocol since 1998, banning commercial mining and other measures, and reviews of both still a long way away in years 2041 and 2048 respectively, have created environmental assurance. Presently, with broad commitments from countries towards the preservation of the pristine environment, Antarctica seems to enjoy a sort of tranquillity in the form of a well-settled, peaceful and cooperative world governance through the ATS. This may not be so in view of the changing situation in the Arctic, mainly due to the effects of global warming.

6.12.1 *Changing Environment in the Arctic*

The stark contrast between the Arctic and the Antarctic as regards the structure, climate, human inhabitation and wildlife has been seen in earlier chapters. However, their importance to the human race, in terms of their potential for natural resources, depleting elsewhere in the world, remains the same.

The Arctic contains about thirteen million square kilometres of sea-bound area of the Arctic Ocean (nearly the same area as that of Antarctica), surrounded by twenty-one million square kilometres of land area. Five littoral countries, namely the US, Canada, Russia, Denmark and Norway ring the Arctic Ocean. As per estimates of the Geological Survey of US, the region may contain about twenty-three per cent of the world's undiscovered and recoverable oil and natural gas reserves. The region is warming roughly twice as fast as the rest of the world. Fram Strait between Greenland and Norway is 3.5°C warmer than what it was a century ago, and chiefly as a result of this, the Arctic now has less sea ice. It is expected that the Arctic Ocean may be largely ice free in summer, sometime between 2020 and 2050 (*The Economist, Arctic Politics*, 2012).

All this would make Arctic shipping a reality in the near future. This would make the sea route through the Arctic, between Yokohama (Japan) to Rotterdam (Holland) at 12, 894 km, as opposed to the existing route through Malacca and Suez at 20, 921 km, much shorter, changing economic and trade equations. The Arctic has also become a popular tourist destination. In 2010, about 50,000 tourists sailed its unspoiled waters, previously thought unthinkable (Shyam Saran, 2012). The scope of this aspect is very wide to deserve a separate book. It is certain that much more will be heard on this subject in the near future. The changing scenario is clearly seen in Figure 6.3.

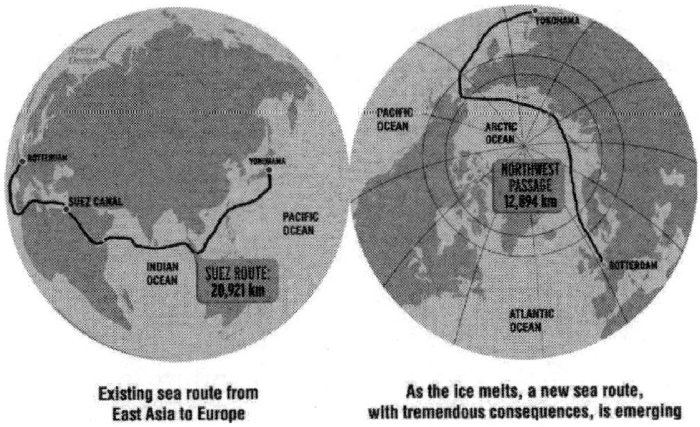

Fig. 6.3: The Position in the Arctic Due to Summer Snow Melting

6.12.2 *Arctic Council and Governance*

Five littoral countries mentioned above, along with Greenland, Iceland and Sweden, formed an Arctic Council in 1996. They claim sovereignty over the entire Arctic area together, but have conflicting territorial claims among themselves. Till 2010, six non-Arctic countries, namely France, Germany, the Netherlands, Poland, Spain and the UK were granted observer status. No country can become an observer unless it accepts the sovereignty of the Arctic Council countries over the area. In May 2013, six more countries namely, China, India, Singapore, Italy, Japan and South Korea were granted observer status. Norway had earlier openly opposed China's efforts. In 2010, in a speech to the Chinese peoples' Political Consultative Conference, Rear Admiral of the Chinese Navy, Yin Zhuo, had declared, 'The Arctic belongs to all the people around the world and no nation has sovereignty over it.' (Shyam Saran, 2012). It seems that China has changed its policy on paper at least, otherwise it would not have got observer status. Maybe, developments in the situation in the Arctic in the next few years require a close watch.

6.12.3 *Likely Effects of the Changing Arctic Situation on ATS*

The opening of Arctic sea routes, possibility of exploitation of natural resources, changes in the global trade and commercial situation and differing perceptions as pointed out in the previous paragraph may cause serious changes in geopolitics in the near future. This may also affect the present situational tranquillity with respect to the ATS, the extent of which may become clear in a few years' time.

6.13 In Conclusion

The uniqueness of all the natural aspects of the Antarctic environment and their stark contrast with those in the rest of the world are truly reflected in anthropogenic aspects like the system of world governance through the Antarctic Treaty (AT/Treaty) and its instruments known as the Antarctic Treaty System (ATS).

It can be seen that exploration voyages to Antarctica followed the usual pattern of any new land discovery, with curiosity, locating renewable resources (seals, whales), exploitation till near extinction, pressing ahead to new areas (sub-Antarctic islands) and, finally, reaching the mainland (ice shelves surrounding the continent). But the similarities end here. In the case of Antarctica, it was found that, even if discovered and its sovereignty proclaimed, it couldn't be colonised. The continent was totally isolated by the fierce Southern Ocean and unapproachable, except about two months a year, due to pack ice. On the continent proper, there can be no permanent human or animal inhabitation owing to the enormous ice cap and forbidding environmental conditions. Till the possibility of the existence of non-renewable mineral resources was seen, it was felt, as James Cook had said in 1772–74 while retreating in the face of pack ice just 121 km (it turned out so) short of the continent, that even if a great continent exists, it would be of no use to human race.

Still, though colonisation and physical occupation was not possible, territorial claims were made and remained on paper as diplomatic fights and minor nuisance conflicts. With the end of WW II, attention came back to the continent. As part of the probing efforts to assert authority, IGY was observed in 1957–58 and its outstanding success proved that scientific cooperation was indeed possible despite political differences. By the end of the IGY, not only were the harsh and unique continental realities realised or reconfirmed, but the environmental interdependence of Antarctica and the world and the subsequent catastrophic effects should that balance be disturbed, were also grasped in the true sense. All this led to the negotiated birth and signing of the Antarctic Treaty in 1959–61 and the subsequent enactment of various instruments for the protection of flora and fauna, including seals, whales, penguins, birds and other life-forms, finally culminating in the Madrid Protocol in 1991–98 for comprehensive environmental protection. Commercial exploitation of minerals is since banned and no military activity is permitted, thus, preserving the pristine environmental conditions. Territorial claims, too, are frozen. Looseness of and ambiguity in language, and ground realities, have proved to be a binding force for the success of the ATS.

As of now, over thirty countries maintain about seventy research stations and are engaged in scientific research beneficial to the whole world. Issues like the ozone hole depletion would never have come to light without such arrangements. ATS remains an example of peaceful international coexistence and can act as a model to tackle differing views and perceptions in case of international discords. Even IAATO, with all its commercial interests in Antarctic tourism, accepts the need for environmental protection and abides by the measures of the Treaty. The waters of the Southern Ocean, after a century of wild life slaughter, seem to be mostly non-threatening to marine life. It seems that countries are happy with the present situation and would like the status quo to remain till 2048 when the Madrid Protocol will be up for review. The situation then will depend on many factors, including world population, effects of global warming and the mitigation measures for it, the response of the world community to the depletion of non-renewable natural resources and development of renewable resources, the state of the Arctic Ocean and uncertainties in the geopolitical situation. It is anybody's guess.

CHAPTER 7

Antarctic Logistics

"The line between disorder and order is logistics."

Sun Tzu (544 BC–496 BC)
(Chinese general, military strategist, philosopher, thinker and author of The Art of War)

7.1 Historical Perspective and Development

The Oxford English Dictionary defines logistics as, 'A branch of military science having to do with procuring, maintaining and transporting material, personnel and facilities.' Logistics was, typically, applied to the art of movement and supply of material to a body of troops engaged in military operations. The volume, range, distance, duration, modes of transport, communication, terrain and the nature of the operation of a body of troops, emphasised the need for a skilful use of the art for a successful military operation. In ancient Greek and Roman Empires, there were military officers, entitled *Logastikas,* specifically responsible for logistics. The overwhelming importance of logistics can be illustrated by the examples of Napoleon and Hitler who launched their campaigns against Russia on the same day, different years: 22 June in 1817 and 1941 respectively. Though the operations were 124 years apart, one of the key reasons for the failure of both was the same – untenable logistics.

As advanced technology started changing the world in every sphere, logistics' parameters, too, changed. Over the years, the interface between technology and logistics became critical. As logistical needs became more complex, the tools required to achieve them became modern, in keeping with the advancement of technology. In today's world, one can say that logistics has moved away from its traditional reference to military movement, to operations in any field. Management of logistics has become an integral part of the efficient and effective supply chain which plans, implements and controls the forward and reverse flow and storage of goods, services and related information between the point of origin and the point of application or consumption, in order to meet the desired customer and legal requirements.

If there is one place on earth which can test the limits of logistical skills, it is the continent of Antarctica and the vast Southern Ocean, separating the continent from rest of the world, the extreme environment replacing the enemy, and a world governance through the Antarctic Treaty System (ATS) and other loose agreements, posing a formidable challenge to logistics in the area. While low temperatures, heavy winds, a thick layer of ice covering over ninety-eight per cent area, blizzards and the absence of any indigenous resources required for inhabitation characterise the continent, a very difficult sea with circumpolar currents, pack ice and other obstacles are the features of the ocean. All these create near-unsurmountable obstacles in reaching and exploring Antarctica. In few spheres of human endeavour are the demands upon a logistics organisation as onerous as the ones in Antarctica. Despite this, tremendous progress has been made in Antarctic exploration, which only underscores the human urge, tenacity and technological progress made over centuries, giving credence to the adage 'necessity is the mother of invention'. Logistics here, an important tool of management, has been transformed from merely an art form to a combination of art, science and governance.

Though scientific inquiry remains the principal goal of any exploration, for its execution, it will remain an adjunct of the logistical support at hand. In other words, if modern techniques are to be used in Antarctic exploration and scientific research, they have to be matched by simultaneous technological advancement in logistics. Consequently, all essential components of modern-era logistics, namely, transportation (ships, aircrafts and vehicles), food, clothing, shelter, fuel, energy, scientific and other equipment, communications, medicines and others, have undergone significant changes with time.

Efforts made by scientists on heroic expeditions between 1901 and 1915 added enormously towards knowledge-gathering, but the time and effort necessary to travel and survive put constraints on the range and scale of scientific research. Skiing, man-hauling and dog-sledging had been the chief modes of transportation for explorers till motor vehicles took over. Journeys were confined to short summer months when the weather was favourable or less harsh. Here again, tactics to reduce distances, to use dogs skilfully, to apply logistics' acumen in preparation and execution, and some luck, played a major role in making Amundsen the first person to reach the South Pole in 1911 and live to tell his tale.

Introduction of vehicles with internal combustion (IC) engines changed everything dramatically. Though, Britain's Earnest Shackleton and Robert Scott failed to use motor vehicles, their potential was never in doubt. Australian Douglas Mawson intended, in 1912, to fly an aircraft over Antarctica, but the plane had crashed in Australia itself, and was used as a tractor in Antarctica for a while. Later attempts to use motorised transport were increasingly successful. US Richard Byrd used cars with outstanding success and also became a pioneer in the use of aircrafts (*Reader's Digest*, 1985).

Despite all these developments, surprisingly, dogs remained the mainstay of Antarctic exploration until the 1950s. As use of motor vehicles became more prevalent, the presence of dogs gradually waned to a mere ceremonial symbol at most stations, mainly out of nostalgia. Subsequently, the Treaty nations dispensed with dogs from the continent altogether in 1994 due to environmental concerns (McGonigal, 2008). At a personal level, whenever I was on convoy and visited Novo, the magnificent Laika was always there to escort me to see the chief of the station, Dr Yevgeny Nikolayevich Uranov, and sit in during the meeting. I am not very fond of dogs but seeing friendly recognition in her eyes was touching and I always felt disturbed while leaving.

The International Geophysical Year (IGY) 1957–58 was a water shed in Antarctic logistics. The sheer scale of IGY enabled logistics planning and development in a way never seen before. If prior to IGY, expeditions were private ventures, they now became essentially government-funded national ventures in line with a country's political aspirations. The cost and complexity grew with the maintenance of year-round research stations, government association and supporting bureaucracy. For the first time, during the IGY, a large number of ships, aircrafts and tractor/trailers chains were used. Twelve countries established about sixty stations with a personnel population reaching 5000, underscoring the growing international attention, which ultimately paved way for the signing of the Antarctic Treaty (AT/Treaty) a year later, succeeding secret/intense negotiations.

Major improvements in all aspects of logistics were achieved by the end of IGY and have continued since then. Vivian Fuchs's trans-Antarctic expedition on four Snow Cat vehicles, covering over 2000 miles in ninety-nine days, saw the effective use of tractors (Trewby, 2001). Fifty-five years later, in 2010, an Indian expedition with four Arctic trucks travelled nearly the same distance, from Maitri to the South Pole and back, in just nineteen days. (*NCAOR Through the Years,* 2010). The use of large, ski-fitted, long-range aircrafts opened up a new era of transportation to and from Antarctica. The landing of an aircraft to rescue a cancer-affected person (herself a surgeon in the wintering team) at the Amundsen-Scott station on the South Pole in the winter of 1999 signalled what could be be achieved in such severe conditions. Dr Jerri Nielsen died in 2009 after a remarkable post-Antarctic active life till the very end (cnn.com, *Health, 2009*).

7.2 Management of Antarctic Logistics within the ATS

The overwhelming importance of logistics in Antarctica has been acknowledged and duly catered for within the ATS. The first symposium on Antarctic logistics and operations was held in 1962. Till 2008, a total of thirteen symposia were held. Since 1986, the Committee of Managers of National Antarctic Programmes (COMNAP) has been organising these through its Standing Committee on Antarctic Logistics and Operations (SCALOP). It is a forum where people from different countries share their expertise and thoughts for the benefit of the wider Antarctic programmes' community. Since 2010, the symposium includes all matters relevant to the activities of the National Antarctic Programmes (NAPs), along with logistics and operations. Reports from these symposia form a very valuable reference material (*COMNAP Symposium on Antarctic Logistics*, 2012).

7.3 Broad Logistics Requirements of an Antarctic Base/ Station

Extending logistics support, despite the developments, remains a big challenge in Antarctica. To understand the complexities of the logistics support needed by any station, an understanding of the salient points of a generic deployment season is helpful (Reed, 1995):

- Antarctica is inaccessible to ships for most of the year due to pack ice, except during the few austral summer months. An austral summer extends, approximately, from September to February and winter from March to August. With the exception of a few adjoining months on either end, the austral summer and winter months are characterised by twenty-four hours of light and twenty-four hours of darkness respectively. The period of accessibility to the continent also varies due to the changing conditions of sea ice from year to year.
- Ships must arrive in Antarctica in austral summer to take full advantage of the period, for pack ice is at its minimum and ships can be berthed at ice shelves or other ice-free points, winds permitting. As the continent is surrounded by the Southern Ocean, ships have to undertake a long and difficult sea journey, manoeuvring through varying pack-ice conditions and icebergs before berthing at ice shelves or the few coastal oasis areas.
- A base is established over a period of two to three austral summers, depending on the availability of working days, unloading conditions, transportation and construction issues, and periods free of blizzards and other environmental uncertainties.
- Once established, Antarctic bases must generate their own fresh water and electricity.
- Every item, including vehicles, helicopters and equipment, spares, food, fuel, building and maintenance material, medicines and other items for survival must be brought to the continent, and all

the waste generated shipped back to the mother country.

- During this period, the scientific work of the summer team and preparation by the new winter team to take over from the existing winter team is in full swing. By February, the stations are handed over to a small cadre of personnel, in both logistics and research, who remain there during the winter months.
- The ship must leave for its return journey by end-February and March-beginning to avoid getting trapped in pack ice. There are many examples, in early expeditions, of ships getting stuck in pack ice and having to experience a forced winter stay.
- Once the ship leaves, the winter team is totally isolated till the return of the ship next summer around December, with a new team to continue the programme.
- Due to the range and complexities of various logistics tasks involved with any Antarctic programme or expedition, many tasks are done by external organisations on a contract basis. This is known as third-party logistics and it includes various forms of outsourcing.

The Antarctic programme of most countries followed the above cycle till recent times. Now, advanced aviation, better landing facilities and availability of airports, all have resulted in vastly improved conditions for physical travel. This has reduced travel time and made more time available for both scientific and logistics tasks. It has also made possible short visits for those who do not have extended free time in their schedules.

Most of the above aspects are discussed in details later.

7.4 Development of the Individual Elements of Antarctic Logistics

There has been tremendous improvement in all elements of Antarctic logistics, in keeping with the correlated technological developments elsewhere in the world. A discussion on elements like shelters, ships, aircrafts, helicopters, vehicles, communications, food, clothing, medicines and others, with reference to their individual development and subsequent impact on Antarctic logistics as a whole, makes for interesting reading. Overall, technology has made Antarctic logistics and by extension, polar exploration, far safer, faster and better in every sense. The only unfortunate fall out of it is the redundancy of an important element of yesteryears, a creature loyal and dedicated to the human race in its polar exploration, the dog. Though dogs have outlived their utility, polar explorers still get nostalgic about their erstwhile companions.

7.4.1 *Ships*

In terms of size, displacement, power, carrying capacity and other parameters, ships have undergone huge changes since the days of early exploration. For example, *L'Astrolabe*, in 1811, was a sail-powered polar ship with a size of 31 m × 8 m and a displacement of 325 tons and no icebreaking capacity. Compare this with the polar ship, *Yamal*, launched in 1992 with a size of 150 m × 30 m, a displacement of 24,000 tons and power of 75,000 hp, and one can see the extent of the development. Whereas earlier, ships were mainly built for sealing, whaling and fishing, today, modern ships are custom-built for operations in pack ice (*Cool Antarctica*, 2012).

The first ships to sail on the Antarctic ice-bound waters were Captain James Cook's *Resolution* and *Adventure* in 1772 (*Reader's Digest*, 1985). The early ships had wooden hulls and were not equipped to travel through thick pack ice. Consequently, many were crushed by pack-ice pressure and sank. Amundsen's *Fram* had a saucer-shaped hull, designed to ensure that ice pressure would squeeze it upwards. In contrast, Shackleton's ill-fated *Endurance* had a steep-sided hull that caved inwards as the moving ice pressed against the ship and crushed it in the Weddell Sea pack ice. Gradually, ships started to adapt themselves to polar conditions. Steel hulls have replaced wooden ones, giving rise to ice-strengthening and icebreaking capabilities. With strengthened hulls and powerful engines, icebreakers carve a path through thick pack ice to access the Antarctic coastline. The first icebreakers to be used in the Southern Ocean were the *North Wind and Burton Island* by the US during Operation High Jump in 1946–47. The speed and icebreaking capacity varies from ship to ship. For example, Britain's *James Clark Ross* can be driven at a steady speed of two knots through level sea ice one metre thick. As they cut through ice, compressed air systems are used to roll the ship back and forth, preventing ice from squeezing the ship. When the ice becomes too thick to cut, ships moor at the edge and the cargo is unloaded for over-snow transport.

As of now, over a hundred ships visit the Antarctic waters each year. They are mostly expedition vessels for national Antarctic programmes. Cruise ships, research vessels, commercial fishing boats, ships of tourism operators and private yachts also frequently visit Antarctic waters.

Some of the old ships are still preserved in museums around the world. When I visited Fram Museum (in Norwegian, *Frammuseet*) in Oslo with my wife, we were amazed to see how well the ship, *Fram*, had been preserved. Seeing the ship, its exterior, the bow, deck, engine and other parts and their primitive nature, I understood Amundsen's gratitude to the skills of old sailors, when he said, 'We must always remember with gratitude and admiration the first sailors, who steered their vessels through storms and mist and increased our knowledge about the land and ice in the south.'

7.4.2 *Aircrafts*

The first aircraft to fly above Antarctica was a balloon named *Eva*, piloted by Robert Scott, on 4 February 1902 when he surveyed the landscape from a height of 240m and noted the progress of his sledging party on the move. The first successful flights over Antarctica were made by Carl Eielson and Hubert Wilkins in 1928–29, in a Lock Heed Vega monoplane. The first real aircraft taken to Antarctica, but which could not be used as such, was Douglas Mawson's air tractor described earlier in the chapter. Richard Byrd made a number of aircraft-supported expeditions over uncharted ice fields in the 1930s and led Operation High Jump, the first large-scale deployment of aircrafts, in Antarctica in 1946–47.

The landing of aircrafts in inland Antarctica was demonstrated when Lincoln Ellsworth and Herbert Hollick-Kenyon did an aerial crossing of the continent in 1935. The first intercontinental flight to Antarctica from Christchurch, New Zealand, to McMurdo Sound was made in 1955–56. Many runways are now built on the periphery and inland polar areas for such flights.

Helicopters were used for the first time in Antarctica during Operation Windmill (sequel to Operation High Jump) in 1947–48. They have now become an indispensable part of Antarctic logistics. Both ships and land-based helicopters are used to transport parties to remote field camps. They can carry large loads of equipment and supplies, sometimes in underslung positions due to the unwieldy size of the payload. Their low-flying and hovering abilities make them invaluable for search and rescue operations and for use on unstable sea ice.

The use of both small and big aircrafts and helicopters is now very common. In the United States Antarctic Programme (USAP), large air-transport operations are carried out by Hercules, Globe Master and Star Lifter planes, which carry equipment and personnel from Christchurch each summer to the Italian, American and New Zealand bases. Smaller ski planes such as Twin Otters are used to ferry parties and supplies to deep field camps out of the range of helicopters.

7.4.3 *Dogs*

An historical and often nostalgic symbol of human presence in Antarctica, dogs were an essential part of any inland expedition team. Driving a team of dogs was a skilled job. Dogs helped to explore, study and map the continent. On their successful journey to the South Pole, Raold Amundsen's party took four teams of thirteen dogs each, also killing the animals for food, on schedule, as they went. In contrast, Robert Scott sent his dogs back to base and his team man-hauled their sledges over the horrors of the polar plateau.

With their thick pelts, the Arctic-bred sledge dogs or huskies are suited to Antarctic conditions. Very tough and vicious fighters, they can haul great loads through strong winds and blizzards, travelling about twice as fast as a team man-hauling a similar load. A team of dogs travelled with the Commonwealth Trans-Antarctic Expedition to scout routes up stiff glaciers for the tractor team led by Edmund Hillary during the IGY. From the 1960s onwards, motorised vehicles gradually replaced dogs (McGonigal, 2008).

American Colonel Norman Waughan (1905–2005) was a member of Richard Byrd's first expedition to South Pole in the 1930s. A colourful personality, he also led the last expedition with dogs in the 1990s before their presence was terminated due to environmental reasons. I had volunteered to go on this expedition, but Norman, in a handwritten note to me said, 'Your credentials are impeccable but my roster is full.' A nice way to decline.

7.4.4 *Shelters*

In Antarctica, shelter is more vital than food. Intense cold may kill more swiftly than any deprivation, save that of air. The fundamental shelter is clothing which conserves warmth and regulates its escape. It can do no more. In still air, even at temperatures well below zero, body heat is undisturbed because a thin layer of warm air remains close to the skin. But once even a slight breeze stirs, body temperature drops rapidly. To be caught in a wind without adequate clothing and with no shelter, is a sure recipe to perish. To survive, a traveller must have food, rest and some sort of refuge, which must be found, built or innovated.

Shelter in any form is protection from the elements and essential for survival. It can be permanent in the form of huts, like those of the early explorers or the modern-day permanent bases, or temporary, like a transit refuge created during a journey or at field camps in the form of an igloo, snow cave or tent.

7.4.5 *Permanent Buildings or Huts*

The early explorers' huts were built mainly of wood with several layers of boards and some insulating material, often seaweed, in between, to protect against the cold. They were, essentially, a winter refuge and a starting point for field parties during summer months. A hut, typically, housed members of an expedition, provided cooking, eating and sleeping space, offered facilities for development of photographs, preparation for field excursions, classifying and storage for collections and other activities. Although stoves burned constantly, it was difficult to keep indoor temperatures above freezing. Blizzards could force large amounts of snow through gaps in walls and could very well bury the hut and its occupants under large snowdrifts. Some huts were doubled with occasional lofts. Dogs were either tethered on the lee of the hut, beyond the drifts, or given shelter in a lean-to.

Since IGY 1957–58, bases have become more complex, with multiple scientific functions, each requiring own building and facilities. Modern buildings are usually made from prefabricated panels, insulated by plastic foam cores and silvered metal surfaces or honeycomb partitions with reflective foil. These panels are bolted together precisely to make the joints wind- and drift-proof. Most bases have a central power station to provide electricity for heating, lighting, cooking and scientific instrumentation. The interior of the buildings is kept at a comfortable room temperature.

Even with all the commonalities, designs of the buildings built during IGY varied considerably. The British and Norwegian expeditions used conventional timber-framed buildings, whilst the French used prefabricated pressed-steel huts. Australia, US and the then-USSR built huts on rock sites with similarity in construction. All used prefabricated aluminium or plywood insulated panels clamped together to produce flat-roofed, box-like buildings.

Now that it is accepted that the Antarctic research stations are going to be permanent features, giving expression to their countries' active and influential presence in the continent, the second major change from earlier days has been the consolidation of buildings. Since the bases are to be permanent, major rebuilding has had to take place to provide widely spaced blocks housing many functions, instead of separate huts. These save movement, time and labour, especially in a blizzard. They are also fireproof, quieter, vibrate less and easier to heat than a lot of the smaller self-contained buildings.

There is also another key difference between a shelter created on an ice shelf or polar plateau, and one built on an ice-free oasis area. Whereas the former is bound to be buried under snow at some stage, the latter does not face that danger. Even on ice, the extent of snowdrifts and wind conditions vary considerably from location to location. Various countries have developed different methods of rebuilding their stations to suit location requirements.

A radical experiment was carried out at the Australian base, Casey, in the 1960s. The units were connected in a line, across the prevailing wind, on a raised, open-iron framework. Men could move through the station in all-weather conditions and the free movement of air beneath the structure prevented snowdrift. However, vibration was a problem and now Casey, rebuilt in the late 1980s like most Antarctic stations, consists of independent two or three storey buildings (*Books LLC*, 2010).

Most countries now, while erecting new bases or replacing old ones, use mechanical or hydraulic devices (stilts or skis) to raise the level of their structures or to make them mobile, particularly those erected on ice, in order to increase their lifespan and guard against snow burial after a few years. This aspect is discussed while examining the NAPs of some countries later in the chapter.

7.4.6 *Temporary Shelters for Field Parties*

- *Caves:* These occur naturally at the base of glaciers or icebergs when moving ice pushes against an obstacle, creating a space in the lee of the obstacle. Snow caves have frequently sheltered marooned parties. Spurred by necessity, Scott's northern party excavated an ice cave in a location in snow drift calling it inexpressible island for its six members to spend the winter of 1912. The ice walls of the cave were lined with snow blocks and on its floor was laid a layer of pebbles topped by seaweed. A tent cloth was spread on the seaweed.
- *Igloos:* Dome-shaped shelters made from blocks of ice, igloos are traditionally used by Eskimos in the Arctic region. Like ice caves, they provide good protection from the Antarctic elements. Field parties are trained in how to construct and live in such shelters.
- *Tents:* Tents are widely used in field camps. These are the famous polar pyramid tents derived from the Eskimo skin tent, which was

originally made to be thrown over four or five strong bamboo poles spread in position. Mawson's description of tent construction, written in 1911, can be applied to any tent construction in a sledging expedition virtually to the present day (Mawson, 1915). To facilitate tent erection in perpetual winds, the skin was sewn onto five bamboo poles instead of being thrown over them. A man inside had to do considerable jugglery to spread it out like a conical umbrella, while others worked outside to erect it. When pitched, the tent was held down by blocks of snow or ice, helped by spare food bags. Ventilators had to be discarded due to the incessant drift. During Scott's 1910–13 expedition, a new type of double tent consisting of an inner and outer lining, spaced a few centimetres, was used for the first time. Humidified air from the occupants could pass through the porous inner lining and condense as frost on the main tent. As the vibration caused by the wind dislodged the frost, it was directed to the perimeter of the tent instead of falling on the occupants. This has become the standard design for many tents used on expeditions today.

7.4.7 *Medicines*

Modern transport systems, stations equipped with every facility, availability of communication, food and medicines have greatly reduced or eliminated the ailments suffered by early explorers.

Medical problems facing explorers about a century ago were about the same as those faced today, but early explorers were not as well informed or well equipped to cope with them. Hypothermia or cold exposure, frostbite, forms of food poisoning, scurvy and psychiatric problems were experienced in a number of early groups. Many of these can be explained now with better medical knowledge. Serious illness befell Douglas Mawson and Xavier Mertz in the Australasian Antarctic Expedition in 1912 while dog-sledging. Mertz became delirious and died and both men lost a large amount of skin. It was only forty-five years later that research suggested that their cause of illness was hypervitaminosis or Vitamin A poisoning caused by eating the livers of their huskies. Scurvy affected several explorers, disturbing a large numbers of expeditions. Investigations revealed that it was due to the lack of fresh fruit and vegetables in the diet, which comprised primarily of preserved foods, and a deficiency of Vitamin C.

Scott's doctor inspected every tin of meat for ptomaine (an alkaline substance formed by decay in vegetable or meat, causing poisoning) in 1901. The symptoms of scurvy disappeared in the second year when the fresh meat of seals and birds was consumed.

The most common medical emergencies on modern-day expeditions are wounds or injuries and other minor ailments. Research stations of today have good facilities and accomplished surgeons and doctors, and most medical problems are well taken care of. In 1961, a physician with the sixth Soviet expedition was forced to operate on himself for acute appendicitis with the help of two co-workers, holding retractors and mirrors (Trewby, 2001).

Antarctic expedition members of today are screened for pre-existing medical problems and mortality and morbidity rates remain very low. Most of the serious health risks in modern Antarctica are not caused by environmental extremes, which are now well understood and guarded against, but by hazardous activities such as engineering tasks and use of vehicles and aircrafts. Thus, in the twenty-first century, the Antarctic medical practice is mainly concerned with the treatment of occupational injuries. The British Antarctic Survey (BAS) offers a good example of the current medical practice in Antarctica. Its contemporary medical unit combines the traditional approach, that of doctors residing on Antarctic bases, with the modern approach of providing paramedical training to all staff, electronic access to specialists in Britain and telemedicine. As a result, BAS doctors spend only part of their time on medical duties and concentrate the rest on conducting high-quality research, thereby, maintaining and perpetuating two hundred years of their Antarctic tradition (Wikipedia, BAS, 2012).

7.4.8 *Food*

In bitterly cold Antarctic conditions, humans need a high-calorie intake to maintain energy levels. This is provided by food high in carbohydrates and fats. Presently, aircrafts and frequent ships ensure fresh supplies of vegetables, fruits and the like, every year; fresh bread is baked daily, and many bases have greenhouses that grow hydroponic produce. Emergency food depots have now become a thing of the past. Dehydration is a serious threat in the arid environment and a high fluid intake is essential.

Modern packaging and freezing or drying techniques allow field parties to have a balanced diet and still travel light. A person camping out in the field needs about 3500 kilocalories per day, about double the amount needed in warmer regions. Field party food includes basics such as freeze-dried meat, dried soup and vegetables, rice, pasta, crackers and cheese, water, sugar, porridge, milk powder, plenty of snacks and chocolates.

Excellent modern-day rations are a far cry from what the early explorers had to contend with. Many of them did not have enough to eat and were preoccupied with hunger. Their rations were uninteresting, unappetising and not very nutritious. Problems of size and weight influenced the choice of food. Rations had to be light, not too bulky and easy to prepare on a primus stove, using small amount of fuel. In addition, food had to be standardised in easy-to-handle man-daypacks. Pemmican, originally used by Native Americans, was a staple food for sledging journeys. Made from dried and crushed meat, it was mixed to a paste with fat and berries. It made a hot, nourishing and filling meal, traditionally called a 'hoosh' porridge when mixed with dried biscuits and water brought to a boil, and served hot.

To supplement dried rations, parties killed their own animals or lived off wildlife. Amundsen even calculated the eating of his sledging dogs on schedule. The men and the surviving dogs alike consumed them.

Scott's comment in 1902 on the inadequacy of rations even with all the preparations sums up the situation in early days, 'There can be little question, therefore, that polar sledging ranks an easy first as a hunger-producing employment.' (Crane, 2005). Scott's northern party, mentioned earlier, survived in the cave, on mainly penguin and seal meat. Little was known in the early exploration days about the role of vitamins and it showed. Research revealed that it was a lack of

Vitamin C and a depletion of Vitamin B in Scott's sledging rations that probably contributed to the eventual tragedy.

7.4.9 *Communications*

Communication in Antarctica is important for the exchange of information, weather forecasting and to allow contact between bases, field parties, home countries and families. The remoteness and natural environment of Antarctica has often made communication difficult. The first use of the telephone was by members of the 1910–13 British Antarctic expedition to transmit a message from Cape Evans to Hut Point, a distance of 24 km. Ships, visiting the continent during the austral summer months each year, were relied upon for postal services in the early days.

The first radio link from Antarctica was established during the 1911–14 Australian Antarctic expedition, via a relay station on Macquarie Island. The use of radio in Antarctica was greatly expanded by Richard Byrd, who used it to communicate with field parties and different bases and to link up with the rest of the world. Floyd Bennet made what, at that time, was believed to be the first flight over the South Pole in November 1929. Its radio signal was picked up by a ham radio operator in New York and broadcast in Time Square.

At one stage it was believed that Antarctica might be a useful location for ground relay stations for global radio transmissions. However, communication satellites have filled this role more effectively. The southern aurora belt made communication between some parts of Antarctica by radio difficult, but this problem has also been resolved with the use of satellites.

Today, communication in Antarctica is organised by consultative parties in conjunction with the World Meteorological Organisation (WMO). A network of low-earth orbit satellites allows voice, data and fax services between Antarctica and rest of the world, and most of the bases now have Internet access, telephone and fax facilities.

Used in Antarctica since the 1970s for mapping, communications and research, a number of satellites continually sweep across the skies. Polar-orbiting satellites, which monitor changes in the sea-ice cover and the ozone hole, are extremely valuable in climate-change studies. Knowledge of the distribution of sea ice is also important for navigation. By 1980, the Global Positioning System (GPS) enabled scientists to use satellite signals to fix a position within a metre. This technology has been used widely in Antarctica. For instance, stakes positioned on the surface of a glacier are fixed at different times in order to calculate the speed at which the ice is flowing. Satellites are also used for monitoring the location of fishing vessels in the Southern Ocean.

India has experienced considerable changes with respect to communications between Antarctica and the mainland. Most of the reports were sent by HF (Morse Code) communication. During the last wintering at DG, 147,000 words were transmitted by HF and the telex was used for 444 minutes (Khadilkar, 1990). Just over two decades later, it seems primitive, what with the free use of video conferencing, satellite connectivity, instant data transmission and Internet access for members, in addition to telephone facilities (*NCAOR Through the Years*, 2011).

7.4.10 *Waste Disposal*

Looking back, one can say that until the 1980s, waste-disposal practices had no or little regard for the environment. In a way, it was free for all. Station locations were selected to suit political choice, urge for continental presence, convenience for sea, air and ground logistics and facilitation of scientific experiments mostly in that order. Not much thought was given to the pollution caused by power generation, fuel and other toxic spills, and waste disposal created a very unhealthy situation around most of the stations.

As most waste decays very slowly in Antarctica's arid climate, a legacy of debris and toxins remains for decades if not attended to. Bases such as the McMurdo Station once burned garbage. Waste was either dragged under the sea ice and into the Southern Ocean or piled up on ice or ice sheets and blown up into the ocean. This waste also had an extensive collection of refuse, including toxic polychlorinated biphenyl. For a long time, fuel drums, chemical waste and garbage from Argentina's Marambio Station were thrown down a slope into the sea, leaking toxins unabated into the soil, streams and ocean. Leftover building material such as wood, iron and debris from abandoned bases had littered many parts of the continent around the stations. Waste disposal of fuel and leaks has caused problems as well. Large waste dumps at the US McMurdo were an eyesore before massive efforts were made to clean the area. Many people, like Eirik Sonneland (of Norway, who with his colleague Rolf Bae walked from their station Troll to Ross Sea, a distance of over 3800 km unsupported in 2000/2001), have used unprintably harsh words to describe conditions at the McMurdo Station. A review of the aspect of waste management was done in a special workshop in Hobart (Australia), and many countries revealed their actions on the clean-up drives for their stations (*COMNAP Workshop on Waste Management,* 2006).

All this has, however, changed dramatically after pressure from environmental groups and the enforcement of the Madrid Protocol. Most bases now have environmental officers, and staff and visitors undergo environmental awareness and safety training. Stations established recently, like Concordia (France, Italy), Kunlun (China), Neumayer (Germany) and Bharati (India), have been built only after a Comprehensive Environmental Evaluation (CEE) was made and implemented in accordance with the Madrid Protocol. For example, only after going through all the stages, like the initial proposal, selection of a proper site with due evaluation, circulation of a draft CEE to all treaty members, and the approval of a final CEE by Committee for Environmental Protection (CEP) and Antarctic Treaty Consultative Meeting (ATCM), was India able to build its third station Bharati in the Larsmann Hills area of East Antarctica (*NCAOR, CEE, 2010 and NCAOR Through the Years,* 2010).

7.4.11 *Vehicles*

Machines now provide the main form of surface transport in Antarctica. They have taken over from skiing, manual hauling and dog-sledging. It all started as an innovation on Ernest Shackleton's Imperial Antarctic Expedition of 1907–09. He tried a 12–15 hp, four-cylinder, air-cooled Arrol-Johnson petrol car with modifications like special oil, a reinforced frame and a recycled exhaust system to heat the engine, to commence the era of mechanical surface transport in Antarctica. This car successfully travelled for a distance of 48 km at a speed of 15 to 24 kmph, laying food depots to support the expedition. Here, unfortunately, it disappeared into a crevasse. It was rescued but could not be used again. Scott benefitted from Shackleton's experience and took three specially built tracked motor sledges on his expedition. Unfortunately, one broke through the sea ice while unloading, disappearing into the ocean, and mechanical failures accounted for the other two. As mentioned earlier, in 1911, Mawson tried a wingless airplane as tractor sledge, which too had to be abandoned due to mechanical failure. In 1929, Richard Byrd used a Ford Snowmobile, which broke down after travelling about 120 km. Through all these, motor transport kept developing its systems, like engine, transmission, exhaust, suspension and wheels/tracks, to survive the severe environmental conditions of Antarctica.

The first successful utilisation of mechanical transport over snow in Antarctica was achieved when Byrd used three modified Citroen cars on his 1933 expedition. The rear wheels were converted into tracks and the front wheels were clamped to short metal skis.

Despite this, dogs remained the mainstay of sledge hauling in Antarctic exploration up until the 1950s.

WW II and the period after, led to numerous advances in the construction and engineering aspects of vehicles and this was reflected in the tractors used in Antarctica. Long journeys are now supported by a variety of low-ground pressure, wide-track tractors developed by many countries. Tractor-sledge trains have become the main form of transport for major trips on the polar plateau. Each tractor usually hauls a number of sledges; the bulk of the load carried is often drums of fuel which, when empty, can be welded together for route-marking for the return journey. The remainder of the load consists of living caravans, scientific and mechanical workshop vans and food for those aboard. People travelling in a convoy can talk to one another by radio. Vehicles always travel in pairs for mutual support, except in an emergency.

A tractor or sledge train on a journey from the coast may meet a number of hazards. Sea ice may be one to two metres thick overall, but it can be thinned by strong currents near islands or narrows, and may also be crossed by cracks that have refrozen and are now weak. Large areas may break and drift out to sea with no warning at all. The junction between land and sea, as also the zones where glaciers and ice sheets move down to become ice shelves, are hazardous to vehicle movement due to crevasses and miniature rift valleys.

The initial slopes of the plateau may be steep enough to force parties to work in relays, making it necessary to winch the vehicles up steep slopes. Often katabatic winds near the coast scour the snow from the surface, leaving slippery blue ice on which vehicles have little grip. Ice on the coastal plateau is often thin as compared to the high areas towards the Pole where it may be 3000 m thick and can be severely crevassed where it moves over the bedrock beneath. All these problems make the coast probably the most hazardous area for vehicular movement. Marker poles or drums need to be checked and replaced frequently to show safe routes through dangerous sections even on very familiar routes.

On the plateau, problems are different but no less significant. Extremely low temperatures, often minus 50° to minus 60°C and lower, make metal brittle and cause it to snap. A sastrugi, two to three metres high, like frozen surf, can cause vehicles to roll like ships at sea, straining at every joint.

Standard toboggans or snow scooters are used for work and travel near bases and in the mountains.

Vehicles are very difficult to start in Antarctica in the low temperatures and with snow filling everywhere. This has now greatly improved with superior batteries, special fuels and lubricants that remain fluid at temperatures down to minus 60°C, and cabin/sump heaters.

Navigation in Antarctica is little different from the techniques used at sea but compasses become unreliable near the South Magnetic Pole. Celestial navigation was used since early expeditions but this has now been replaced by satellite navigation.

The development of air transport has greatly reduced the work of motor transport. Still, stations like Vostok (Russia), Concordia (France/Italy) are maintained by tractor-chain convoys from coastal stations like Mirny and Dumont de Urville respectively, over a distance of 1000 km on polar ice plateau (*Books LLC*, 2010).

The Indian Antarctic Programme (IAP) used tractors, snow scooters, cranes, snow cutters and sledges for snow clearance, local logistics tasks and convoy-running at DG till 1990. Maitri, situated about a 100 km away from the ice shelf, does not have a problem of snow clearance, but requires to run frequent convoys to the coast. This aspect is discussed in details while discussing the IAP in the next chapter.

7.4.12 *Harbours and Airports*

As seen earlier, the nearly 18,000 km coastline of Antarctica consists of ice shelves, ice walls, ice streams/outlet glaciers and promontories. Except at two places, natural harbours do not exist on the continent. One natural harbour is at the US McMurdo Station and the other, horseshoe shaped, is at Australian station Mawson. Most coastal stations have offshore anchorages, and supplies are transferred from ship to shore by small boats, barges, and helicopters. At a few stations where ships can berth at the shelf, direct unloading can be done by the ship's crane, wind permitting.

A network of airstrips supports air operations in Antarctica. Airstrips are constructed on permanent rock surfaces in ice-free areas or on sea ice or compacted snow, or on hard blue ice. There are about thirty available airports in Antarctica but none of these are developed for public or commercial landing facilities. Of these, five are equipped for landing aircrafts with wheels, nineteen for aircrafts with skis and the rest with skis and wheels both. The length of the runways varies greatly from less than 1 km to 3 km and more. Details of these airports are given in *Appendix IV* (*COMNAP – Main Antarctic Facilities,* 2009 and Wikipedia 2012, *Research Stations in Antarctica*).

Argentina, Britain and Chile have constructed rock airstrips on the Antarctic Peninsula near their stations.

Operating an airport in Antarctica is a very different and complex ball game as compared to operating a conventional airport. This complexity is best understood when we examine the three airstrips at the USAP's McMurdo Station: one each on sea ice, shelf ice (Williams Field) and on the polar plateau (Pegasus Field) (*Books LLC,* 2010). These three represent the types of runways that are found in Antarctica.

The annual sea-ice runway for the wheeled aircraft is USAP's principal runway due to its proximity to the McMurdo Station. It is used from October to early December at the start of summer operations. After this period, sea ice begins to break and operations are moved back to Williams Field. Pilots landing C-17 Globe Master III cargo aircrafts on the sea-ice runway report that the surface is stable, not unlike landing on a concrete runway. However, the similarity with land bases ends when the jet aircraft rolls to a stop. The nearly 50,000 pound weight of the plane, including cargo and passengers, causes it to sink into the ice, albeit only for a matter of inches. A laser light is trained on the aircraft to measure the settlement rate. The $200 million aircraft is moved to a new location on the six-foot thick sea ice as a safety measure if the ten-inch red line is reached, indicating the permissible limit of sinking.

Williams Field is USAP's principal airfield in Antarctica on shelf ice. It is a snow runway located on approximately 8 m of compacted snow, lying on top of a nearly 80 m thick patch of visible ice shelf, floating over water. About 12 km from Ross Island, it also serves New Zealand's Scott Base. In addition, it is a major airfield for continental aircraft operations in Antarctica. It operates from December to February (end of summer operations) and its groomed snow surface can support the landings of ski-equipped aircrafts only. The extraordinary conditions encountered at this airfield include the fact that it is in a continuous slide towards the sea along with the ice shelf on which it is constructed. The seaward movement of the floating McMurdo ice shelf has forced Williams Field to be relocated three times since its construction.

Pegasus Field is the southernmost of the McMurdo Station's three airstrips and is 65 km from the station. This white-ice runway is used for wheeled aircrafts that cannot land on the busier Williams Field runway. It is used in August and December through February of each season.

Argentina and Chile with their proximity to the continent (only Drake Passage separates South America from the Antarctic Peninsula) maintain a number of stations and demonstrate a very strong presence in Antarctica. At the Chilean base Frei, a 1300 m airstrip supports fifty intercontinental and 150 intra-continental flights each season, serving as a means of support to many bases around. The airfield at the main Argentinean base of Marambio is the first airfield in Antarctica and used very frequently, supporting over a hundred intercontinental flights each year. The base and the airfield are very well developed in terms of facilities and communication.

Britain maintains an airfield at their main station, Rothera. It has a 2500m crushed-rock runway, with an associated hanger and other facilities to support their stations.

In East Antarctica, Australia has developed its own network of air and shipping for its Antarctica programme.

All the airfields mentioned above represent the airfields existing near the stations of some countries. Whereas countries like the US, Britain, Argentina, Chile and Australia and a few others, have developed their own logistics network, other countries have their own limited networks, plus they utilise the services of commercial organisations developed over the years.

One such example is the specially created airfield at Troll, owned and operated by the Norwegian Polar Institute, which caters to the requirements of many stations in collaboration with the Antarctic Logistics Centre International (ALCI) and a consortium formed by eleven countries, including India, known as Dronning Maud Land Air Network (DROMLAN) since the last few years.

A personal experience of an airfield on glacier is worth mentioning. On one of the convoys, I went to visit the glacier airfield near Novo. They had just done lot of work there to smoothen the satstrugi-affected areas and the airfield looked smooth, white and beautiful. I decided to take a look around on a snow scooter belonging to the

Russian on duty. Now, riding a snow scooter at DG in summer was easy as the snow there was very soft. Not realising that the surface here was extremely hard, with no friction, unlike at DG, I increased my speed. As the scooter went out of control, I applied the brakes hard and, to my horror and the concern of all those watching from a distance, my scooter slid and violently turned a full circle. Fortunately it did not come crashing down. On that hard glacier airfield that day I learnt a huge lesson –not to try anything in Antarctica without making a prior study.

7.5 Organisations Engaged in Antarctic Logistics

7.5.1 *Antarctic Logistics Centre International (ALCI, 2012)*

The ALCI was formed due to an increase in the activities of Antarctic operators and agencies utilising Cape Town as a gateway to the continent. It drew its origin from the International Arctic and Antarctic Research Initiatives Ltd (INTAARI), formed on the premise of the Arctic and Antarctic Research Institute (AARI) at St Petersburg (Russia) and Cape Town (South Africa). They have experience in the field of Antarctic logistics since 1990s. ALCI became fully operational in 2001. In 2004, one more company joined the corporation – The Antarctic Company (TAC). TAC specialises in providing logistics support to countries, non-governmental organisations (NGOs) and tourist groups. This combination of experience, knowledge and resources forms the foundation of ALCI.

ALCI aims to professionally service the specific requirements of all Antarctic operators, as well as to establish the necessary infrastructure and facilities to provide, maintain and service specialised Antarctic vehicles and equipment in Cape Town.

ALCI provides a full range of services for flights from Cape Town to Antarctica. These include chartering suitable aircrafts or helicopters when necessary, providing any additional equipment required for aircraft safety, rescue equipment and all other operational, administrative and international logistics services like forwarding, custom clearing, warehousing, as well as procurement and leasing.

In collaboration with a leading South African waste-management company, they also undertake all waste management/ecological projects in accordance with the following:

- Madrid Protocol, 1991.
- The Hazardous Substance Act (Act 15 of 1973) for the handling and conveyance of hazardous substances.
- Acting in accordance with both the above to obtain a 'certificate of safe disposal'.

As of now, ALCI has become a premium organisation in providing logistics services in Antarctica to all those countries who have not been able to develop their own individual logistics network.

7.5.2 *The Dronning Maud Land Air Network (DROMLAN, 2011)*

The Norwegian DROMLAN is a coordinated project between eleven countries with bases in Dronning Maud Land (DML) in East Antarctica, to create coordinated logistics services among countries. The multination consortium has opened an entirely new era in internationally coordinated logistics support in East Antarctica with the likelihood of deeper repercussions in the future.

In the 1990s, the Norwegian Polar Institute took an initiative to establish DROMLAN. The Troll Airfield was opened on 11 February 2005 by Queen Sonja of Norway and the country's new research station, Troll, the following day. The airfield is 6.8 km from the station. It consists of a 3300 m × 100 m runway on a blue-ice glacier on the Antarctic ice sheet.

Through meetings and discussions, an agreement was reached and eleven national Antarctic operators active in DML formally established DROMLAN as an international project at the XIV Committee of Managers of National Antarctic Programmes (COMNAP) meet in Shanghai in July 2002.

The founder members of DROMLAN are the NAPs of Belgium, Finland, Germany, India, Japan, the Netherlands, Norway, Russia, South Africa, Sweden and the UK. DROMLAN is administered by a steering committee, comprising an executive director and a support group with a representative from each member country. The committee takes decisions, by consensus, on financial matters, operating guidelines and annual programmes. The locations of the stations of DROMLAN's national partners and the approximate distances between them are shown in Figure 7.1 and Table 7.1 (DROMLAN).

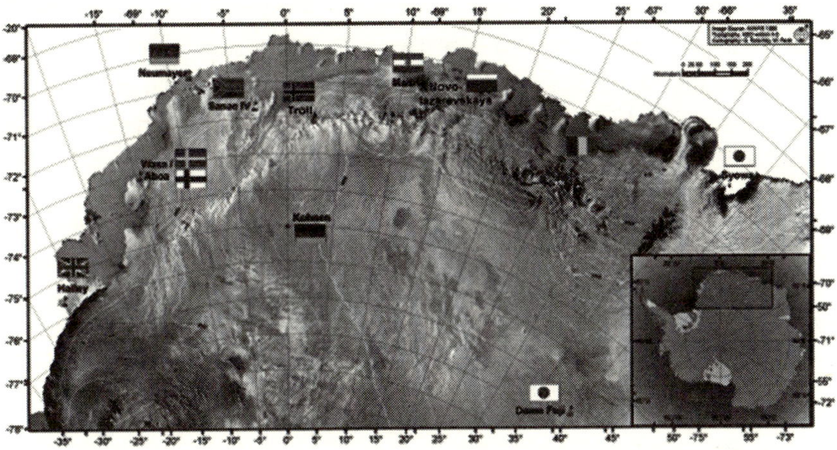

Fig. 7.1: Stations in the DROMLAN Network

Table 7.1: Approximate Distances between Stations Covered By DROMLAN

Miles / Kilometres	Cape Town	Novo	Aboa / Wasa	Dome Fuji	Halley	Kohnen	Neumayer	SANAE IV	Sør Rondane	Syowa	Tor	Troll
Cape Town	—	2630	—	—	—	—	—	—	—	—	—	2690
Novo	4200	—	550	680	810	370	450	330	250	670	160	220
Aboa / Wasa	—	890	—	950	290	310	220	260	740	1180	390	360
Dome Fuji	—	1090	1520	—	1020	660	990	850	480	580	720	750
Halley	—	1290	460	1680	—	460	500	530	950	1380	650	610
Kohnen	—	600	490	1060	740	—	340	240	490	920	240	210
Neumayer	—	720	350	1584	800	550	—	140	680	1120	310	260
SANAE IV	—	530	420	1360	850	380	230	—	540	980	180	130
Sør Rondane	—	410	1180	770	1520	780	1100	880	—	440	370	430
Syowa	—	1080	1180	930	2210	1470	1790	1570	700	—	810	860
Tor	—	260	630	1150	1040	380	500	280	590	1300	—	60
Troll	4300	350	580	1200	980	340	420	210	690	1380	100	—

As the Indian station Maitri is situated 15 km from the Novolazarevskaya (Novo) airfield, it is not shown here separately. Novo/Maitri are about 4200 km from Cape Town from where the DROMLAN network starts.

The aim of DROMLAN is to provide an intercontinental air-link from Cape Town to destinations within DML to any member country of the Committee of National Antarctic Programmes (COMNAP) and SCAR (Scientific Committee on Antarctic Research) for science-related activities, including logistics. This regular air-link has improved accessibility and extended the length of the summer season for all activities of national operators, except tourist activities.

The DROMLAN cooperation includes the following activities:

- Maintain, improve and operate two airfields in DML close to Novo (Russia) and Troll (Norway) stations for intercontinental flights from Cape Town.
- Organise intercontinental flights with an appropriate aircraft to transport personnel and cargo between Cape Town and the airfields at Novo and Troll.
- Organise connecting flights with small ski-equipped aircraft to all stations and field destinations in DML, including further options such as Vostok (Russia), Amundsen-Scott (US) and the stations of the East Antarctic coastal region.
- Organise the necessary support services, such as weather forecasting, provision of fuel and accommodation at stations in DML.

This new air network facilitates communication and transportation of scientists and equipment between Cape Town and DML, and between the scientific stations and field locations within DML. The network connects the 3000 m ice runway at Novo air base, close to the Russian Novo Station, to the Cape Town International Airport by an intercontinental flight. This flight, in an Ilyushin 76TD cargo/passenger aircraft, is operated by the ALCI. The Novo runway acts as a hub from which feeder flights by ski-equipped Antonov-2, Dornier 228, Twin Otter or Basler (BT-67) aircraft services from the Russian, German and British Antarctic programmes, can connect to other stations and field locations within DML. From mid-December to end-

January, the Novo runway suffers from extensive surface melting and has to be closed. An alternative 3000 m ice runway at the Norwegian Troll Station is currently being prepared that will not suffer from surface melting and will be able to operate throughout summer.

Since the 2002–03, flight operations in the framework of DROMLAN have been established by annual contracts between national operators and ALCI to organise intercontinental and inter-DML area flights. National operators support feeder flights to various stations and provide other necessary services and facilities at their stations. Further progress was made by the establishment of a detailed and customised weather-forecasting service at the German Neumayer Station to improve the safety and reliability of intercontinental and feeder flights within the DROMLAN network.

DROMLAN facilities are not available for commercial or private flights.

7.6 A Broad Review of National Antarctic Programmes (NAPs) and Logistics Networks

Out of fifty-three signatory countries, twenty-nine (twelve original signatories and seventeen who have demonstrated their interest by carrying out substantial scientific activity in Antarctica) enjoy the right to participate in decision-making and are known as consultative parties (CPs). As discussed earlier, of these, seven had made territorial claims before the Antarctic Treaty (AT/Treaty) was signed. The claims of Argentina, Chile and UK overlap. There have been many disagreements but are contained while the Treaty is in force. Germany had made a claim but it has been since abandoned. Japan has given an undertaking that it will not make any claim. Brazil has declared it a zone of interest but is not a claimant country. Peru, South Africa, Russia and the US have formally reserved the right to make a claim. The position as of now is that all the claims are frozen and no new claims can be made while the Treaty is in force. Also, no mining can be undertaken till the Madrid Protocol is in force. This means that any review of the current situation can be done only in 2041 for the Treaty and 2048 for the Protocol. Matters concerning the Convention for the

Conservation of Antarctic Marine Living Resources (CCAMLR) and the International Whaling Commission (IWC) keep popping up now and then for a resolution within the parameters of these agreements at the Treaty or International Court of Justice (ICJ).

The abovementioned background highlights the strategic importance that individual member countries attach to their presence in Antarctica.

The way in which Antarctic research and presence is organised differs considerably between countries. A major difference is whether an organisation is highly centralised, integrating logistics and science, or whether it is dispersed between different agencies. A second important feature is whether those who determine policy and administer programmes are actively involved in research. Most of the countries use a national Antarctic committee of some sort to report to SCAR on their present and future activities, but the way in which projects are selected and supported is varied. There appears to be a mixture of government, including military, and university personnel. As has been brought out earlier, though the Treaty prohibits all military activities, it allows the participation of military personnel in support. Many countries use their military personnel and equipment to meet logistics demands. The diverse and costly nature of Antarctic logistics support always ensures that a variety of other official bodies are also involved. Politics and national strategy play a major part in Antarctic affairs and political considerations inevitably influence the number of stations and their locations, as well as the structure, selection and execution of scientific projects in Antarctic programmes.

The Antarctic programmes of some countries with an effective and influential Antarctic presence, represented to obtain a clear picture in this regard, are discussed below.

7.7 The Antarctic Programme of Argentina (Wikipedia, 2012, *Argentina: Antarctic Programme,* Trewby, 2001)

Argentina is geographically the closest country to Antarctica. It is one of the seven who have made territorial claims as also one amongst the twelve original signatories to the Treaty. As a country, it demonstrates a deep involvement in all matters concerning Antarctica.

The purpose of the Argentine Antarctic Programme is to support, strengthen and increase the Argentine sovereign claims over the portion of the Antarctic continent and surrounding seas from 25°W to 74°W longitude and from 60°S latitude to the South Pole.

The Argentine Antarctic programme is composed of several organisations at the highest level of the national government. The Argentine Army, under the Defence Ministry, provides the logistics, whereas the Instituto Antartico Argentino (IAA) of the Ministry of Foreign Affairs rules all other activities. The scientific priorities are set according to the national Antarctic policy to investigate, understand and preserve natural resources, protect the environment and to maintain historic monuments. The IAA is divided into scientific departments grouped into four major scientific areas related to life, earth, ocean and atmosphere.

Argentina maintains six permanent or year-round stations, Orcadas, Jubany, Esperanza, Marambio, San Martin and Belgrano II. In addition, it maintains seven seasonal or summer bases. It has the maximum number of stations for any country and all the stations consist of a number of permanent buildings and facilities. All these stations are located within its claimed area. The locations of these stations are seen in Figure 6.3.

Orcadas is the first permanent Antarctic base since 1904. Esperanza happens to be the birthplace of the first human born in Antarctica – Emilio Marcos Palma – in 1979. Over a thousand tourists visit this base every year.

There is an airfield at Marambio Base. It is suitable for wheeled landing and remains one of the most frequently used airfields in Antarctica. There are over a hundred intercontinental flights every year.

All the supplies are taken to Marambio Base first and then distributed to other stations. All the bases are exclusively maintained by personnel of the armed forces. The Argentine Navy operates a large icebreaker (*Almirante Irizar*), an oceanographic vessel (*Puerto Deseado*) and several smaller patrol vessels. Air operations from South America and within Antarctica are provided by the Argentine

Air Force through several C-130 Hercules planes. A small DHC-6 Twin Otter plane year-round is based on Marambio Base. Bell 212 helicopters are deployed at Marambio Base during summer.

The Argentines treat their Antarctic bases as an extension of their country. They had declared Marambio Base the provisional seat of government in August 1961 and President Lastiri had spent a month there with his cabinet to assert sovereignty. It has regional economic interests, notably in fishing in the South Atlantic sector of the Southern Ocean. Argentina also benefits from the expanding interests in Antarctic tourism, Ushuaia being the main departing point for several excursions of the International Association of Antarctic Tour Operators (IAATO) and other organisations.

Due to its proximity, ease of operations and the highest national priority, Argentina seems to be self-sufficient in terms of its logistics structure in Antarctica.

7.8 The British Antarctic Survey (BAS) Programme and Bases (Trewby, 2001, Wikipedia, 2012– BAS)

By 1917, UK had laid claims to seventeen per cent area of the continent and the sub-Antarctic islands of South Georgia, South Orkneys, South Shetlands and South Sandwich, calling it the British Antarctic Territories (BAT) as a part of the Falklands Islands Dependencies (FID). Further in 1923, it claimed Ross Dependency and, in 1933, made claim to nearly thirty per cent of the continent and ceded the same to New Zealand (160°E to 150°W) and Australia (45°E to 136°E and 142°E to 160°E) respectively. In 1962, BAT, consisting of the mainland section of Graham Land, South Shetlands and South Orkneys islands, was split from FID which now includes South Georgia and South Sandwich Islands.

During 1943–45, Operation Tabarin was launched to establish a permanent British presence on the Antarctic Peninsula to counter the territorial claims made by Argentina and Chile and to keep a watch on German submarines. Three research stations were established. At the end of World War II, the area was renamed Falkland Islands Dependencies Survey (FIDS). In 1962, FIDS was renamed British

Antarctic Survey (BAS) which now runs the British Antarctic programme.

All the above amply illustrates Britain's deep involvement and strategy so far as its interests in Antarctica are concerned.

As of now, BAS operates five permanent research bases in BAT. While Rothera and Halley are year-round (as shown in Figure 6.3), Signy (research) and Fossil Bluff and Sky Blu (both for logistics facilities) are manned during the Antarctic summer only.

BAS has a well-developed and organised logistics cycle for all its stations. Rothera is on a rocky promontory and has a wharf and a 2500 m crushed rock or gravel runway. It is an important station of BAS on the peninsula. Flights link it to the Falkland Islands. A very large complex of many buildings houses about 130 persons in summer and twenty in winter. Fossil Bluff and Sky Blu are fuel dumps and blue-ice runways going south to Halley.

The Halley research station is on the Brunt ice shelf. Since 1956, four stations have been buried by snow and abandoned. Various methods were tried, from unprotected wooden huts to steel tunnels. In Halley V, the main building is on steel platforms that are raised annually to keep them above snow surface. Halley VI is designed by Faber Maunsell Hugh Broughton Architects, winner of the design competition, and built by a South African consortium. It is a structure which, like Halley V, is jacked up on legs to keep it above snow. But unlike Halley V, there are skis on the bottom of these buildings which allow them to be relocated periodically. One of the reasons for its location is that it is under the aurora oval, resulting in frequent displays of Aurora which are easy to see during the 105 days when the sun does not rise above the horizon.

7.8.1 *BAS Ships*

BAS operates two ships as part of the British Antarctic programme. The *James Clark Ross* replaced *RRS John Biscoe* in 1991 and *Earnest Shackleton* succeeded *RRS Bransfield* in 1999. Whereas the former is primarily an oceanographic vessel, the latter is mainly a logistics vessel. Both vessels depart UK in September/October each year and return the following May/June. Both undergo the refit/dry dock during the Antarctic winter, but are also used elsewhere for scientific work or for chartered commercial survey work.

The two BAS ships are complemented by the capabilities of the Royal Navy's ice-patrol vessels that operate in the same waters. *HMS Protector* replaced *HMS Endurance* in 2011. Two Lynx helicopters on these ice breakers enable the BAS staff to go to remote field sites that BAS aircrafts cannot access.

7.8.2 *BAS Aircrafts*

Five aircrafts operate to support BAS's Antarctic programme. The aircraft used are all products of de Havilland Canada and comprise four Twin Otters and one Dash 7. The planes are maintained by Rocky Mountain Aircraft Ltd from Canada. During the Antarctic summer they are based at Rothera. Since Antarctic winter precludes flying, they return to UK.

The larger Dash 7 aircraft undertakes regular shuttle flights between either the Port Stanley airport on Falkland Islands, or Punta Arenas in Chile, and Rothera. It also operates to and from the ice runway at the Sky Blu base. The smaller Twin Otters are equipped with skis for landing on snow and ice in remote areas and operate out of the bases at Rothera, Fossil Bluff, Halley and Sky Blu.

7.8.3 *Life on BAS Bases*

Life on the BAS base Halley is representative of life on any Antarctic base. The very few over-winterers are scientists, mostly technical specialists who are required to keep the station, its life-support systems, like heating, snow melting, ventilation, sewage, food preservation and ongoing scientific experiments, running. One of the winterers each year is designated as the base commander and sworn in as a magistrate – a job that is carried out in addition to his normal duties for a small additional allowance.

Life in Antarctica is dominated by short or pointed summers and long or coreless winters. The base is supplied by sea, by *RRS*

Shackleton, during the two weeks from end-December, a period marked by hectic unloading activity. The summer season is defined as the period between the first call when the ship arrives and the second call in February when it returns to collect everyone who is not staying for the forthcoming winter (basically, the old winter and current summer team). Significant dates in winter are sundown (the last day when the sun can be seen), midwinter day (21 June) and sun-up (the first day when the sun rises after the winter). Traditionally, the oldest person on the base lowers the tattered flag on sundown and the youngest raises a new one on sun-up. On midwinter, a week's holiday is given and a member of the winter team is chosen to keep the old flag. With minor changes, in line with individual customs and traditions, all countries here follow a similar schedule.

7.9 The Antarctic Programme of China (Chinese Arctic and Antarctic Administration, Books LLC, 2010, Wikipedia, 2012)

Like India, China, too, is a late entrant on matters Antarctic. It acceded to the Antarctic Treaty on 9 May 1983 and was accepted as a Consultative Party on 7 October 1985. The Polar Research Institute of China had begun research in Antarctica in January 1980 when it sent two scientists to the Australian Casey station. The first major Chinese expedition to Antarctica was in November 1984.

Beijing-based Chinese Arctic and Antarctic Administration (CAA) organises, coordinates and supervises China's scientific programmes in both the Arctic and the Antarctic and provides logistics support to polar expeditions. Its responsibilities include developing an integrated national polar research strategy, a policy to guide national polar research activities, a five-year plan to implement national policy and an annual updating of the national polar plan. On ground, it plans and conducts polar research expeditions and coordinates polar research programmes with other agencies in the country. At an international level, it coordinates all Chinese activities concerning polar areas with other countries and international organisations.

Presently, China maintains three research stations in Antarctica. It established the Great Wall Station on King George Island in the

Peninsula in 1985 and Zhongshan Station on Larsmann Hills of Prydz Bay in 1989. Both are year-round stations on ice-free oasis areas, accommodating about 14 to 25 persons in winter and 40 to 60 persons in summer. The bases are for research on marine, glaciological, geological and atmospheric sciences. Incidentally, in the Prydz Bay area, research stations from Australia (Davis), Russia (Progress), Romania (Rocovita) and India (Bharati) are also located along with Zhongshan.

China's third station, Kunlun, established on Dome A or Dome Argus on 27 January 2009 indicates its ambitious plan of an influential Antarctic presence in the years to come. Located on the polar plateau at 80°22'S and 77°21' E, Kunlun is midway between Lambart Glacier and the South Pole which are 2400 km apart. Dome A is considered to be the highest point in the East Antarctic Ranges and underneath its 2400 m thick ice sheet is the Gamburtsev mountain range. The coldest air temperature recorded at Dome A during 2005 was minus 80.5°C. The lowest air temperature of minus 89.2°C on the surface of the earth was at the Russian Vostok station in July 1983, which is 600 m lower in elevation than Dome A.

A Chinese team first visited Dome A in January 2005 after a traverse of 1228 km from Zhongshan and located the highest point of the ice sheet (4093 m above sea level) and established one Automatic Weather Station (AWS) there. Another AWS was established at the midpoint between the summit and Zhongshan. These are operated as part of a collaboration between China and Australia. The Polar Research Institute of China deployed a robotic observatory called PLATO at the Dome in January 2008. The advanced observatory is equipped with instruments from various institutions in Australia, US, China and UK.

With elaborate preparations, Kunlun was established in 2009, as mentioned above. Presently, it is a seasonal or summer-only station but may become a year-round station in the near future. Its major interests are exploration of deep glacier ice cores and mountains under the Antarctic ice, astronomical and terrestrial magnetic observations and a study of the effects of extreme weather conditions on human beings. Kunlun is only one of the five stations established deep inside on the polar ice plateau, the other four being Amundsen-Scott (US), Vostok (Russia), Concordia (France/Italy) and Dome Fuji (Japan). These stations face extreme cold temperatures (even summer temperatures rarely rising above minus 20° to 30°C) and extremely difficult logistics.

China purchased an icebreaker from Ukraine in 1993. Named *Xuelong*, it has made twenty-four expeditions to Antarctica and three to the Arctic and is responsible for logistics support to the Chinese stations. Presently, China is building a new icebreaker ship of its own and plans to enhance its Antarctic aviation capability, too, for its massive expansion plans in the continent (*The Age*, 2012, 'China Widens Antarctic Operations').

A recent incidence in the Antarctic waters points out China's increasing interest, strength and enhanced polar capability. A Russian ship, *MV Akademik Shokalskiy,* was chartered during the summer of 2013–14 for an Australasian Antarctic expedition with seventy-four people on board (including twenty-two crew members) to celebrate the centenary of Douglas Mawson's last-century expedition. The ship got trapped in thick pack ice in Commonwealth Bay on 25 December 2013 for two weeks. A massive rescue operation was launched with three icebreaker ships – *Aurora Australis* (Australia), *L'Astrolabe* (France) and *Xue Long* (China). Fifty-two passengers were evacuated by helicopter and boarded on to *Xue Long*. In the process, the rescue ships were also trapped in ice. Finally, the Antarctic wind freed all of them when the pack ice broke on its own before the arrival of American heavy icebreaker *Polar Star* (Wikipedia, *Akademik Shokalskiy,* 2015).

It can be said that China has a well-coordinated plan in place to make up for its late entry and is set to establish an effective and influential presence in Antarctica.

7.10 The Antarctic Programme of Australia (*Trewby, 2001, McGonigal, 2008, COMNAP, 2009,* Wikipedia 2012, AAD)

The Australian Antarctic Programme, like that of the UK, US, Russia, Argentina and a few other countries, has come a long way. After participation in the British Antarctic Expedition (1907–09), Douglas Mawson led the Australasian Antarctic Expedition (1911–14) and the British-Australian-New Zealand Antarctic Research Expedition (BANZARE) (1929–31), gained prominence and charted the East Antarctic coastline from Enderby Land to George V Land. BANZARE prepared the way for a British territorial claim which was then placed under Australia in June 1933. The Australian Antarctic Territory (AAT), as they call it, is the largest claim in Antarctica, covering about 6.5 million sq km in two wedges, comprising over thirty per cent of the total area of the continent.

The formation of the Australian National Antarctic Research Expeditions (ANARE), its effective participation in IGY 1957–58 and creation of the AT, as well as locating COMNAP and Convention for Conservation of Antarctic Marine Living Resources (CCAMLR) in Hobart indicate Australia's deep involvement with the continent. Its refusal to ratify the Convention for Regulation of Antarctic Mineral Resources Activities (CRAMRA) in 1989 was one of the principal reasons for its being put in cold storage and the beginning of intense negotiations, leading to the signing of the Madrid Protocol for environmental protection in 1991.

The Australian Antarctic Division (AAD) is an agency of the Department of Sustainability, Environment, Water, Pollution and Communications of its government. The AAD undertakes science programmes and research projects to contribute to an understanding of Antarctica and the Southern Ocean. Under its charter, the AAD is responsible for conceiving, planning and organising all activities of the country's Antarctic programme and maintain its permanent stations with the objective of an influential and effective presence in Antarctica.

The AAD is based in Kingston, Tasmania, just south of Hobart. It houses laboratories for science, electron microscopy and instrument workshop, a krill research aquarium, a herbarium, equipment stores, communications and other operational and support facilities. Along with COMNAP, CCAMLR, the krill research centre and a library, AAD is an excellent source for Antarctic information.

7.10.1 *Australian Antarctic Stations*

The AAD maintains three permanently manned year-round stations on the Antarctic continent and one on Macquarie Island in the sub-Antarctic. The stations are shown in relation to one another in Figure 6.3 in the earlier chapter.

Mawson is Australia's oldest Antarctic station and was established in 1954. It is also the oldest continuously inhabited Antarctic station of any nation south of the Antarctic Circle (66.7°S latitude). Its site has a large natural harbour (Horse shoe Harbour). The earlier station consisted of small prefabricated huts. Through a major rebuilding programme in the 1970s, the station now consists of large, steel-framed modular buildings. Along with scientific activities, its use of wind power is a unique feature. Its wind generator provides for over seventy per cent of the power needs of the station, and saves over 600,000 litres of diesel fuel.

Built on Vestfold Hills in a remarkable ice-free oasis area, Davis is the busiest of all the Australian Antarctic stations. Sensing that the Russians might build a station in the area, the Australians built a station to pre-empt the Russians in 1957. Mawson originally sighted the area in 1931 during the BANZARE expedition. However, the first recorded landing in the area was by the Norwegian whaler Captain Klarius Mikkelsen with his wife Caroline and seven others in 1935. They erected a rock cairn to mark the first recorded Antarctic landing by a woman. It was seen by an ANARE team in 1960 but lost again for many years. Caroline Mikkelsen was still alive in 1996 to hear about the rediscovery of the cairn and the old flagpole. Davis was temporarily closed from 1965 to 1969 to make resources available for rebuilding Casey Station. Presently, in the same area are the stations of Russia, Romania, China and India's third station, Bharati.

Casey is close to the now-abandoned Wilkes Station of the US established during IGY 1957–58. It, too, was rebuilt in the 1960s and 1980s. As stated in an earlier chapter, it was the first station on stilts, raising the height of the structure to counter snow accumulation.

AAD also maintains a permanent year-round research station on Macquarie Island which is roughly mid-way between Australia and its three Antarctic stations mentioned above. The Island, however, forms part of the state of Tasmania and not the Australian Antarctic Territory (AAT).

AAD has established an elaborate and systematic logistics network to support its own Antarctic stations and a few of other countries. It also promotes limited and organised tourism.

The relative locations and distances of the Australian stations are shown in Figure 7.2 (http:/www/aad.gov.au. History of Australian Antarctic stations).

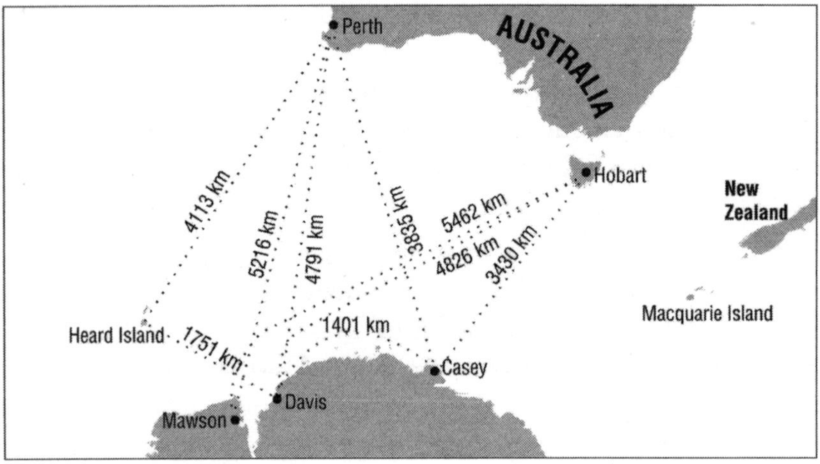

Fig. 7.2: Locations of Australia's Antarctic Stations

7.10.2 *Aviation*

The AAD utilises an air-transport system for both transport to and from Antarctica and within the continent. An Airbus, a 319–115 LR, operated by Skytraders provides services to and from Antarctica between November and February. It operates from Hobart, Tasmania, to the Wilkins ice runway, located 65 km from Casey Station. The inaugural flight landed in Antarctica on 9 December 2007.

Services within Antarctica are provided by a mix of fixed-wing aircrafts and helicopters. Two fixed-wing aircrafts (CASA 212–400) equipped with skis are operated by Skytraders. These aircrafts are operated from the Wilkins ice runway at Casey and from smaller snow runways at each of the three permanent stations. They also provide services to field camps where flat areas of snow and ice are available. Helicopter services are provided by three Eurocopter Squirrel helicopters, operated by Helicopter Resources (*Australia's Antarctic Aviation*, 2012).

7.10.3 *Shipping*

The AAD uses icebreaker *RSV Aurora Australis*, a multipurpose research and resupply ship. It was specially built by Carrington Shipways in Newcastle, New South Wales, and launched in 1989. This ship is the flagship of Australia's Antarctic activities. Various other ships from commercial companies, namely, *L'Asrolabe*, *Orion*, *MV Eastern Voyager* and *Akademic Shokalskiy* are also used for Australian and other countries' expeditions and limited tourism.

A very elaborate aircraft and shipping schedule is worked out and implemented with a commercial angle. The timings and dates for aircrafts are, however, uncertain due to the very uncertain weather conditions in Antarctica.

From every aspect, the AAD logistics network is totally self-reliant and developed with a long-term view to effectively cater for the protection of Australia's strategic interests.

7.11 United States Antarctic Programme (USAP) *(NSF Publication 4.0, 1997,* Trewby, 2001, McGonigal, 2008*)*

7.11.1 *Background*

Though the American sealer Nathanial B Palmer saw the Antarctic Peninsula from his ship *Hero* and by chance met Russian Bellingshausen (the first person to site the continent of Antarctica) in 1820–21, the efforts of the US in Antarctic exploration were individual and sporadic. While seven countries made official territorial claims and a few others expressed their specific interests, the official US position was neither to make any claim nor recognise the claims of the other countries in line with the Hughes Doctrine (1924), amounting to a non-policy towards Antarctica.

US Antarctic activity in the last century began with Richard Byrd's hugely popular, privately financed expeditions during 1928–30 and 1933–35 in which he flew over the South Pole (the claim is still doubted) and considerable addition was made to the geographical knowledge of the continent. He pioneered the successful use of tractors and airplanes which truly revolutionised logistics in the years to follow. This background prompted President (1933–45) Franklin D Roosevelt to authorise a re-examination of the US position towards Antarctica. That is when the real US involvement in the continent started. Based on secret instructions, Lincoln Ellsworth dropped brass canisters in East Antarctica with a document of US sovereignty inside. A US government expedition was mounted in 1939–41, established two stations and did considerable survey and other scientific work.

In the Cold War climate after the conclusion of WW II, US military strategists thought that next major global conflict may be in the Arctic. Military manoeuvres and practice operations conducted in the polar climate were viewed as essential ingredients for a healthy state of battle readiness. Accordingly, Operation High Jump (1946–47) was launched. It was on a very large scale, involving 4, 700 military personnel, 24 civilian scientists, 55 observers, 13 ships, 23 planes and helicopters, assorted ground transportation and construction material. Over 50,000 photographs were taken, maps were drawn, land features were spotted and various scientific experiments were carried out. All this created a solid base, along with earlier activities, for territorial claim-making if required. This was followed by Operation Windmill in 1947–48. All these activities, plus an effective and leading participation in IGY and AT and the establishment of permanent stations form the background of the USAP.

7.11.2 *Current Programme*

The National Science Foundation (NSF), through USAP, manages US scientific research in Antarctica. USAP was established in 1959, immediately after the IGY.

NSF is an independent federal agency whose management role in the USAP is reconfirmed through a memo in 1982 by President

Reagan. The USAP is a part of NSF's Office of Polar Programmes (OPP). NSF not only manages the USAP, but NSF's director also chairs the Integrated Arctic/Antarctic Research Policy Committee (IARPC).

The USAP's goals are to understand the Antarctic and its associated ecosystems, the region's effects on and responses to global processes, one such issue being the climate, and to use Antarctica's unique features for scientific research that cannot be done anywhere else.

Among the scientific disciplines encompassed by this broad mandate are astronomy, atmospheric sciences, biology, earth sciences, environmental science, geology, glaciology, marine biology, oceanography, and geophysics.

While the US Department of State leads the US delegation to the annual Antarctic Treaty Consultative Committee Meeting (ATCM), NSF leads the US delegation to the ATCM's Committee on Environmental Protection (CEP) established to oversee the implementation of the Madrid Protocol.

7.11.3 *Logistics Support to USAP*

About 3500 personnel are involved each year in USAP's research and logistics activities. Women constitute about thirty per cent of the total work force. Every year more than 800 scientists and their support teams conduct research in the Antarctic environment.

7.11.4 *US Research Stations in Antarctica*

The USAP maintains three permanent year-round stations, namely McMurdo, South Pole and Palmer which are far from one another. An elaborate and comprehensive logistics support network has been developed over the years. Detailed information on various aspects is available through various books and on public domain sites.

McMurdo Station: This forms the principal facility and logistics hub of USAP to support all the three stations and numerous field camps established from time to time. It was opened on 16 February 1956. It is built on the bare volcanic rock of Hut Point on Ross Island, the most southerly solid ground that is accessible by ship for about two months every year in Antarctic summer. It is located just twenty miles south of Mt Erebus, an active volcano that steams continuously and erupts frequently but not violently.

Robert Scott's Discovery Hut built in 1902 still stands adjacent to the harbour, now declared an historic site under the Treaty.

The station has developed over the years and has a harbour, landing strips on sea ice, snow (ice shelf), and hard white ice (glacier). It can be used at different periods of the year and it has a helicopter pad as described in details earlier. It has eighty-five or so buildings, varying greatly in size, from a small radio shack to three-storey structures. Repair facilities, dormitories, administrative buildings, a firehouse, power plant, water distillation plant, wharf, stores and warehouses are linked by aboveground water, telephone and power lines. The water and sewer lines are heat-taped and insulated. About four feet of drifting snow can accumulate here every year, although the station becomes snow-free in summer.

The Albert P Crary science and engineering centre of the station consists of various laboratories with a working area of about 46,000 sq ft. It is a staging area for a range of scientific disciplines, like earth sciences, meteorology, snow and ice mechanics, ice-core studies, geology, geophysics, atmospheric sciences, and so on.

The station had activated a nuclear power plant but it had to be decommissioned as the experiment failed miserably as described earlier.

Amundsen-Scott Station: Americans have occupied the geographic South Pole continuously since November 1956 in the form of this station. In a real sense, it signifies America's central position in matters Antarctic. As it is located on glacier ice, it drifts at the rate of 10 m per year. Due to snow accumulation, the original station, Old Pole, was abandoned in 1975 and replaced by a second one known as the Dome. Dome was decommissioned in 2003 due to snow burial. The new station, officially dedicated on 12 January 2008, includes a modular design to accommodate more people, and an adjustable elevation to prevent the station from being buried in snow. The building faces the wind and the lower portion of its wall is sloping. The angled wall increases the speed of the wind as it passes above, keeping the building from being quickly buried in snow. During a jack-up operation, a new height will be added to the existing columns and jacks will pull the building up to a higher elevation.

Research at the station includes glaciology, geophysics, meteorology, upper-atmosphere physics, astronomy, astrophysics, and biomedical studies. There is a 10 m telescope among other equipment. A greenhouse at the station produces a variety of fresh vegetables. Between February and October, the station personnel are totally isolated.

The station has a runway 3658 m long. Between October and February, the ski-equipped LC-130 Hercules aircraft makes several flights from McMurdo to supply the station. Resupply missions are collectively termed Operation Deep Freeze.

Palmer Station: Built on Anvers Island, off the Antarctic Peninsula, Palmer station is located north of the Antarctic Circle. It is built on solid rock and it consists of two huge and three small buildings, two large fuel tanks and a dock. Established in 1965, it is not inaccessible and does not have winter isolation like the two other US stations. Any ice-strengthened ship can transit to and from Palmer any month of the year after crossing Drake Passage from South America. *Polar Duke*, built in 1983, is an ice-strengthened ship under charter to the NSF since 1985 and it resupplies Palmer Station, provides research support and performs research in collaboration with the station. The station is superbly located for conducting a biological study of birds, seals and other elements of the marine ecosystem. Meteorology, upper-atmosphere physics, glaciology and geology are also pursued.

All the three US stations are complementary and signify a strong and influential US presence in sub-Antarctic (Palmer), Peninsula (McMurdo) and polar plateau (Amundsen-Scott).

7.11.5 *Yearly Logistics Cycle using Aircrafts and Ships for US Antarctic Stations*

By February, the bases at McMurdo and South Pole are turned over to a small cadre of personnel who remain during the austral winter months.

7.11.6 *Aviation Support*

NSF has a fleet of seven LC-130 aircrafts (the polar version of C-130 cargo plane). They also charter other planes (Twin Otter) and helicopters when required. In August, the 109th Airlift Wing of the NY Air National Guard (till 1999, the US Navy was doing this duty) fly six to eight missions using the ski-equipped LC-130 to transport the additional personnel needed to augment the winter overstaff in preparing McMurdo for summer deployment.

In October, the summer deployment begins. In phases, personnel and supplies are flown to McMurdo to land on a runway constructed on sea ice. Wheeled LC-130, US Air Force C-17 and C-130, flown by the New Zealand Royal Air Force, conduct round-trip flights. Aircrafts remain in McMurdo just long enough to offload and refuel and return to Christchurch, New Zealand. On occasion, Air Force C-5 will fly a few turn-around flights. Concurrently, the ski-equipped LC-130 aircrafts fly to Amundsen-Scott with replenishments of fuel and stores from McMurdo Station. LC-130, and HH-1N helicopters are used to provide supplies to scientists working at camps outside the McMurdo area. They also undertake local search and rescue missions.

By the first week of January, the upper crust becomes very soft and the sea-ice runway becomes unusable and operations are shifted to Williams Field on the ice shelf and Pegasus Field located on glacier ice, as explained earlier.

7.11.7 *Sea Route Support with Icebreaking and Cargo Ships*

The final approach to the US McMurdo station is called McMurdo Sound and its ice-clogged waters extend to about 55 km long and equally wide The Sound connects the Ross Sea to the north and the Ross ice-shelf to the south where the US station as well as New Zealand's Scott Base are located. The adjacent Ross Island is dominated by the active volcano Mt Erebus (13,202 ft).

Every year, in early January, two research vessels with icebreaking capability cut a channel in McMurdo Sound, leading to McMurdo Harbour. Usually, the ships used are *Polar Star* or *Polar Sea* from the US Coast Guard as well as a chartered ship, *Nathanial B Palmer*. They also carry out other logistics tasks. About two weeks after the channel is cut, a chartered civilian fuel ship arrives and spends five days pumping fuel to storage tanks at McMurdo Station. By early February, the annual commercial cargo ship arrives escorted by icebreakers. It comes from the homeport of Hueneme, California, with the majority of food, supplies and equipment needed for the upcoming winter and summer seasons. The amount of fuel and cargo supplied is huge, for example, recorded figures for unloaded fuel and cargo during the summer of 2004–05 are 8.5 million gallons and 14.2 million pounds respectively.

Even if chartered ships carry out the duties to open the channel, a coast guard vessel remains on standby to assist if required.

All these operations are done under the military sealift command, *USNS Green Wave,* or some other ship on charter. Throughout the deployment, the naval support unit or coast guard units at Christchurch, New Zealand, provide stevedoring duties (loading and unloading of vessels) and coordinate all staging activities for the joint and combined operations of the Navy, Army, Air Force, Air National Guard, Coast Guard, Royal New Zealand Air Force, and merchant ships. Add to this the weather and sea-ice uncertainty and one can visualise the logistics complexities till the conclusion of the first phase. Logistics requirements for Amundsen-Scott and other stations have to be catered from here.

In February, all personnel redeploy to their homeport except the winter teams staying back. During the homeport season (March through September), the NSF and all the agencies prepare for the next deployment. In some cases they plan and order items needed for deployments two or three years away.

7.11.8 *Review of the Present USAP Resupply System*

Due to the very tricky and difficult icebreaking operations, the present system has a single-point potential for failure. It has been felt so particularly in the last few years due to greater iceberg calving and drift in McMurdo Sound. With this in view, studies initiated by the OPP (Office of Polar Programmes) of the NSF are underway to determine an alternative or a combination of alternatives for resupplying the stations, including the continued use of icebreakers. The outcome of these studies may be interesting and far-reaching in the present logistics scenario.

The detailed analysis as above underscores the vital importance of logistics for the USAP and efforts involved to evolve a fool proof logistics network in Antarctica.

7.12 The Russian Antarctic Programme (*Russian Antarctic Stations: An Overview*, 2012, *Books LLC*, 2010)

If there is one country which has gained tremendous competence in polar conditions over the centuries, it is Russia. As mentioned earlier, Bellingshausen discovered Traversy and Alexander Islands, circumnavigated Antarctica and became the first person to site the continent in January 1820. The erstwhile USSR was involved in whaling and krill fishing in the past, took part in IGY 1957–58 and became one of the twelve original signatories of the Antarctic Treaty. After USSR broke up, Russia assumed its place as a strong supporter of the ATS. Like the US, though it has very strong grounds on Antarctic exploration, it has neither made nor recognised any territorial claims but reserved its right to do so.

The Arctic and Antarctic Research Institute in St Petersburg (AARI) is responsible for the Russian Antarctic Programme. Founded in 1917, it is one of the oldest of its kind, with a considerable collection of information (*Arctic and Antarctic Research Institute*, 2012). During my visit to Moscow and St Petersburg, I regret missing the chance to visit AARI but while talking to some people, I could sense their deep involvement in both the polar areas.

At present, Russia maintains five permanent year-round bases including Mirny, Vostok and Novolazarevaskya (Novo). Vostok is located at the South Magnetic Pole. In one of the biggest achievements, they drilled 3769.3 m down and hit the sub-glacial lake below Vostok on 5 February 2012 (*Nature News and Comment*, 2012).

In terms of vehicles, aircrafts and ships, Russia is extremely competent and truly self-reliant for all its logistics needs. For a long time, the Russians have maintained Vostok by tractor/sledge convoys over a distance of over 1400 km from Mirny in the most severe conditions, bearing testimony to their logistics expertise. Russia also has a dominant presence in DROMLAN as seen earlier. Its Novo and Progress stations are located very close to the Indian stations Maitri and Bharati respectively.

When one sees the map of Antarctica marked with various stations, it is seen that Russia is the only country which maintains its presence all around the periphery as also deep inside the polar plateau. Alongwith its proximity and influential position in the Arctic, the Russian presence in Antarctica clearly indicates its global strategic preferences.

7.13 The Antarctic Programmes of Other Countries

In terms of logistics skills, scientific programmes, quality of shelters, programmes in scientific disciplines, logistics and other aspects of an effective and influential permanent presence in Antarctica as part of AT, countries like Chile, Germany, France, South Africa and Japan deserve mention. New Zealand also is quite active (Peat Neville, 2007 and *Ice Age,* 2007) and has completed more than fifty years of its Antarctic presence. Their Antarctic Centre at Christchurch is one the prominent tourist attractions in New Zealand. They have built the world's coolest wind farm to supply energy to their Scott Base and to the US McMurdo Station to greatly reduce carbon print (Meridian, 2009). They claim to meet a hundred per cent of the energy needs of Scott Station from wind energy. In terms of international institutions, too, New Zealand participates very actively (Samson Lou, 2008).

The position, participation and expertise of Norway and South Africa in Antarctic logistics have been described earlier.

It can be said that the seven countries who had laid territorial claims along with the US, Russia, China and all the other countries mentioned above, will be certainly be in the forefront of matters Antarctic, in terms of their presence and involvement. Many more would like to join their ranks. The position of India is examined in the next chapter.

7.14 Antarctic Logistics as Military Operation other than War

In peacetime, countries all over the world use field training as opportunities for military operations other than war, to train their armed forces, which are potent instruments of projecting national power. These operations are varied from country to country, depending upon the size, geographic position and the geopolitical strategic objectives of each country. The operations range, depending on the support of national and state governments, from disaster relief, national assistance, drug interdiction, peacekeeping, support for insurgencies and counterinsurgencies, non-combatant evacuation, to policy enforcement and other operations outside of the gamut of any declared war.

One of the elements in which expertise can be gained in any of the operations mentioned above is through the effective maintenance of a body of people in a very challenging environment and at a long distance and duration. Thousands of kilometres away from the rest of the world (except that of the Antarctic Peninsula from the tip of South America), with a hostile Southern Ocean, non-accessibility for most of the year due to pack ice and environmental factors that preclude any normal human habitation, Antarctica will always pose the ultimate challenge from a logistics and survival point of view. It can hence be characterised as an excellent peace time military operation other than war.

A review of some of the countries' Antarctic programmes shows that many countries have looked at their operations from a 'military exercise' point of view. Other than individual sealers and whalers in the early days, exploration teams were governed by military organisations and personnel precisely because of their expertise in logistics and their influential position in national security matters. Through Operation High Jump and Operation Windmill, the US trained its personnel and tested its equipment in the polar environment and the Navy continued to provide logistics support to the NSF for USAP for years. Military operations other than war, usually involve a combination of air, land, sea, space, and special operations forces in a complementary fashion. This is certainly the case of USAP where the Navy, Air Force, Air National Guard, Coast Guard and Army jointly provide logistics support. Additionally, it is a multinational effort, with the Royal New Zealand Air Force providing airlift support to the Scott Antarctic station in exchange for fuel and transportation services. All the air services are operated from Christchurch (New Zealand). Such services were provided to the bases of other countries, namely Italian Terra Nova, French-Italian Concordia and Russian Vostok. All this is an excellent execution of Memorandum 6646 of 5 February 1982 from President Reagan, reaffirming that 'the USAP shall be maintained at a level providing an active and influential presence in Antarctica designed to support the range of US Antarctic interests.' The memorandum further stated that the department of Defence shall continue to provide, on a reimbursement basis, the logistics support requested by the NSF (Reed, 1995).

Similarly, there are recorded examples in the form of the Antarctic programmes of Argentina and Chile where the institutionalised use of military is resorted (*Reader's Digest*, 1985). Other countries with a prominent and effective presence in Antarctica, like Australia, UK, and Russia, enjoy a developed logistics structure, and countries like China will achieve this in the very near future.

Obviously, all this points to the realisation among the world community that should a conflict situation arise in Antarctica, a well-developed, self-reliant, powerful and effective logistics structure may be the answer to guard a country's strategic interest.

All this is well reflected in the Treaty document itself. Article 1 of the Antarctic Treaty states, 'The establishment of military bases and fortifications, the carrying out of military manoeuvres, as well as the testing of any type of weapons is prohibited.' This does not preclude the use of military personnel or equipment for peaceful purposes.

The Treaty argues that no nation should ever consider Antarctica worth fighting for. But UK and Argentina fought fiercely in 1982 over sovereignty rights in the nearby Falklands or Malvinas, as Argentina calls it. Maybe the breakdown of the Treaty and the possibility of an armed conflict in Antarctica is a remote scenario now but nations have probably given it considered thought while drafting the Treaty.

Military involvement in Antarctic logistics provides any country institutional knowledge, skill, expertise and continuity, while outsourcing logistics tasks can have long-term adverse effects (Koltz, 1990). Some of these can be summarised as under:

- The armed forces of a country may lose an opportunity to train in polar conditions. The valuable Antarctic expertise within the defence department of the country will be eliminated in the long run.
- The country may lose an institutionalised cadre of experienced polar hands.

A considered use of armed forces in Antarctica for peaceful purposes may be considered an excellent and beneficial by-product to train an army in the logistics of polar environment while carrying out scientific tasks, within the accepted parameters of the Treaty.

The above discussion may be very useful in the Indian context. Therefore, the relevance of this aspect as regards India and the IAP is discussed in the next chapter.

7.15 In Conclusion

Logistics has come a long way from its earlier and exclusive reference to military operations to operations in any field. It is an important management tool which plans, implements and controls the forward and reverse flow and storage of goods, services and related information between the point of origin and the point of application. In the case of Antarctica, today's logistics are a far cry from the days of James Cook, Bellingshausen, Amundsen or Scott. In some areas like communications, logistics has come a long way in the last three decades since the advent of the Internet. Technological developments in all the other fields have also been reflected in elements of Antarctic logistics and tremendous changes have taken place in shelters, transportation (aircrafts, vehicles), food, medicines, waste disposal, communications and others. Even then, despite all these developments, Antarctic logistics remains and will remain in the foreseeable future, a daunting task due to its distance, isolation, the Southern Ocean, low temperatures, continental ice cap, absence of permanent inhabitation and the requirements of the AT and Madrid Protocol.

The technological boost from WW II and the highly successful conduct of the IGY paved the way for an apolitical and non-military presence, primarily aimed at permanent scientific research stations in the continent in line with the provisions of the AT and its instruments. Over seventy permanent research stations from thirty countries exist in vastly different environmental conditions and logistics requirements. With all this, various forms of Antarctic logistics have also evolved over the years. With the excellent quality, longevity and comfort level of structures, a revolution in air transportation due to the availability of airports and capable aircrafts, icebreaking ships, vastly improved snow vehicles, proper food, medicines and communication evolved over the years, the difficulties faced by the early Antarctic explorers have become a thing of the past. Still, establishing and maintaining Antarctic stations remain a matter for detailed planning, execution and resources management. The requirement to preserve the pristine Antarctic environment, too, in accordance with the Madrid Protocol adds to logistics parameters, responsibilities and complexities.

While some countries, like the US, Russia, UK, Australia, Argentina and few others, have developed their own logistics structures, countries like China are engaged in massive efforts to develop and enhance their capabilities to bridge the gap. Those who do not possess their own logistics capability are resorting to combined efforts through international organisation like ALCI and consortiums like DROMLAN.

It seems that an effective and influential presence in Antarctica, in terms of the number of stations, scientific research contribution through ATS and efforts to preserve the pristine Antarctic environment, will remain a matter of strategic importance and prestige for countries in the world. To achieve this, the development of matching logistics based on self-reliance, will always be a matter of necessity. The aspect of Antarctic logistics as a military operation other than war, therefore, will assume importance if a conflict situation were to arise in the future. No country can be immune from this consideration.

CHAPTER 8

The Indian Antarctic Programme, Environment and Logistics

"All of you have brought enough warmth from India to beat both the cold and the wind."

Jagadish P. Khadilkar
(The book's author, welcoming the new team as it alighted from the helicopter, when asked how, in shorts and without a headgear, he was so comfortable in such conditions)

8.1 IAP and the Overwhelming Importance of Logistics

Geological time-line reveals that at one point, long ago, India formed a part of the supercontinent Gondwana, with Antarctica as a hub and Africa, South America, Australia, New Zealand as its other parts. However, due to long colonial subjugation, India could not make any contribution to Antarctic exploration, unlike its Gondwana counterparts and other countries, and was a late entrant in matters Antarctic. India first expressed its Antarctic interest to the United Nation (UN) in 1956–58 (Chaturvedi, 2005), but it was persuaded to withdraw its proposal as negotiations for the Antarctic Treaty (AT/Treaty) were underway. A few Indians had participated in Antarctic expeditions as individuals – Dr. G.S. Sirohi and Dr. P.S. Sehra were part of the US and then-USSR expeditions in 1960 and 1971 respectively and were helped by the Indian embassies there (Sehra, 2003). Apart from attempted proposals at the UN, a couple of early non-aligned summits and the participation of people like Sirohi and Sehra, in their individual capacity, India had been too busy after Independence to give attention to Antarctica. However, by the end of the 1970s, India decided to join the group of nations actively involved in Antarctic affairs through the AT which had come into force in 1961.

Once decided, India took decisive and rapid steps to establish its presence in the continent. Initially, two short summer expeditions were launched in 1981 and 1982. Along with scientific tasks, they made a reconnaissance for establishing a permanent year-round research station in Antarctica. It was during the course of their third summer expedition in 1983 that the first Indian scientific research station, Dakshin Gangotri (DG), was built on an ice shelf off Princess Astrid Coast in Central Dronning Maud Land (DML) in East Antarctica. India acceded to the Antarctic Treaty on 19 August 1983 and within a month, on 12 September 1983, was accorded Consultative Party (CP) status. With this, India started its Antarctic journey. India is also a member of important AT instruments like the Committee of Managers of National Antarctic Programmes (COMNAP), Convention for the Conservation of Antarctic Marine Living Resources (CCAMLR) and the Madrid Protocol. It is also a member of the Scientific Committee on Antarctic Research (SCAR). It enjoys a prestigious position in these organisations due to its active participation.

The life of any station or structure built on the Antarctic ice shelf or polar plateau is limited. Stations like Halley (UK), Neumayer (Germany), Sanae (South Africa) and Amundsen-Scott (US), built on ice, have been replaced a number of times at regular intervals due to snow burial (*Books LLC*, 2010). DG, too, had to be abandoned. Still, in what is a testimony to its tenacity, DG lasted twice the duration of its expected life and became uninhabitable only in 1990 after six Antarctic winters of national service.

To continue India's Antarctic presence, a new site about 85 km away from DG was selected. On the hilly ice-free area of Schirmacher Oasis, a new station named Maitri was built during two austral summers between 1988 and 1990 near two already-existing stations, Novolazarevskaya or Novo (Russia) and Georg Forster (then GDR). With no danger of snow burial and conditions vastly better than at DG, the location assured safety, longevity and scope for a greater range of scientific research activities on the ice shelf, oasis and polar plateau. Maitri has been serving the country with a year-round presence continuously since then. The area near DG now functions as a base for supplies unloaded from ships at India Bay and taken to Maitri by ice tractor-trailer convoys throughout the year. Probably, after over twenty-five years of service, Maitri, too, is ageing and may have to be rebuilt in the near future.

After about thirty yearly expeditions and the experience of a year-round presence in the continent, India has recently built its third Antarctic station, Bharati. Located in the area of Larsmann Hills of Prydz Bay Area of East Antarctica, it is over 3000 km away from Maitri, about 7 km from three stations, Progress (Russia), Racovita (Romania) and Zongshan (China) and 70 km from Davis (Australia).

To maintain its permanent stations and scientific programmes, India has progressed with matching bureaucratic support and institutions to deal with all aspects of its presence in Antarctica as a member of the AT and its instruments.

As pointed out earlier, in few spheres of human endeavours are the demands upon a logistics organisation as onerous ones as in Antarctica. To ensure both survival and scientific pursuits, supply procedures need to be as fool proof as possible. Polar history is full of hardships caused by inadequate logistics expertise and experience. Problems are minimised when experienced teams and tools with advanced technology are available. The International Geophysical Year (IGY) 1957–58 marked the initiation of a new phase for many nations much before India arrived on the scene. Most of these countries also gained from their centuries-long association with Antarctic exploration. Even then, leading countries keep revaluating and revising their logistics plans for better performance, and India, therefore, has much to catch up with to match the Antarctic capability of many leading nations.

Logistics-wise, Antarctica can be divided into four different types of areas – two on ice and two on ice-free zones. Areas permanently under ice cover are the ice shelves and continental glaciers or the interior polar plateau. Ice-free areas, known as oases, consist of coastal promontories and some hilly areas near the coast. It is, by coincidence, that DG, Maitri and Bharati cover three area types, with a considerable coastal distance between Maitri and Bharati. Even if DG lies buried deep in the snow, the area near the decommissioned station serves as a supply depot to Maitri about 85 km away. As far as Bharati is concerned, the approach for berthing and transportation of stores is only over a short distance and fairly simple. Maintaining two stations widely spaced on the Antarctic coast with different approaches, terrains and parameters has certainly added a new dimension and complexity to the logistics network, requiring an integrated approach. There are very few stations on the interior polar plateau, like Amundsen-Scott (US), Vostok (Russia), Concordia or Dome C (France/Italy), Dome Fuji (Japan) and Kunlun (China). Keeping all other parameters equal, they require a more exacting and advanced logistics network due to the long distance from the coast to the polar plateau. With no station on the polar plateau so far, India is still a long way from this scenario.

The success of IAP in pursuing a perceptible, effective and influential Antarctic presence will depend on how it meets the present and future challenges – with viable scientific programmes and well-developed logistics network geared to maintaining the present Antarctic stations and adding new ones.

8.2 National Centre for Antarctic and Ocean Research (NCAOR)

Since the inception of the IAP in 1981, the Department of Ocean Development (DOD), Government of India, had acted as its logistics and scientific organiser in New Delhi. The Antarctic Study Centre (ASC) established in 1988 was renamed National Centre for Antarctic and Ocean Research (NCAOR) in 2000 and was established in Vasco da Gama in Goa. Now it is the NCAOR that functions directly under the Ministry of Earth Sciences, Government of India, as a nodal agency for all matters concerning Antarctic logistics and research.

NCAOR is organised into eleven major divisions to service all its functional requirements.

Five major groups attend to all matters concerning polar and ocean research: National Antarctic Programme (NAP), National

Arctic Programme (NAP), Antarctic Logistics and Operations (ALO), National Antarctic Data Centre (NADC) and Oceanographic Surveys and Services Group (OSSG). Two key programmes are for complementary tasks: Legal Continental Shelf (LCS) programme and the Exclusive Economic Zone (EEZ) programme. Three operational support groups or divisions are: Research and Development Group (RDG), Programme Planning and Coordination Division (PPCD) and the Information and Communications Technology Division (ICTD). The last major group, Management Services, provide all administrative and IT services to the organisation.

Presently, NCAOR has a strength of twenty-two scientists and thirty-two managerial and administrative staff. Apart from that, around thirty-five research scientists and research fellows are working at NCAOR. Excellent laboratory facilities for ice core, microscope, marine stable isotope, polar biology, cryobiology, polar environment and polar remote sensing have been established in the last few years (*NCAOR Laboratory Facilities, 2010* and *NCAOR Research Activities,* 2010).

8.3 Mission Mandate of NCAOR

'To plan, promote, coordinate and execute the entire gamut of polar science and logistics activities of the country in order to ensure a perceptible and influential presence of India in Antarctica and to uphold India's strategic interests in the global framework of nations in the southern continents and the surrounding oceans.' (*NCAOR Annual Report, 2013–14*)

Appearing prominently in all their annual reports, the mission mandate is also announced in bold brass letters at the entrance of the NCAOR complex in Goa.

8.4 Indian Scientific Research Stations in Antarctica

To get a general idea of the relative positions of DG, Maitri and Bharati, see Figure 6.2 in Chapter 6, where their locations are shown, along with that of the other stations in Antarctica.

8.5 General and Logistics Requirements of the Indian Antarctic Research Stations

For five years, from 1982–83 to 1988–89, only one station, DG, existed on the ice shelf at Dronning Maud Land, about 15 km from the berthing point named India Bay. In 1989–90, a second station, Maitri, was established about 85 km on the Schirmacher Oasis area. After one year of co-existence, DG, was abandoned due to snow burial. However, India Bay and the area near the now snow-buried DG, continue to serve the IAP as berthing point and supply area for goods before they are transported to Maitri by a helicopter-lift during summer and by vehicular convoys all-round the year. After over twenty-five years of national service, outlasting its designated life of ten years, the ageing Maitri will have to be rebuilt or replaced in the near future. NCAOR has already made its intentions clear and actions on this score may start soon. This aspect is discussed in details later in the chapter.

Now that there are two permanent stations, Maitri and Bharati, 3100 km away from each other on the Antarctic coast, the following major administrative, logistics and scientific requirements will have to be catered for, coordinated and executed by NCAOR, to effectively maintain them:

- Perspective planning consistent with national strategic missions, goals and objectives.
- Execution of the above, mid-course reviews and corrections.
- Planning, coordination and organisation of all aspects of Antarctic expeditions.
- Coordination with the Antarctic Treaty System (ATS) and other national and international organisations.
- Working out detailed requirements for ships, aircrafts, helicopters, vehicles, major and minor equipment, their indigenous procurement or import, chartering and other actions required for yearly expeditions and continuous year round maintenance of the two stations when 10 out of 12 months in the year they are isolated from rest of the world.
- Selection of personnel and projects both for scientific activities and logistics and the necessary pre-expedition training and post-expedition reports and publications.

- Organisation of the supply and delivery of items like food, clothing, medicines, communications, spares and other requirements.
- Transportation of personnel, vehicles, stores, etc., from India to Maitri, Bharati and vice versa.
- Liaising with Dronning Maud Land Air Network (DROMLAN), Antarctic Logistics Company International (ALCI) and other organisations and the creation of an integrated logistics network for both stations.
- Catering for emergencies and unforeseen situations.
- Periodical review of the position of the IAP in relation to India's avowed perceptible and influential position in Antarctica vis-a-vis its strategic interests as mandated by the mission statement.

8.6 A Review and Functioning of the Indian Antarctic Stations – DG, Maitri and Bharati

Though all three stations are located in East Antarctica, they represent different eras, areas, environments and logistics parameters. A detailed analysis of the three highlights the logistics structures supporting them and provides suggestions for developing an integrated logistics network in the future for them, or for any new stations, that will help consolidate the country's position in the continent.

8.7 Dakshin Gangotri (DG)

8.7.1 *History and Background*

Irrespective of previous experience, a first visit to any area in Antarctica is of an exploratory nature and the selection of a site is invariably marked by trial and error methods. For instance, Russia (then USSR) constructed its third station, Lazarev, in February 1959 in the area of Princess Astrid Coast of DML in just three weeks' time (Russian Antarctic Stations Overview, 2012), the first two being Mirny (1956) and Vostok (1957). The station names were derived from ships commanded by Admiral Bellingshausen and Mikhail Lazarev during the Russian expedition led by the former in 1820, sighting the continent for the first time. Lazarev consisted of a large house (radio station, cabins, kitchen, store rooms, bath and laundry and power station), an aero-meteorological pavilion and a power generation and radio station. Those were the initial Cold War days of competition between the US and erstwhile USSR, and, in a bid to demonstrate one upmanship, it seems that the Russians did not give much thought to site selection. Lazarev had to be abandoned just after three years in 1962 due to snow burial and the danger of crevasses. In the meantime, the Russians had also carried out a study of Schirmacher Hills, an ice-free oasis area of about 35 sq km and 85 km away from Lazarev, which formed the hinge point or the strand-crack area where the polar glacier meets the floating ice shelf. The area has a number of lakes of varying sizes. In 1961, they constructed a new station, Novolazarevskaya (or New Lazarev popularly known as Novo), on Schirmacher Hills. Even the new location was found to have a number of drawbacks, like a non-representative meteorological ground, difficult water supply and the melt-down of an adjoining glacier, causing frequent heavy floods on the station. So the station was shifted to a new location 600 m away in 1979–80 (Uranov, 1990). As of now, it is functioning smoothly in close cooperation with Maitri.

The first Indian expedition to Antarctica was with twenty-one members who stayed for just eleven days in January1982, carrying out a few scientific experiments (*NCAOR, TP No 1*). Apart from a brief mention of the Russian abandonment of Lazarev and showing its location in the expedition report (*NCAOR Expedition Report TP No 3* and Sharma 2001), the initial Indian expeditions make no mention of the above history or if any other information regarding Lazarev was considered while searching for a location for a permanent Indian station. Anyway, the Indians landed in an area near the Russian Bay, renamed it India Bay and carried out a reconnaissance for the establishment of their first permanent station.

Selection of Site for **DG**: During the period of the second expedition and the initial period of the third expedition, a total of five locations were examined in detail, leading to the selection of the final site as shown in Figure 8.1 (Sharma, 1997 and 2001).

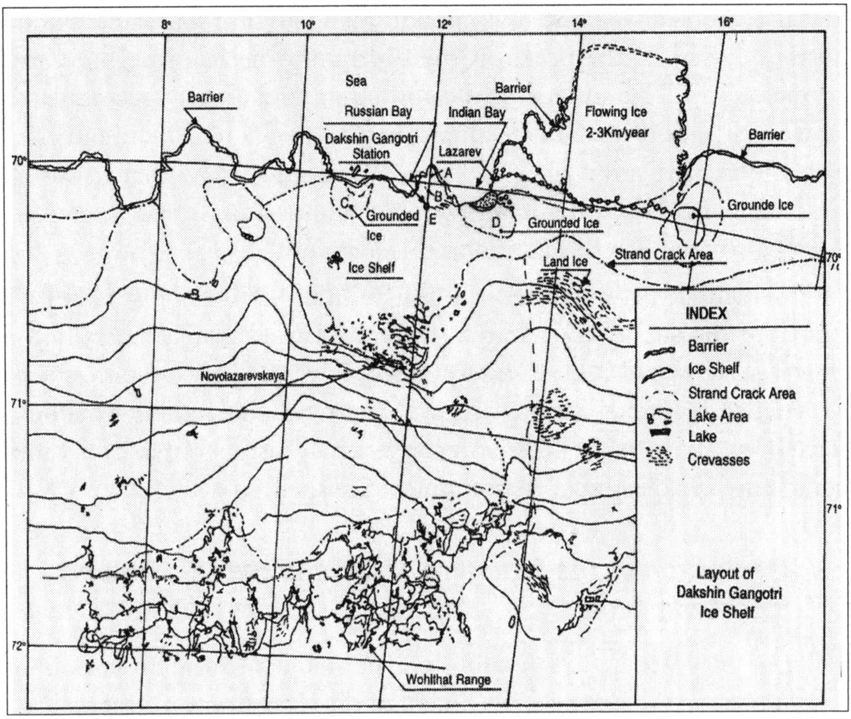

Fig. 8.1: Selection of the Site for DG

A comparative analysis shows that Sites A and B were just one and five km away respectively from the edge of the ice shelf. The area was full of cracks and, being at the edge of the shelf, the wind was more severe. Proximity to the shelf edge also heightened the chances of the area calving into the ocean as an iceberg. Also, at the farthest points from the hinges or strand-crack region where the shelf starts floating, the maximum movement during ocean swells and instability would be experienced.

Sites C and D were on high ground on either side of sites A and B. Intermittent patches of white ice on wind-swept slopes of blue ice indicated crevasse propensity and enhanced wind activity even more than sites A and B. Sites C and D also indicated a less safe manoeuvring space for the inhabitants if any structure were to be built there.

As compared to the above, Site E presented a better picture. It was 20 km south of the ice shelf edge, flat and there were no signs of crevasse proneness. There was enough manoeuvring space and a good all-round view of the ocean in the north and the Schirmacher Oasis in the south. The surface at the top strata had freshly accumulated snow of up to 40 to 50 cm, a hard icy layer of 5 to 10 cm and then a medium hard snow layer of 15 to 20 cm and below that it was all hard ice. This was suitable for laying a foundation. It was, therefore, decided to establish the station at this location.

Incidentally, the selected site for building DG was just about 10 km from the site where Lazarev remained buried many meters under the ice. DG would join it under the ice six years later. Despite lack of experience in matters related to the Antarctic, more thought should have been given to the Russian experience of just a couple of decades ago before establishing a station on the ice shelf.

8.7.2 *Selection of the Structure of DG and the Life-Support Systems on It*

Once it was decided to establish a permanent research station in Antarctica, on the ice shelf, the design of the station was conceptualised and bids for procurement were invited. Since no agency in India had any experience of the polar regions, especially with respect to construction technology and life-support systems, emphasis was laid on considering offers from international firms with past experience in similar work. Accordingly, after due consideration, an offer from M/S Structa Ply Ltd of the UK was accepted. This company had supplied a similar station to the British Antarctic Survey (BAS) – Halley IV on the ice shelf. A team from the Corps of Engineers and Corps of Electrical and Mechanical Engineers (EME) of the Indian Army was sent to the UK and Germany in the summer of 1983 for obtaining training on structure construction, commissioning of various life-support systems and maintenance. The team was closely associated with the packaging and sequential loading of items in the ship, *MV Finpolaris,* at the port in UK before its departure to pick up the expedition and construction teams from India. This was essential to avoid any confusion and delay during unpacking, structure construction and commissioning of life-support systems by the trained teams.

8.7.3 *Design of DG and Its Life-Support Systems (Structa Ply Ltd., 1983)*

Two double-story blocks, 'A' and 'B', linked by a narrow single-storeyed passage, termed as 'C' block, formed the station with a total floor area of 800 sq m. The structure was built on a raft-type foundation in order to ensure an even distribution of load to obviate unequal sinking in snow. Prefabricated panels were made of specially treated timber. Both 'A' and 'B' blocks were each made of eleven portal frames of 6 m height, each made of wooden panels, with suitable attachments to foundation beams. The foundation raft, with beams of 60 cm and 30 cm deep placed in a grid-type arrangement, was positioned on the hard-ice layer which was obtained by digging 1 m deep trenches in ice. In order to further ensure proper levelling of the ice surface, as also giving it a higher bearing capacity, the ice layer was covered with corrugated galvanised iron (CGI) sheets, of which a large number were lost during a helicopter accident during unloading. The sidewalls were made of 100 mm thick prefabricated timber panels of marine plywood sheets 9 mm thick, attached to the two sides of a timber load-bearing frame. The space between the two faces was filled with a thermally insulating material, Styrofoam. The outside face of the structure was protected by cladding, consisting of a 25 mm thick layer of insulation material with bolted metallic sheets. The indoor finishing varied from area to area, depending on utility. The radio room had acoustic tiles, while the generator and boiler rooms were provided with asbestos lining. The kitchen and the bathrooms had melamine-faced plywood. In all other rooms, hard board was used.

Block 'A' housed essential services, including generators, fuel supply system and various workshops on the ground floor; storage space was on the first floor. Block 'B' had laboratories, a kitchen, dining room and lounge on the ground floor, and the first floor had two-man bunk rooms, radio room and toilets. In all, there were six bunk rooms to accommodate twelve members of the wintering party. An office provided earlier was later modified to serve as an office-cum-living accommodation for the base commander. In addition, some space was left for extra bunk rooms if required. In subsequent years, two more bunk rooms were created. These were used during the last wintering as, due to enhanced needs, the team had sixteen members.

The station was energised by electrical power generated by three 62.5 kva three-phase generators, of which one was for running, one for undergoing maintenance, and one ready to run as a stand-by. The peak load of the entire station was 40 kva initially and, later, it increased to 50 kva. In addition to the main generators, there were two 12.5 kva Kirloskar generators and six 1.5 kva generators to assist the construction party. A weekly fuel storage tank of 4500 litres was provided within the station from which fuel was pumped to daily storage tanks for the generators and boilers.

Boilers were provided for central heating as well as for the snow-melt plant to melt ice for water supply. To maintain a steady temperature of 15°C, heating was provided through close-circuit hot-water pipes and radiators. All waste water was discharged into underground twin drains. Arrangements were also made for the circulation of fresh hot air in living spaces and, to that end, exhausts were provided at suitable locations. A modern kitchen was made, with facilities for cooking on both power and gas. The station was also equipped with a small operation theatre and a medical room with drugs.

The water supply was made available by a snow-melting plant of 1000 litre capacity, installed on one side of the 'B' block, which had a hatch from where snow could be shovelled from outside. After shovelling, one had to wait for about half an hour for the snow to melt. Snow was melted by the circulation of hot water mixed with an anti-freeze in the boilers. Once the snow-melt tank was full, water had to be pumped into a storage tank provided in the bathrooms.

The toilets were of the type using humus, where night soil got converted to air and a negligible quantity of soil waste which could be taken out once a fortnight. After each use, one had to add some humus (soil conditioner) in the toilets and churn the night soil through a handle provided for the purpose above the seats. Continuous circulation of hot air was essential for the smooth operation of the toilet system.

A satellite communication system was provided, with an operating terminal inside the radio room, and a receiving and transmitting dome on the 'C' block. This system, which was linked to International Maritime Satellite Organisation Network (INMARSAT), offered telephone and telex links on a global basis and also had facilities for a telefax and a slow-scan TV transmission. The station was designed to house laboratories on the ground floor in 'B' block, where meteorology, microbiology and glaciology laboratories, as well as facilities for atmospheric studies, were planned.

Additional storage facilities were provided in the loft, created by using the space between the trussed roof panels and the ceiling panels of the first floor. The entire building was designed to withstand 80 knots of wind velocity. The stability of the building against the wind was safeguarded by a deposition of snow on the 1 m extended foundation all-round the structure. Additional safety was ensured by anchoring the entire building by means of straps put over the roof with their lower ends buried in suitable anchors in ice. As the station started sinking, this picture totally changed, bringing in new worries due to accumulating snow pressure from all sides.

The floor plans of the station are shown in Figure 8.2 (*NCAOR Expedition Report TP No. 3*).

8.7.4 *Construction of DG*

After the ship was berthed at India Bay, unloading of vehicles and stores on the shelf, construction of a helipad for helicopters (two MI-8 and two Chetaks) near the station site, establishment of a temporary camp for the construction party, and other activities, started in earnest. January and February 1984 were marked by continuous work, using sledge vehicles and helicopters for land and air transportation, for the completion of the station in a virtual race against time. There were interruptions due to blizzards, ship movements, difficulties in construction on ice and snow and a major MI-8 helicopter accident (the helicopter fell into sea, was rendered out of use but the crew escaped miraculously) and other hiccups, but somehow the station was built and all the life-support systems were installed and made

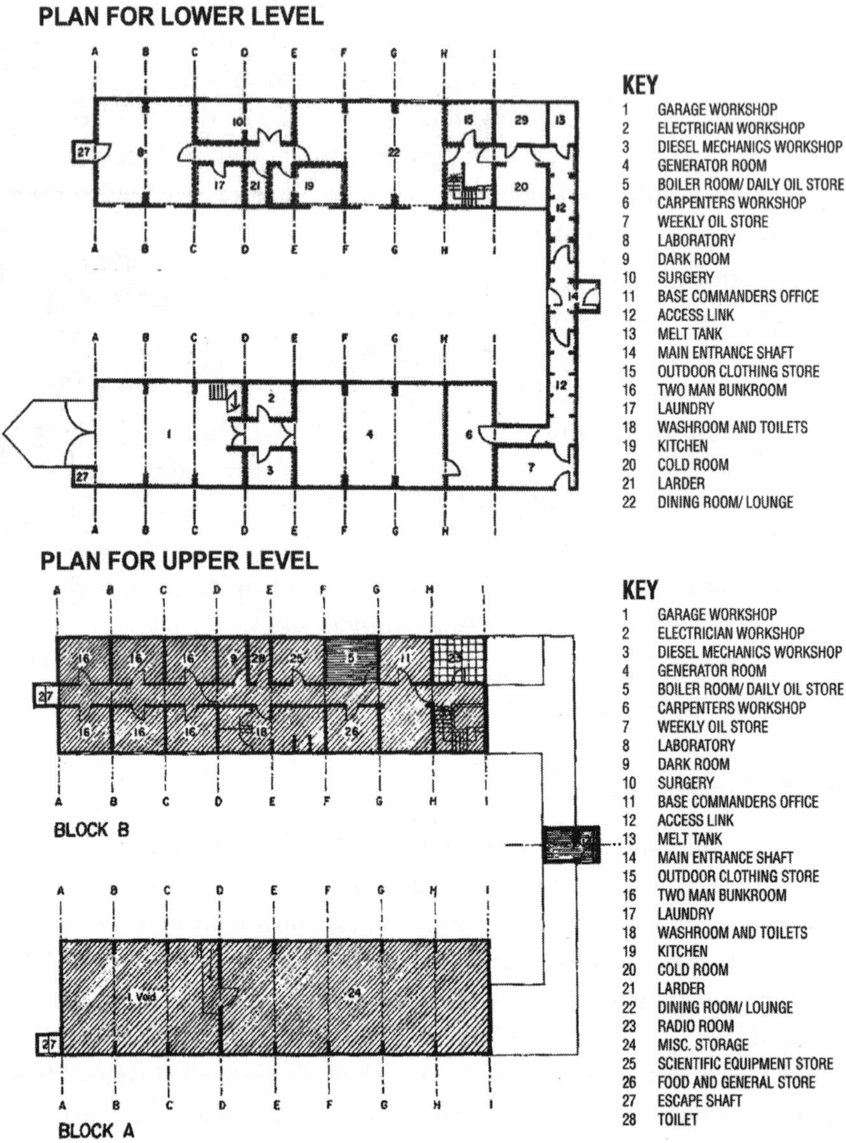

Fig. 8.2: Floor Plans of DG

operational. Once the structure and facilities were duly tested, the ceremonial moving in, by the first Indian wintering team of twelve personnel, took place on 25 February 1984. With the achievement of this major task, the ship, *Finpolaris*, could leave Antarctica for India on 1 March 1984 (*NCAOR TP No 3*). As the sea starts freezing around Antarctica very rapidly once summer is over, ships must leave around this date. There are many examples in the past of ships getting stranded for 8 to 10 months till next summer if they fail to get out in time.

8.7.5 *Six Years of DG in IAP*

Being on the ice shelf, the fate of DG was known from day one. One has to accept the inevitability of snow burial, like that of all other stations on the ice shelf, continental glaciers or polar plateau, and prepare accordingly. India did this well. While continuing its Antarctic programme from DG, it had also started a detailed reconnaissance of Schirmacher Oasis for building a station near two pre-existing stations – the Russian Novo and the East German Georg Forster (*NCAOR Expedition Reports TP No 4 and 5*). GDR had also completed twenty-five years of presence in Antarctica that year (*GDR,* 1984).

However, the responsibility of the six winter teams to keep the station running while carrying out the assigned scientific tasks, was becoming more and more daunting as the years went by. The sixth and last wintering team had the toughest time and the dangerous station had to be abandoned at the end of their winter stay. The danger of technical breakdowns and the necessity to leave the main structure always looms on any structure built on ice. For example, if the power supply fails, the liquid in all the pipes of the life-support systems freezes, causing permanent damage. The quality of material used and the workmanship of the structure, however, were excellent in the case of DG. Instructions for the maintenance of the structure and the systems were given in great detail and self-explanatory, a fact instrumental in carrying out the tasks effectively by the successive winter teams (Structa Ply Ltd, 1983).

A new complex based on four prefabricated containers was also established both as an emergency shelter for winter teams as also a refuge for summer teams. A half round-shaped big hanger (Christian Nielson) was constructed for vehicle repairs during the third wintering (1985–86) but it got buried in ice. Despite all efforts, the structures kept sinking and the snow level in the surrounding area kept rising, ultimately, taking the station with it. After the first two winters, the entrance and exit of the station was through shafts created with a 30 ft tall ladder inside and a long slanting path outside joining two ladders. Every year the height of the shaft had to be increased. With successive winters, the task of snow clearance also got tougher. For example, the last winter team had to stay in the totally snow-buried station with only a part of the triangular roofs visible from all sides in good, blizzardless conditions. With the general snow level much higher, the station used to be in a bowl. During a blizzard, the bowl would get filled, with the roofs becoming totally buried. Considerable vehicular and shovelling skills were required to clear the roofs and surrounding areas of the snow. A number of modifications were also carried out for strengthening the walls, trusses and roof and for overhauling all the life-support systems. After an overhaul in the last few months before closing the station, the snow-melt plant worked much more efficiently than at any time before. The area outside was also similarly affected. Fuel sledges and barrels were mounted on temporary ramps made with planks on barrels. Food and supplies' containers were similarly mounted. These, along with the repair hanger, the DRDO complex (summer refuge), survival and other portable cabins, Dobson's hut based on a container for ozone experiments, and a large number of vehicles spread over a large area around the main station, had to be cleared of near-snow-burial after every blizzard. In the last two winters, there was also a need to run a number of convoys to Schirmacher Oasis to prepare the area for summer tasks and for building Maitri. This created a need for expertise in running convoys in the crevasses-ridden ice-shelf area, marking of routes and other essentials for logistics support. Separate skills and routes are required for running convoys in winter and in summer due to different ice/ crevasse conditions.

No two objects should be stationed on Antarctic ice near one another, and a distance of at least 50 m should be maintained between them. When DG was in its last winter, around the totally snow-buried main station, there was an emergency complex of four containers, a hanger, a Dobson's hut (meaning a container), two portable cabins, nine vehicles of different types, five snow scooter parked on top of containers, five big sledges to be towed with vehicles on convoys, two big fuel tanks and thirty tanks on sledges filled with fuel and two containers with supplies – all placed in a widely dispersed area of over 10 sq km. Every item required to be cleared from snow burial and the containers mounted on barrels needed frequent repositioning with the help of crane vehicles. Indians called the crane Hanuman. This was a constant battle in the area due to frequent blizzards.

All tasks, like clearing objects of snow, starting a vehicle, running a convoy, releasing a balloon and other jobs, while staying in a totally snow-buried station under constant threat of caving under the weight of snow, particularly during the sixth and last winter of the station, were unbelievably difficult and taxing, both mentally and physically. Massive snow clearance operations were the most important, demanding and exacting tasks during the last winter occupation.

It was to the credit of the designers, fabricators and eighty-seven winterers over six years that, when the station was finally abandoned on 25 February 1990, exactly six years after its occupation by the first winter team, it had provided safe shelter all along and was in excellent shape. But it could no longer fight the elements. Many useful stores, tools, minor equipment, vehicles, Dobson's hut, portable cabins, library, kitchen items and the communication dome and set-up were duly shifted to Maitri for continued usage, through convoys. The quality of the mechanical and electrical items was so good that samples of about hundred and fifty of such items, along with the description of their usage, was taken back and handed over to DOD by the last station commander of DG after the decommissioning of the station for future guidance (Khadilkar, 1990).

A shelter accommodation was also constructed in the hanger during the summer of 1990 for future convoys during summer. As the last station commander of DG, I had strongly recommended that this not be done as the hanger was totally buried under the snow and, instead, suggested creating a mobile-supply base founded on available containers and mounted on sledges whose positions can be changed and snow burial avoided. It was not accepted and a new shelter-accommodation arrangement proved to be totally unworkable from the beginning (*NCAOR Expedition Report, TP No. 6, Khadilkar 1990*).

Thus when I, the last station commander of DG along with the second station commander of Maitri, left DG for Maitri on a convoy of seven vehicles and seven sledges on 25 February 1990, after decommissioning and abandoning DG, it marked the conclusion of DG's service to the country.

The site of DG has since been marked as a historical site – HSM 44, and the area now serves as a supply base for Maitri (*NCAOR,* CEE for Bharati, 2009).

8.8 Life of DG as a Structure on the Antarctic Ice Shelf

The transformation of DG from the time of its commissioning to its abandonment, a period of six years, as well as its two-storey building, which housed a total of eighty-seven winterers, is shown in Figure 8.3 (*NCAOR TP No 3 and 5*; Sharma, 2001 and Khadilkar, 1990). It is said that a picture is worth a thousand words. These six photographs should tell the entire story of DG's life on ice.

8.9 The Second Indian Antarctic Station – Maitri

When India decided to establish a new station, Maitri, in Antarctica, on Schirmacher Oasis, to replace its first station, DG, which was heading towards snow burial and decommissioning, the DOD floated a global tender in February 1987. The Research and Development Establishment (Engineers) (RDEE), Dighi, Pune, had been actively involved in providing engineering support to Indian Antarctic expeditions since 1983. The support included strengthening of DG's structure against pressing ice during its gradual sinking, designing and establishing a container-based shelter on ice near DG for winter emergency situations as also for the stay of summer parties, development of sledges for Antarctic snow vehicles, fuel tanks and

Dakshin Gangotri under Construction

Dakshin Gangotri after 1 year

Dakshin Gangotri after 3 year

Dakshin Gangotri after 5 year

Dakshin Gangotri after 6 year

Fig. 8.3: Transformation of DG from Commissioning to Abandonment due to Snow Burial (The last photograph taken on 25 February 1990 just after leaving the station. The author is standing in the middle)

a variety of other engineering tasks. Their personnel were part of all the expeditions to execute the tasks. The organisation had received valuable experience in providing engineering support in harsh Antarctic conditions, thanks to the station gradually sinking in ice.

With this background, RDEE decided to take a plunge and responded to the tender with an indigenous proposal to build Maitri. The technical subcommittee of the DOD awarded the prestigious task to RDEE and, as a result, Maitri, the second Indian Antarctic station, designed and built indigenously, was established over two consecutive austral summers between 1987 and 1989.

8.9.1 *Site and Sequence of Construction*

As pointed out earlier, conditions on the ice shelf and those existing at the Schirmacher Oasis, though only 85 km apart, are vastly different. In the general area where Maitri was to be constructed, two permanent stations, Novo (Russia) and Georg Forster (then GDR), had been in existence for many years already. They were merely 5 km away. Unlike DG, there was no danger of snow burial, lakes were available for fresh water and Maitri could be planned for national service on a permanent, long-term basis.

However, work in Antarctica to build Maitri imposed several unchangeable restrictions. Being over 10,000 km away from India, only 30–40 persons for station construction were available during two months of austral summer. About 1200 tons of construction stores needed to be transported by helicopters from the ship berthed at the ice shelf. Often that was not possible due to strong winds and the ship had to be anchored in the ocean more thana 100 km away. Unpredictable winds, blizzards and low temperatures add to difficulties in construction activities. The structural components have to be of lightweight material and easy to handle by four men at a time. For easy airlift by MI-8 helicopters, the maximum load per pallet cannot exceed 2 tons and the dimensions of each pallet cannot exceed 1.5 m × 1.5 m × 3 m.

In view of the above restrictions, it was decided to build the station over two austral summers–the construction of foundational columns (telescopic stilts) and some portion of the station during the seventh expedition (1987–88) and the remaining work during the eighth expedition (1988–89).

Accordingly, the construction of the foundational columns of the entire station and the construction of block 'A' of the building was completed during the seventh expedition (*NCAOR TP 7)*). While this was on in Antarctica, the fabrication of the remaining structure of the building and the station's life-support systems were in full swing in India at RDEE, Pune. All the stores were made ready and rigorous training was imparted to selected army members on the construction of the building and the installation of the various life-support systems.

With this preparation, the expedition ship *MV Thuleland* left Goa with the eighth Indian Antarctic expedition team in November 1988.

All the planning and prior preparation a year earlier paid off handsomely. The remaining stores for the station construction (about 600 tons), fuel, food and other stores were transported from the ship to Maitri location by helicopters. As per the record, a total of 295 and 245 sorties were made by two MI-8 and two Chetak helicopters during the available sixty summer days and construction was duly completed. Twenty-six members of the wintering team stayed back in the newly constructed station (*RDEE*, 1991).

This was also the first time when two Indian winter teams were present in Antarctica – one at DG in its sixth and last wintering and one at Maitri in its first wintering. Both stations were a mere 85 km from each other, but existed in totally different environmental conditions.

8.9.2 *Structural Design and Original Layout*

The foundation of Maitri's main building is laid on adjustable telescopic stilts or columns (the main block and 'B' block) to safeguard it from the boulders and uneven surface of Schirmacher Hills. 'A' and 'C' blocks stand on relatively level ground. The stilts can negotiate a relief difference of up to 1 m to give a level top surface on which grillage timber beams are fitted for superstructure. The columns are especially grouted into the foundation pit with the help of bolts and special cement concrete. While designing the structure, factors such as dead load, snow load, a wind speed of up to 200 kmph (the maximum recorded wind velocity in the area till then was 104 knots), a temperature of minus 40°C and the permafrost condition of the soil were considered. The transfer of heat from the centrally heated building to the steel framework in the foundation is controlled, so as to not affect the permafrost level of the ground.

The design of the structure consists of standard modules of 3 m × 2.4 m and a few larger rooms. The panels of the modular construction are made of lightweight timber framework, 12 mm thick marine plywood cladding and sandwiched polyurethane foam as insulating material. The panels are interconnected by wooden connectors. A prefabricated structural steel system of columns, bracings and beams forms the foundation of the superstructure. The modules have been lined internally by a fire-retardant material or gypsum board and externally by plastic-coated galvanised iron (GI) sheets. The building has been well anchored by means of steel anchorages. The entire station is single-storey high except the main block, which has a loft and two bays at the centre raised to a double storey to house the satellite communication system and communication dome.

Originally, of the four blocks, the main block housed the living accommodation, medical facilities, communication control system and laboratory. Block 'A' accommodated the power supply system and workshop. Block 'B' housed the central heating system, water-storage tanks, snow-melt plant, kitchen, dining hall and cubicles of chemical toilets. Block 'C' provided accommodation for incinerator-type toilets. Adequate storage facility was provided in the loft above the main block.

8.9.3 *Present Layout and Facilities*

Maitri was originally designed for ten years of operational service when it was constructed in 1988–89. It was the first totally indigenous effort and it is a matter of great credit to the designers, fabricators and manufacturers, and all the wintering teams, that the station is functioning well even after twenty-five years. However, due to the severe Antarctic climate, the structure and life-support systems have needed structural and technical reviews followed by strengthening, readjustments, repairs, replacements and modifications from time to time over the years. New facilities for smoother functioning during summer and winter and for the growing activities of IAP had to be created around the station. When DG and Maitri were constructed, environmental considerations were voluntary and did not attract the attention they deserved. The signing of the Madrid Protocol in 1991 and its coming into force in 1998, enforced new responsibilities on all member countries for the protection of the pristine Antarctic environment and mandated new measures.

With passing time, considerable changes have taken place and today Maitri has become a sprawling complex around the original structure of four blocks. The old structure is ageing and may require a complete makeover in the near future.

Schematic layouts of the main station, the overall complex and a few photographs showing the station, lake and glacier are shown below in Figure 8.4 (*NCAOR TP No. 11 and 21 and NCAOR Through the Years,* 2010) to give a general idea about the Maitri complex as it looks today.

An understanding of the complex development, described after the layout, is necessary to appreciate how a permanent station in Antarctica survives over the years.

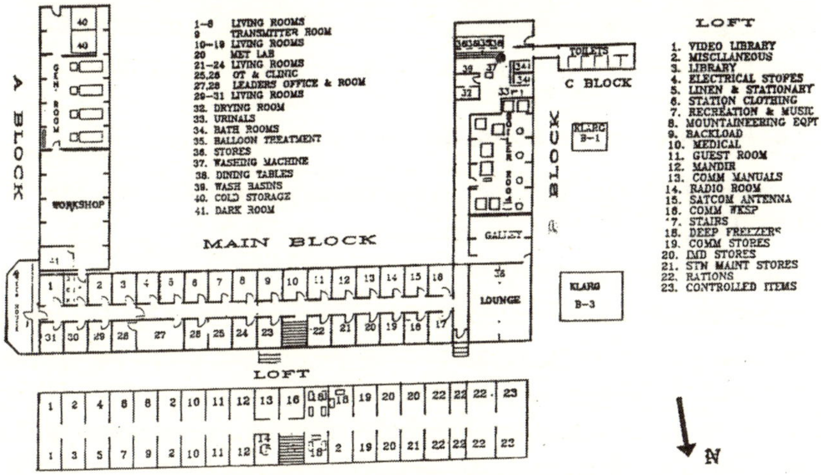

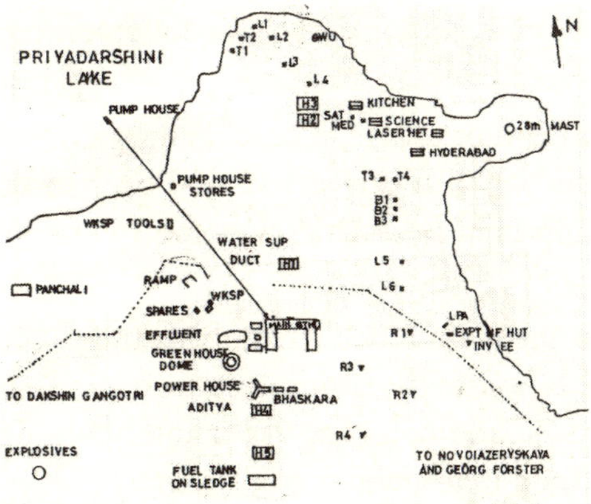

Fig. 8.4: Floor Plans, Schematic Layout and Aerial Views of Maitri

8.9.4 *Facilities in the Main Station (Maitri Manual, 1988)*

The main block and 'B' block are constructed on stilts, whereas 'A' and 'C' block are on level ground. There are eighty-four steel stilts of 2 m height under the main block.

The main block is located east-west and is 51.6 m long, 8.08 m wide and 4.75 m high. It consists of twenty-five single-occupancy living rooms, a lounge-cum-dining room, station commander's office, a computer room, reception, hospital, meteorology laboratory and a post office. The central corridor and passages lead to the 'A' and 'B' blocks. There are staircases for the loft, main entrance and emergency escape.

At the eastern end of the main block, an experimental greenhouse was constructed in 1990. It was abandoned later and converted to an environmental-impact analysis (EIA) laboratory in 2003–04 (*NCAOR TP No. 21*).

The loft over the main block strengthens the structure and creates additional space. It houses the communication rooms, communication stores, temple, music room, audio, video and books library, stores for the kitchen and food items, and medical stores.

'A' block, at the eastern end of the main block, is 26.4 m long, 7.51 m wide and 3.76 m high. The block has 12 timber frames containing 11 bays. It is constructed on metal beams (box girder using structural steel angles and flats) resting on level ground. Initially, it housed the power-supply system which was later moved out. The length of the block was extended by two bays in 1997 to accommodate two walk-in deep freezers for food and drug items (*NCAOR TP No 15*). The block houses a gymnasium, sports area, guest rooms, medical store, operation theatre, electrical items stores and a carpenter shop.

'B' block is 24.76 m long, 8.08 m wide and 3.70 m high. Along its length, the block has10 bays with 11 timber frames. It is constructed on stilts similar to the main block and has 40 columns underneath. The block consists of a kitchen, boiler room with four boilers, snow-melting plant, washing area for washing and drying equipment, wash basins, bath modules, urinals and a saloon.

'C' block is 9.24 m long, 2.69 m wide and 3.70 m high. Its construction is similar to the 'A' block on level ground. It consists of five toilet modules which are of the dry-incinerator type.

8.9.5 *Facilities Located Outside the Main Building of Maitri*

In the sprawling complex around the main building, many facilities have come up over the years. They include three power houses with generators, a workshop, two effluent treatment plants, incinerators, a fuel dump, a container complex for the summer team, a Brewer spectrophotometer and a geomagnetism observatory, a seismological observatory, communication facilities, a pump house in the water-supply lake and an environmental observatory. The lake is named Priyadarshini Lake and the powerhouses are called Aditya, Surya and Bhaskar. All other facilities, too, are named keeping in line with Indian culture.

8.9.6 *Life-Support Systems at Maitri*

With the advancement in vehicular technology and other logistics aspects, the hardships and dangers faced by early explorers have become a thing of the past. However, harsh environmental conditions and the continent's total isolation necessitate fool proof life support systems for year-round permanent stations, both for the survival, safety and health of the people and for the continent's environmental protection. Some of the essential systems are electrical power generation and distribution, water supply, central heating and waste disposal.

Electrical Power Generation and Distribution: Initially, four 62.5 kva and one 30 kva generator sets were installed in Block 'A' for the main power and for fluctuation-free communication. Shortly afterwards, a centralised new power station with 2 × 125 kva and 4 × 62.5 kva generating sets was created outside the main station but it was destroyed in a fire in 1991. So, in 1992, a new powerhouse, Aditya, was created. It has 4 × 62.5 kva sets and a control room in three knock-down type containers, arranged in a Y shape. To augment the power-supply system, additional powerhouses with two and four same-capacity generating sets in standard ISO containers were installed in 1994 and 1997 respectively and named Bhaskara and Surya (*NCAOR TP No. 12, 15*). All the containers have a strengthened floor for the direct mounting of the sets on the floor and with anti-vibration mountings. Diesel engines have been given cold-starting devices. Each generator set is provided with a control panel near the set and a remote control panel installed at the control room for ease of operation and observation. With this arrangement, 'A' block was reorganised to house facilities as mentioned above.

Each room and all the corridors of the four blocks of the main station building have been suitably illuminated. Efforts have been made to select the best-suited cables, switchgears, exhaust fans and luminaries for low-temperature conditions. Instead of going in for heavy and bulky PVC/rubber cables, Teflon silver-plated copper wires have been used for the complete system. Light miniature breakers with built-in protection against overload and short circuits have been used instead of the conventional heavy iron-clad switches, distribution boards and fuses. Miniature earth-leakage circuit breakers have also been used. Silver contact switches have been used for longer life.

Water-Supply System: Water supply to Maitri is ensured by pumping water from Priyadarshini Lake, about 255 m from the station, through a specially designed system. During winter, the top layer of water in the lake is frozen to about 2 m or so and, therefore, an elaborate system is required. A pump house has been constructed in the lake, at a distance of about 80 m from the edge and a submersible pump has been lowered in the water. The pump is enclosed in two concentric stainless steel jackets which are heated electrically to ensure the continuous pumping of water even at low temperatures of up to minus 40°C. To further ensure a continuous flow from the pump to the boiler room in the main station, the entire copper pipeline is enclosed in a rectangular air-tight insulated duct, supported on a steel structure. The duct is made of marine plywood and a 100 mm rigid polyurethane foam.

The daily requirement of water is pumped from the lake by the submersible pump through a 40 mm diameter pipe. The duct houses two additional pipes of 25 mm diameter which carry hot water mixed with anti-freeze media (monoethylene glycol) to maintain the duct temperature in the pipeline. The water fills two stainless steel tanks of 2500 litre capacity, located in the boiler room and distributed to the utility points through laid-down pipes and circulation pumps. A control panel with indicators and safety systems is fitted in the boiler room for observation and operation. A standby centrifugal pump has been installed in the pump house.

To avoid any contamination of lake water in case of leakage, an electrical trace-heating system was designed. The trace-heating tape was installed in 1997 on the fresh-water pipeline. Temperature sensors installed in the duct indicate the temperature on the control panel in the boiler room to avoid any malfunctioning.

A snow-melt plant of 1000 litre capacity has been provided in the boiler room as a standby for use in winter.

***Central Heating System*:** The station building has been centrally heated to provide comfortable living and working conditions. The central heating system consists of sixty-four radiators of 1000 kcal/hour capacity, which are heated by circulating hot water and an anti-freeze mixture at about 65°C to 75°C temperature. Four hot water generators of 2 lakh kcal/hr capacity, each connected by a pipeline to the radiators, meet the entire heat load of the station. Out of these only one is used at a time, the second is switched on to meet the peak load, the third being a standby and the fourth undergoes maintenance. Two fuel tanks of 2500 litres capacity each are connected at parallel points from which fuel is pumped to the daily tank (500 litre capacity) of the hot-water generator.

Fire Protection System: One of the grave dangers faced by all Antarctic stations is fire due to the very dry climatic conditions and high-velocity winds. Once a fire starts, it is very difficult to control it in Antarctica. Maitri is provided with the means of self-contained prevention, detection, spread check, fire suppression and evacuation to the refuge area inside the building itself. Suitable fire-protection devices have been ensured in addition to the fireproof gypsum board lining provided in the main block. Fire-check doors and fire-barrier walls have also been installed at various points to check the spread of fire in case of an accident. Manually operated fire extinguishers are also positioned at fire points near the main entrance of the station building.

Communication System: Adverse climatic conditions in Antarctica, like magnetic storms, static electricity, earthing problems and heavy magnetic fields pose a strong challenge to communication engineers. In addition to High Frequency (HF) Morse/Voice communication (between India and Antarctica, with other stations and with ships during voyages), a Very High Frequency (VHF) communication (between camp areas, convoys, air-traffic control, ship, helicopters and the station) has to be provided. Maitri is also linked to the rest of the world via a satellite communication terminal, the International Maritime Satellite (INMARSAT) system. This is mainly used for sending and receiving official messages as well as for making official or authorised private telephone calls and fax messages.

To help the stations perform smoothly their scientific activities and to have global communication, DRDO established ERNET (Education and Research Network) as well as an electronic mail system in Antarctica. This made for reliable and faster communication and exchange of scientific data between scientists in Antarctica and their parent organisations.

A direct satellite communication link between Maitri and NCAOR has now been established to create an archetype in scientific and logistics operations, planning and coordination by providing 24×7 connectivity @ 1 mbps. This facilitates online collaboration, online real-time data transfer, video conferencing, email and web-browsing facility and live television broadcast (*NCAOR Annual Report, 2010–11*).

It can be seen that communication has undergone a sea change in the last twenty years or so due to the Internet, making all the systems and methods used till the 1990s outdated.

Waste-Disposal System: Five incinerator-type toilets are provided in Block 'C'. The incinerator unit situated in the toilet flap consists of an insulated incinerator chamber, a crucible tray for waste collection and an automatic kerosene-fired burner. The gas and fumes are removed from the incinerator chamber through a suitable exhaust chimney. Ash is collected in bags and transported back by ship. Kitchen and other combustible waste are destroyed in an outside incinerator. A proper disposal of solid waste is effected in the kitchen, dining hall, MI room and laboratory. Waste water from the kitchen and bathrooms is carried to the two effluent treatment plants where the bio-disc treatment plants 'Klargesters' are located. The plants treat waste water by a bio-degradation process and the treated water is allowed to flow to nearby pits. The water from the pit is pumped once or twice a year to a far-off location from the station.

8.10 A Comparative Assessment of the DG and Maitri Structures

Both the structures faced entirely different environmental conditions six years apart and their broad performance study makes for interesting reading.

DG was designed and constructed in 1983–84 by Ms Structa Ply of UK who had previous experience in building the Halley Station of UK on the ice shelf in Antarctica. Maitri was built in 1988–89 by RDEE, Pune, on an oasis area with a totally indigenous design and construction. Both the stations were made well before Madrid

Protocol came into force in1998 and did not have to go through the process of a Comprehensive Environmental Evaluation (CEE). Decommissioned in 1990 after six years of service, DG has sunk many meters below the surface without a trace. Maitri, designed and built to last about ten years, has been serving even after more than twenty years. The process of CEE, mandated under the Madrid Protocol, will have to be followed as and when Maitri undergoes any rebuilding or renewal.

8.10.1 *Assessment of the DG structure*

The structure of DG was very compact, well built and strong. Being on an ice shelf, its fate was known from day one and it was only a question of how safe and secure it would be till the moment of abandonment/decommissioning. Periodic inspections and actions to strengthen the structure and in particular the roof trusses, wall panels and other components along with careful snow clearance helped the station's structure to withstand snow pressure while sinking. It can be said that the station endured the harsh environmental conditions on an ice shelf and gradual snow burial in exemplary fashion.

DG lasted six years, twice the period envisaged while designing it. With regular maintenance, repairs and replacement of minor items, all the life-support systems, like generators, boilers, snow-melt plant, central heating and deep freezing, functioned with a high degree of efficiency. On both the days – inauguration and decommissioning (25 February 1984 and 25 February 1990 respectively) – the station was functioning with equal efficiency sans ice burial.

As mentioned earlier in the chapter, the excellent quality of the structure and life-support systems of DG deserves high praise.

8.10.2 *Assessment of Maitri-structure*

The construction of the entirely indigenous Maitri was completed during two austral summers (1987–88 and 1988–89). Even today, in 2015, after over twenty years, it is still functioning well past its designated life of ten years. Its design, initial quality of workmanship and maintenance carried out over the years by successive expedition teams deserve praise.

DRDO representatives have been associated in all the expeditions and have worked closely to maintain and monitor the structure and equip it with additional facilities. Like all Antarctic structures, Maitri too needed constant engineering attention. The first detailed assessment of the station and its life-support systems was carried out during the thirteenth expedition in January–February 1994 by DRDO (Gaikwad, 1996, *NCAOR TP No 11*).

NCAOR organised an independent structural assessment during the twenty-third expedition in January-February 2004 (*NCAOR TP No 21*). It was carried out by two organisations, namely Structural Engineering Research Centre, Chennai, (Raghava, 2010), and Engineers India Limited, New Delhi, (Sinha, 2010), based on the following parameters:

- A comprehensive visual inspection of structural components like columns, beams, bracings, walls, floor and roof panels for the extent of distress and damage.
- Existence of cracks in structural components.
- Sagging of floor panels.
- The collection of samples of structural members and materials for study of their usage under sub-zero temperatures.
- Precautionary measures to be taken related to the structure.
- Relevant points from support facilities created outside the main structure.

Both organisations separately carried out a detailed examination of the entire structure, using techniques like accelerometer, impact hammer technique to determine dynamic sensitivity to wind, ultrasonic thickness gauge for columns and ultrasonic impact wave measurement on marine plywood panels.

Separate comprehensive reports on the above and related engineering aspects were submitted to NCAOR. Both the reports make near-similar observations and recommendations. The major points and recommendations in both the reports can be summarised as under:

- The structure has generally withstood the harsh environmental elements well. No drastic localised reduction of thickness was observed anywhere in the columns, ruling out excessive corrosion. Nine numbers of columns of the main block and one number of columns from the B block are found damaged due to longitudinal cracks requiring strengthening or replacement.
- Excavation of soil reveals good condition of the base plates.
- Additional stiffener plates with cap plates for columns and suitable packing plates for beams should be provided where needed and recommended.
- Structural columns on stilts are of telescopic type and of different sizes, the upper portion being smaller than the lower one and the two are welded at the joint. Though no visual defects are observed in the weld between the two pieces, a through bolt with two outside nuts on either side should be provided in the lower column, in the hole immediately below the joint. This precaution will hold the upper column from sliding into the lower column in the worst condition when the welded joint may give way.
- Ineffective sealing of the flooring has resulted in the leakage of hot air in the station and consequent ice formation has occurred below the flooring panels.
- 'A' and 'C' blocks, on the ground, are in a dilapidated condition despite repair works from time to time. These two blocks need total dismantling and redoing. If rebuilt, they should be on stilts of the same height as the 'Main' and 'B' blocks.
- The facilities created outside the main structure need regular assessment. It is difficult to approach these areas during winds or long blizzards due to an absence of a proper pathway with a hand railing.

The above study reflects positively on the structural soundness of the station. However, it makes adverse comments on the lack of preservation of samples from the original construction materials, poor record-keeping and documentation from the initial work and subsequent maintenance of the structure and facilities. All these are vital for such an assessment as also competence-building in Antarctic conditions in the long run. Already it is seen how last station commander of DG had handed over hundred parts to DOD for preservation along with a detailed report. As of now, neither that report nor the parts are traceable. This point needs very close attention of NCAOR as an apex polar body in the country.

NCAOR plans to build a new structure for the winter team, keep the existing main building for the summer team, reorganise the entire complex and carry out a massive cleaning drive in the near future. The process has already started through a public expression of interest. All this will be subject to the requirements of the Madrid Protocol on environmental protection.

8.11 Madrid Protocol and Statutory Requirements for Antarctic Stations

Madrid Protocol, the agreement on environmental protection came into force in 1998, as part of the ATS. The Protocol lays down binding parameters for the comprehensive protection of the Antarctic environment and sets out rules concerning existing and new structures, station governance and human activities. As DG was decommissioned in 1990 well before the Protocol signed, only Maitri is discussed here in detail. Bharati, India's third station, just commissioned in 2012, is discussed separately.

The Protocol defines Antarctica as a 'natural reserve devoted to peace and stability'. It consists of twenty-seven articles and six annexures. The articles contain general rules governing the Protocol, prohibition of mineral resources activities other than for scientific research and various procedures. Annexures contain specific guidelines, processes and actions concerning the establishment, governance and maintenance of research stations. All these are important to understand as the expected standards from the ATS (*Cool Antarctica*, 2012, *The Madrid Protocol*).

Aspects covered by the annexures are as under:

- Annexure 1: Environmental Impact Assessment (EIA).
- Annexure 2: Conservation of Antarctic Fauna and Flora.
- Annexure 3: Waste Disposal and Waste Management.
- Annexure 4: Prevention of Marine Pollution.

- Annexure 5: Area Protection and Management (Introduces the concept of Antarctic Specially Protected Areas and Antarctic Specially Managed Areas).
- Annexure 6: Liability.

Out of all the annexures, Annexure 6 is still not ratified by all the CPs and, therefore, has not come into force as yet.

Guidelines for any Activity by Member Countries: The Protocol lays down a series of procedural guidelines to be implemented by national Antarctic programmes. It requires an Environmental Impact Assessment (EIA) to be applied in the planning and operation of all activities to be undertaken in Antarctica in the following sequence:

Preliminary stage: Activities with minor or transitory impact can proceed after a national review.

Initial Evaluation Stage (IEE): For activities with minor or transitory impact as mentioned above, *an IEE* must be prepared and reviewed through national procedure before implementation.

Comprehensive Environmental Evaluation (CEE): For activities with the likelihood of having more than a minor or transitory impact, for example building or renovating a station, a CEE must be prepared. It must be circulated for the comments of the Antarctic Treaty Consultative Parties (CPs) and the public. It also must be submitted to the Committee on Environmental Protection (CEP), which advises the Antarctic Treaty Consultative Meetings (ATCM) who will take the final decision.

8.12 State of Environment at Maitri

In view of the guidelines imposed under the Protocol, NCAOR has put into place efforts for evaluating the impact of scientific and logistics activities on the environment in and around Maitri with a view to drawing an effective Environmental Management Plan (EMP). It takes into consideration various environmental components like air, noise, water, land, waste disposal and life in the lakes.

Though some sampling was done by Bhabha Atomic Research Centre (BARC) during the eighth (1988–89) to the tenth (1990–91) expeditions, a detailed analysis of the environment at Maitri was conducted during the fourteenth (1994–95) (Nair, 1998, *NCAOR, TP No. 12*) and nineteenth expedition (1999–2000) (Tiwari, 2004, *NCAOR, TP No. 17*) by the National Environmental Engineering Research Institute (NEERI), Nagpur; the twenty-second expedition (2002–03) (Bhave, 2004 *NCAOR, TP No. 20*) by Veermata Jijabai Technological Institute (VJTI), Mumbai; and the twenty-third expedition (2003–04) (Veerbhadraiah, 2010, *NCAOR, TP No. 21*) by Engineers India Ltd, Delhi. All the studies assessed the impact of various scientific programmes and associated logistics support facilities at Maitri on the fragile ecosystem of Antarctica. All the reports have pointed out that the existing systems at Maitri are functioning adequately but can do better, with improvements and modifications, to meet the desired parameters and requirements of the AT and the Protocol on environmental protection (Madrid Protocol).

The areas covered in the assessment were the four blocks of the main Maitri Station, surrounding structures, systems for power generation, water supply, central heating, waste disposal and other station activities. Salient points from these studies are described below.

8.12.1 *Waste-Management Strategy*

The waste-management strategy regarding Indian stations, field camps and the ship while it is in Antarctic waters, is derived mainly from recommendations adopted at the ATCM and based on the guidelines prepared by the Scientific Committee on Antarctic Research (SCAR). The aim of this strategy is to minimise, within practicable limits, any adverse environmental impact on Antarctica while conducting scientific research for peaceful purposes, in accordance with the provisions of the AT and Madrid Protocol. To achieve this aim, the specific objectives of the strategy are the improvement of the existing waste-management practices or procedures in Antarctica or at sea. These practices include:

- Waste avoidance and minimisation.
- Waste reuse and recycling.
- Waste disposal in an environmentally accepted manner and cleaning up of old sites.

8.12.2 *Monitoring of Air Quality*

Ambient Air Quality Monitoring (AAQM) was carried out at both the permanent (exhausts from generator and boiler houses where continuous operation are done and toilet incinerators which are fired for some time during the day) and temporary (exhausts from vehicles and helicopters used during convoys and summer operations respectively) sources causing air pollution.

The AAQM was done for Suspended Particle Matter (SPM), Sulphur Dioxide (SO_2) and Nitrogen Dioxide (NO_2).

It was found that SPM, NOx and SO_2 varied in the range of 6 to 6.97 mg/cu m and the values of the gaseous pollutants as well as particulate matter are on the higher side at locations on the downwind of generators due to fuel burning.

8.12.3 *Noise Environment*

The sources contributing noise likely to exceed the permissible levels were identified in the generator room, boiler room and radio room. The noise level at all these locations was monitored using precision sound level with octave filters. It was also observed that the use of flat rubber strips for tightening and sealing in the wooden summer huts cause enhanced sound level owing to the high-velocity winds prevalent in Antarctica. Though this causes little disturbance in sleep, no appreciable noise was recorded.

8.12.4 *Monitoring of Water Quality*

In order to assess lake-water quality, samples were collected from nine different lakes located within few kilometres' distance from the main station. Physical parameters like the pH (Potential of Hydrogen and measure of how acidic/basic water is), temperature, conductivity and turbidity were estimated immediately after collecting the samples. The dissolved oxygen was also measured. Samples were preserved for an analysis of various chemical parameters, such as nitrate, nitrite, phosphates, dissolved solids, hydrocarbon, heavy metals and oil or grease as per standard methods. It was found that seepage from an unknown source gets into Priyadarshini Lake during December-January and, thereafter, subsides before finally freezing in February. Though the source of this seepage could not be identified, it must be from the adjacent glacier with no harm.

Water samples from twenty-six locations in the Priyadarshini Lake, used as a source of water for human consumption, were collected for physical (odour, pH), chemical (chloride, nitrate, sulphate) and biological (manganese, iron, and bacterial) examinations. The quality of water from the lake was found fit and within all parameters.

8.12.5 *Waste-Water Treatment*

This aspect is very important from the point of water contamination and its effect on the health of personnel, especially during austral winter when the station and its occupants are totally isolated from the rest of the world.

There are two bio-discs in operation, installed in an insulated room. These bio-discs treat the kitchen waste and grey water separately. It is a complete self-contained waste-water treatment plant which consists of the following:

- A primary settlement zone.
- Biological treatment using rotating discs, and
- A secondary settlement zone.

The treated effluents from both bio-discs are temporarily stored in separate ponds close to the station building. A trace-heated plastic pipe with insulation is connected to the outlet end of the bio-discs. The collected waste water in these ponds is pumped out to a location at a higher contour towards the south-west of the main station. The glacier lake is located on the downward slope at a distance of around 210 m from the discharge point. Samples were collected from three locations of this lake to ascertain any contamination occurring as a result of discharged wastewater. Another glacier lake is free from such contamination as it is situated at a higher contour. Samples were collected from three locations of this lake, to provide the reference level.

The effluent discharge of bio-discs B1 and B3 varies throughout the day. Further, the discharge between 2300 hrs and 0600 hrs is negligible as all supply and delivery pumps are shut down during this time. Hence, the effective supply to the two bio-discs is only for seventeen hours every day. Continuous monitoring of effluent discharge for twenty-four hours is not practicable in the Antarctic weather. In view of these constraints, the discharge rate was monitored only twice, once when the discharge of B1 is lean and of B3 at its peak and vice versa. The daily discharge from both these bio-discs is around 3100 litres per day, whereas the quantity of water pumped from the lake is approximately 3500 litres per day. Composite and grey waste-water samples were collected from the inlet and outlet of both the bio-discs B1 and B3 at different times.

The permissible limits and quantities observed during the studies conducted from the thirteenth to the twenty-third expeditions are given in Table 8.1 (*NCAOR TP No. 21*).

Table 8.1: Actual Outlet Quality

Parameter	*Expedition Wise*										*Desired Quality*
	B1					*B3*					
	13th	*14th*	*19th*	*22nd*	*23rd*	*13th*	*14th*	*19th*	*22nd*	*23rd*	
PH	5.5	4.3	5.1	NA	4.5	6.5	7.9	6.9	NA	7.3	5.5–9.0
TSS	124	112	20	NA	131	30	53	26	NA	51	20
BOD	310	444	454	NA	320	400	430	114	NA	100	30
COD	920	1332	828	NA	548	1100	1296	306	NA	208	250

TSS: Total Suspended Solids.
BOD: Biochemical Oxygen Demand COD: Chemical Oxygen Demand.

The parameters for allowable limits have been taken from the SCAR and COMNAP guidelines and general standards issued by the Ministry of Environment and Forests, Government of India, in 1993 on the subject.

Reasons for not achieving the desired results as per the studies are as under:

- Temperature of B1 and B3 should be between 25°C to 34°C, whereas they are between 4°C and minus 20°C.

- Treatment for both the kitchen and other water should be together and not separately.
- Improper piping, inadequate shelter, lack of awareness and proper monitoring and an improper arrangement for discharge.

All the drawbacks are taken care while formulating the Environmental Management Plan (EMP)

8.12.6 *Soil and Sediment Quality*

Soil samples were collected from six locations adjacent to oil and fuel tanks. The samples were stored in polyethylene bags and preserved at 4°C. The physio-chemical parameters like particle-size distribution, soil texture, water-holding capacity, pH, cation (positively charged ions) exchange capacity, exchangeable cations, and the percentage of the organic content were determined as per the standard methods. Hydrocarbon and oil and grease in soils were also estimated to ascertain the leakage or spillage of oil. Sediment samples from Priyadarshini Lake and waste-water discharge ponds were collected from eight different locations by using an acrylic metallic hand-coring device. These samples were wrapped in aluminium foil and refrigerated at 4°C. They were analysed for total carbon, nitrogen, sulphur, oil and grease and hydrocarbon.

8.12.7 *Biological Environment*

Monitoring of the ecological characteristics surrounding the station is mandatory in terms of Article 5, Annexure I, of the Madrid Protocol.

Water samples were collected from ten different places in Priyadarshini Lake and nine different sites from six glacial lakes for analysis.

The abundance and diversity of algae was recorded in different samples with a total count, ranging from 440 to 880 in Priyadarshini Lake and from 440 to 2640 in glacial lakes. The Shannon-Weaver diversity index values for the two varied from 0 to 2.32, which are low, indicating poor productivity, less numbers and an uneven distribution of species.

The zooplankton is at the top of the aquatic food chain and, hence, plays an important role in Antarctic lakes. The total count ranged from 180 to 540 in Priyadarshini Lake and 20 to 100 in the glacial lakes' samples. While the former represented rotifera, copepod and protozoa, the latter represented these and, in addition, nematode. The diversity index varied from 0 to 2.17, indicating non-productive water bodies.

8.12.8 *Solid-Waste Generation*

The particulars regarding the generation of solid waste are summarised as under:

- Approximately 1000 empty fuel drums were held to cater for exigency and to support remote field camps.
- Around 20 kg food, consisting of cooked-food waste, non-vegetarian stale food or any other unusable food items are collected per day during summer and the quantity gets reduced to half during winter.
- The food and paper waste generated at the station is burnt in an incinerator operated at a high temperature.
- The non-hazardous solid waste is segregated into different categories – plastics, metal, glass, ash, can, etc, are compacted, reducing the volume considerably. Around 240 barrels and 1000 garbage bags of the waste generated during March 1994–December 1995 were brought back to India. Similar actions may have been taken subsequently.
- Around 200 litres of liquid waste, mainly photographic chemicals, unserviceable lubricants, oil, etc, are generated in a year. This waste is brought back to India. Similar actions at other times may have been taken.

8.13 Environment Management Plan (EMP)

Based on the above analysis, an EMP, which would go a long way to ensure that all the statutory requirements of the AT and the Madrid Protocol on Environmental Protection are met, was recommended to NCAOR (Tiwari, 2004, *NCAOR TP No. 17* and Chaturvedi & others, 2005) with the following salient points:

- Inculcation of the EMP and its benefits to expedition members.
- Reduction of the amount of waste of all kinds to minimise its impact.
- Enhancing efficiency of bio-discs by using higher-capacity discs and maintaining proper temperatures for plant operations at all the times during the year.
- Leak-proof lining of a waste-water collection pond for preventing the seepage of treated/untreated waste water into the lake.
- Collection of the waste water of summer-camp urinals and bathrooms in a lined tank and its treatment instead of the present disposal on land in an untreated form.
- Burning of wet-food waste in an incinerator at 600°C, instead of burning it in a half bin.
- Labelling of non-hazardous wastes. Empty Automatic Transmission Fluid (ATF) drums should not be used for the purpose of storage.
- Sterilisation of medical waste.
- Compaction of plastic and metal cans, cardboards, etc.
- Use of spill kits while handling fuel. *In situ* bio mediation experiments at areas near generators and oil storage facilities.
- Use of chemical reagents only in the environmental laboratory and a proper back-loading of the waste generated.
- Increasing the roof level of the generator room.
- Proper placement of fuel tanks.
- Keeping of fire extinguishers in summer huts.
- Keeping an additional 50 mm fire-fighting pipe in the generator room.
- Maintaining an automatic fire extinguishing system in the generator and boiler rooms.
- Replacement of the old pipeline system.
- Use of earmuffs where necessary.
- Coating of the walls of the generator and boiler rooms with acoustic plaster or wood-wool to absorb noise and reverberations.

- Appointment of an Environmental Management Officer as a permanent measure.

With this EMP, it can be seen that the environmental situation at Maitri is quite good and in accordance with all the provisions of the Madrid Protocol. One can safely assume that as and when the old complex undergoes a rebuild or makeover, it will be fully in tune with the Protocol guidelines.

8.14 Third Indian Antarctic Station – Bharati

DG, Maitri and Bharati are situated at three diverse environmental zones of Antarctica. DG and Bharati were both built close to the coast, but while DG existed on an ice shelf, Bharati is on rocky promontory. Maitri and Bharati are both on oasis areas but one is on a the coast and the other is nearly 100 km from the coast. Unlike DG, however, Maitri and Bharati are in no danger of being buried under snow. The coastal distance between them is 3098 km which is a major factor for integrated logistics planning. With such two full-fledged permanent research stations in Antarctica, India has become the tenth country to have multiple bases.

Built in the pre-Madrid Protocol era, Maitri has nonetheless made substantial efforts to match the required environmental standards. With Bharati, starting from its conception to commissioning, the whole process has taken nearly ten years to complete all the mandatory requirements of the Protocol.

Bharati became fully functional from the austral summer of 2013. As of now, a brief review is done based on the limited information available (*NCAOR*, 2010, *NCAOR, Through the Years*, 2011and *NCAOR, Annual Report 2010–11, Final CEE- 2010* and Wikipedia 2013).

The relative positions of India Bay (on the coast near the ice-submerged DG from where Maitri is maintained) and Bharati (also on the coast but about 3100 km away in the Southern Ocean indicate the complexity of the integrated logistics required to maintain them from India. This point is illustrated in Figure 8.5 (NCAOR website).

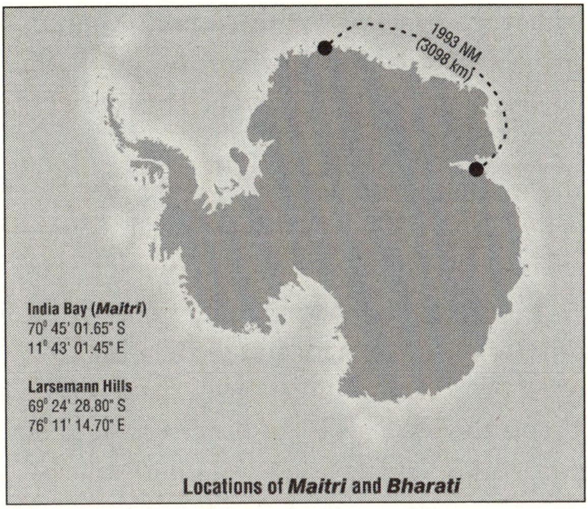

Fig. 8.5: Relative Positions of Maitri and Bharati on the Antarctic Coast

8.14.1 *Background and Selection of site for Bharati*

Once Maitri was fully functional and India was well established as a member of ATS, an expert group recommended the establishment of an additional station. Three areas were suggested for consideration – the Antarctic Peninsula, Filchner Ice Shelf in Weddell Sea and Amery Ice Shelf-Prydz Bay area in East Antarctica. As the peninsula area is overcrowded with existing stations and the Filchner Ice Shelf has considerable logistics constraints, the third option emerged as favourable option.

An expert group on a special mission visited the area between 66°E and 78°E longitude in February 2004 and studied many possible locations. The task force recommended a rocky promontory in the central part of the Larsmann Hills bound by latitudes 69° 24′ and 69° 25′ and longitudes 76° 10′ and 76° 14′ as the most suitable site. The site was selected based on better availability of the following from a logistics point of view:

- Open-sea approach to the site,
- Ease of cargo discharge operations,
- Flat terrain for station construction, and
- Fresh water lakes.

Along with the suitability of the aspects concerning logistics, certain advantages for scientific activities also justify the selection of the site at the Larsmann Hills for Bharati. The major aspects brought out in the final CEE are as under:

- While Maitri is in the sub-auroral zone, the new location is in the auroral zone. This makes the data obtained from both the stations complementary and important from a scientific point of view.
- Magnetic field lines at the new location pass over the Middle East and Europe, close to Hungary and Denmark who have magnetic observatories. The combined data with these would be useful for conjugate area studies. This is not the case with Maitri.
- The earth station at the new location will be very useful for extended coverage over the Indian Ocean and for online transmission of data.
- The Mahanadi Graben, on the Eastern Ghat mobile belt of India, exposes an entire sequence of rock types ranging from middle-late Proterozoic to Permian as a result of the India-Gondwana fit.

Once the exact location was identified, special teams were sent during subsequent expeditions to initiate the area's baseline data collection at the proposed site for a CEE as well as to facilitate the planning of the station infrastructure. A special expedition was undertaken in January-March 2006 to conduct multi-beam bathymetric surveys on the planned approach. Meteorological and oceanographic observations were carried out and topographical maps of the area were prepared.

8.14.2 *Planned Scientific Activities*

The following scientific activities from Bharati are planned:

- Meteorological and atmospheric studies will be done based on real-time data-transmission capability. A high-resolution multi-parameter Automatic Weather Station (AWS) will be installed to acquire data. Other areas include aerosol-radiation-flux estimation using sky radiometer, aerosol-size distribution using multi-channel photometer and the establishment of a geomagnetic laboratory.

- Earth science studies in the area, given the India-Gondwana link, study of Lambert Glacier, sediments records in lakes in the area.
- Supplement detailed ongoing environmental studies at Maitri to evaluate the anthropogenic impact on the Antarctic environment. Studies of the flora, fauna, their population trends and other aspects.
- Biodiversity studies, conservation of terrestrial and aquatic biota and monitoring of the wildlife population and behavioural studies.
- Oceanographic studies in the adjacent Indian Ocean sector and a study of the hydrographical parameters at different locations in the area.

Bharati's close proximity to the research stations of Australia, Russia, China and Romania are likely to have a very positive effect in fostering significant collaboration and cooperation.

8.14.3 *Environment of the Area*

The Larsmann Hills area is marked by persistent, strong katabatic winds that blow from the east to the south-east during austral summer. Daytime air temperatures from December to February at times exceed 4°C. The extreme minimum temperature recorded in the region is minus 40°C. The annual mean and maximum wind speeds recorded at the neighbouring Chinese Zongshan Station are 7 m/sec and 50 m/sec respectively. Precipitation occurs as snow upto 250 mm water-equivalent annually. Pack ice is extensive. The sea ice grows steadily during March-September.

Lakes are mostly saline with low microbial diversity. However, the promontory supporting the station has small fresh-water lakes. The area adjoining the site does not support any population of breeding penguins or seals.

8.14.4 *Preparation of the Comprehensive Environmental Evaluation (CEE), ATCM Approval and Station Planning*

Based on all the available information, a draft CEE as mandated under the ATS was prepared and tabled at the XXX ATCM in Delhi in April 2007. After deliberations, it was endorsed. Accordingly, the final CEE was prepared and circulated to other AT nations.

Simultaneously, planning for the station started in earnest. The expert committee, after a due process, shortlisted four and selected the design offered by a company in Germany, who had the experience of planning the German Station Neumayer III earlier. The design is environment friendly and commensurate with the Protocol for environmental protection. It duly takes into consideration the base-foundation stability, aerodynamic stability, fire protection, energy supply with low-noise generators, heat ventilation and air conditioning, two-stage filtration water supply, recycling and waste-water management, proper lighting and ease of maintenance. It caters for twenty-five people during winter and forty-seven additional people during summer. It has a designated life of twenty-five years and a proper decommissioning methodology. A structural review will be carried out after every five years.

The station building is an elevated three-storey structure supported on columns at the base. The garage is located at back of the building at ground level and is part of the main structure. The three-storey building is aligned in the north-south direction. At the site, the main wind direction is from the east and north-east, which strikes the station transversely. Apart from access through the garage, the building also has two entrances from the eastern and western sides. The station complex comprises a main building, helipad, fuel farm, etc.

8.14.5 *Station Construction*

The pre-construction activities for Bharati commenced in the austral summer of 2009, when the most critical logistics task of establishing the infrastructure necessary for construction was accomplished successfully. All the heavy earthmoving and construction material and cargo were transported from ship to shore either by two helicopters or by vehicles over fast ice. A road about 250 m long from the landing site up to the helipad was carved out using bulldozer, excavators and deflagrating cartridges.

Phase I of construction activities commenced during the austral summer of 2010. Work planned and executed under this phase, as per contractual agreement with M/s Leonhard Nielsen and Sonner AS,

Norway, comprised the following:

- Piling for the foundation of the main building, erection of a garage sump and walls.
- Construction of a helipad.
- Construction of the foundation of the fuel farm.
- Laying of a pipeline for fuel, fresh and waste water.
- Construction of the foundation for a satellite ground station.
- Preparation of approach roads.

Phase II activities involved construction of the main building, life-support systems, laboratories, earth station and other activities. After completion of most of the work, a team of twelve people started wintering from 2012–13. The station was fully functional from the austral summer of 2013.

8.14.6 *Appearance of Bharati and its Surrounding Area*

Though the functional assessment will be available only after a few years, the experience of the last thirty years in running two stations has taught much and the third station, Bharati, is a definite improvement in terms of planning, structural design, life-support systems, sophistication and aesthetic sense as shown in Figure 8.6 (*Final CEE, NCAOR*, 2010).

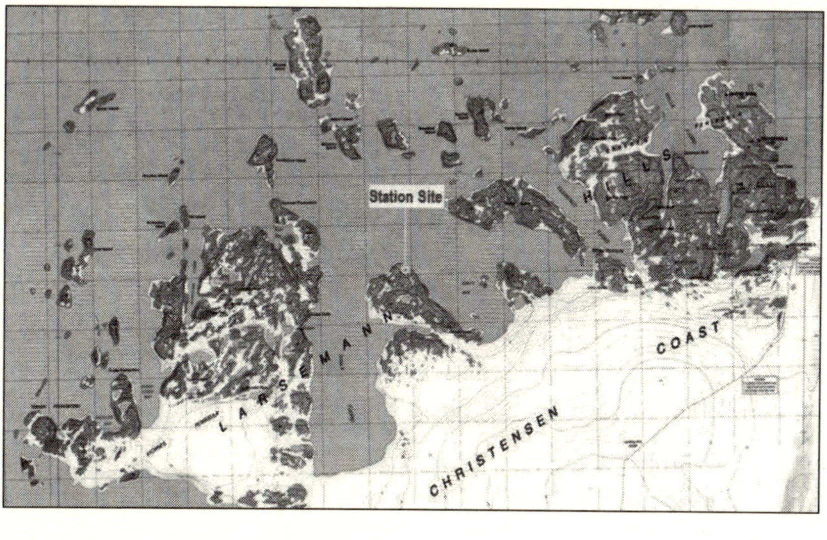

Fig. 8.6: Location, Construction and View of Bharati

8.15 The Logistics Scene for IAP from 2013 Onwards

All activities of Antarctic expeditions fall under two broad categories – scientific research and logistics. Scientific activities consist of those pertaining to projects concerning scientific research and include selection of institutions, projects and personnel/scientists, expedition reports and other scientific tasks. All other activities enabling and sustaining these can be termed as logistics support. Logistics consists of the establishment and maintenance of stations and facilities, provision of vehicles and equipment, selection and chartering of ships and aircrafts, transportation of stores and personnel, needs of expedition members, including training of selected personnel before leaving India, food, clothing, medicines and others, both in India and Antarctica, before, during and sometimes after the expedition, interaction and coordination with organisations, reports and publications and management at the national and international level.

Most countries have developed own institutional framework to manage National Antarctic Programmes (NAPs) to suit their national ethos and political aspirations. In India, the NCAOR, a part of the Ministry of Earth Sciences of the Government of India, acts as a nodal executive agency.

We have already seen the broad requirements of a permanent Antarctic station in the earlier chapter. In the Indian context, overwintering started in 1983–84. Each year, in austral summer, a new expedition arrives. It comprises of a summer and a winter team. The winter team takes over from the existing winter team after a few days of working together. The weary old winter team then returns to the ship after a total Antarctic isolation of well over one year. For the next few weeks till the ship starts its return journey, it either remains on board or assists the summer team or advises the new winter team when required. The new winter team now maintains the structure, life-support systems, such as power and water supply, waste disposal, fire fighting, communications and all other matters of administration for the station which will remain isolated for the major part of the next year. It also carries out or assists in the ongoing scientific tasks.

For the first five winters, only DG (just about 20 km from the edge of the shelf) was operational. During the sixth winter, before it was decommissioned in 1990, both DG and Maitri were operational. During the last few winters at DG, the winter teams had to run a number of convoys to Maitri, about 85 km away, for rendering help in building it. Since then, the old area around DG serves as a supply base to keep stores unloaded from the ship. From 1990 onwards, the winter team at Maitri runs vehicular convoys between the station and the supply base at the DG location to tranship fuel and other stores to Maitri. All these logistics tasks require skill and experience due to the difficult environmental conditions.

While all this continues at Maitri, from 2013 onwards, India has had to cater for the logistics requirements of Bharati, too, about 3100 km from the shelf near old DG location now serving as supply depot to

Maitri. However, unlike Maitri, Bharati is just about 200 m from the rocky berthing point at Larsmann Hills and does not need convoys to take the stores from ship to station.

In all the years since India started its Antarctic expeditions, distinct patterns of logistics support emerged as under:

- From the first to the eighteenth expedition (1981–1998), there wasa uniform pattern. Every year, the ship with the summer and the next winter team started from Goa around end-November. The voyage to Antarctica took about a month or so depending on pack-ice conditions in the Southern Ocean. After carrying out all the logistics and scientific tasks for about two months during the austral summer, the ship returned with the summer and the previous winter teams.
- From the nineteenth to the twenty-fifth expedition (1999–2005), teams would travel by air from Goa to Cape Town in South Africa, from where the sea voyage would start. This reduced the overall travel time by about ten days or so.
- From 2006 onwards, after India joined DROMLAN, the task of moving personnel from Goa to Antarctica was carried out entirely by air, whereas expedition ships would start from Goa some members and bulk logistics stores. Then started the expeditions to Larsmann Hills for the third Indian station Bharati, which entailed the help of two ships for three years – one for the usual annual requirements for Maitri and one with construction material and equipment for Bharati. This combined arrangement of the ships and DROMLAN for transportation of personnel continues till today.
- But once the rebuilding of Maitri takes place in the near future, different arrangements may have to be made, with changed requirements, at least for a few years before a uniform logistics pattern is established for Maitri and Bharati. The logistics cycle may again undergo another transformation if and when India decides to build its fourth Antarctic station. Its location – the peninsula, West Antarctica, inner polar plateau or somewhere else – would make a huge difference to all the logistics calculations.

It would be seen from the ensuing discussion that if India is to succeed in this, the present thinking and execution system may have to undergo considerable change from over-reliance on outsourcing to self-reliance.

- All the above concerning self-reliant logistics is easy said than done. As we have seen in earlier chapters, the logistics support systems developed by the countries intimately involved in the continent are the result of considerable time, effort and money spent based on a long-term national strategy to build institutional polar competence. It seems China has understood the polar game well and it is time for India, too, to do so and do what it takes to be counted among the leading polar nations in the world.

8.16 India's Position in the Development of Logistics

After over three decades of continuous presence and the experience of maintaining two stations on the ice shelf and oasis area, India's third station Bharati is established on a simpler version of the oasis that is a costal promontory. But the new station is over 3000 km from Maitri (which is about 10,000 km from India) which adds considerable complexity in providing logistics support to two stations from India. The next logical step for an effective and influential Indian presence in Antarctica may be the establishment of a station on the inland polar plateau. So it may be prudent to examine the position of major logistics elements such as station structures and life-support systems, ships, helicopters, vehicles and convoy expertise, food, clothing, medicines, communications, travel time, personnel and other elements of the IAP. It will throw light on India's real strengths and weaknesses in Antarctic logistics in comparison to other countries.

8.16.1 *Stations' Structures and Life-Support Systems – An Overview*

The structure of DG was designed and fabricated in the early 1980s by a company in the UK which had previously built the British Antarctic station. A team from the Indian Army, trained in UK and Germany, constructed the station in one austral summer. As the station functioned on the ice shelf in the pre-Madrid Protocol period, the emphasis was on getting a foothold in the continent. Focus on safety and logistical convenience and not on environmental matters was the norm those days. The structural design and life-support systems were of very high quality and the station functioned admirably. It was functioning flawlessly when it had to be abandoned after six winters. In its years in service, RDEE, Pune, was associated with its structural maintenance and modifications to fight the difficulties caused by gradual snow burial, gaining much-needed and valuable Antarctic experience in the process. The last station commander at DG handed over samples of over one hundred components of DG to the Department of Ocean Development as examples of quality for future guidance.

Following limited Indian individual and institutional experience with DG, Maitri was a complete Indian effort by RDEE, Pune, in terms of design, fabrication, construction and maintenance. Built in two phases over two austral summers, it was commissioned in 1989 and though it was designed to last for ten years, it is still in national service. Understandably, a number of changes and modifications have been carried out over the years. The structure and the systems have been ageing for a while now and soon actions will be initiated to rebuild the station. It is a commendable feat and testimony to the expertise displayed by the RDEE, Pune, India.

Bharati, on the other hand, reverts once more to the reliance on overseas outsourcing. It is designed by a firm in Norway and fabricated and constructed by a firm in Germany with their manpower over a period of two austral summers. The construction firm has previously built the German station Neumayer III in Antarctica a few years back. Bharati has been fully operational since 2013. At the outset, it looks very impressive, conforming to requirements of the Protocol. However, thc efficacy of the structure, life-support systems, environmental friendliness and other functional aspects can be reviewed only after few years of its existence. In a way, instead of building further on the national experience and expertise, India has chosen to forgo the advantages while building Bharati outsourcing before full stop.

8.16.2 *Ships*

The details of ships chartered from 1981 to 2010 for Antarctic expeditions are shown in *Appendix V* (Mishra, 2012).

India totally depends on chartering and has used twelve ships from twelve countries over the years. Antarctic expeditions need ice-strengthened or icebreaking ships and India does not have one till today. The recent rescue of a Russian ship from thick pack ice in the Southern Ocean by countries, including China, indicates their efforts to acquire Antarctic expertise.

8.16.3 *Aircrafts and Helicopters*

India uses aircrafts and airfields from organisations such as DROMLAN and ALCI and has not developed any indigenous expertise.

The details of helicopters chartered from 1981 to 2010 for its Antarctic expeditions are shown in *Appendix VI (*Mishra, 2012).

India has so far used nine types of helicopters. Up until the fourteenth expedition (1994), Indian Air Force (IAF) and Indian Navy (IN) helicopters were used; in the eighteenth and nineteenth expeditions, helicopters from the Indian company, Pawan Hans, were commissioned. As of now, IAP depends upon chartering helicopters as in the case with ships. This may be detrimental to building self-reliance in terms of equipment as also for institutional competence within the country.

8.16.4 *Vehicles and Convoy Expertise*

In the first few years of IAP, a number of vehicles were imported. When *DG* was decommissioned in 1990, fifteen snow vehicles, namely five Pisten Bullys (German), two Snow Cats (US) and, one Liebherr Crane (German), one Schmidt Snow Cutter (UK) and one Foremost Crane (Canada), five Snow Scooters namely with - three Scandic and two Bombardier, and six sledges namely with - namely articulated and fixed (German and Indian) were handed over to Maitri. At Maitri, no vehicles are required except those to run convoys and cranes, thus reducing vehicular usage. As of now, the latest version of a Pisten

Bully with 330 hp is the only vehicle in use for convoy purposes. These have made convoy operations between Maitri and the supply depot at the shelf, on the crevasses-ridden route, much smoother.

Till 1993–94, old vehicles used to be brought back for an overhaul at an army base workshop in Pune. A total of eleven vehicles in six years (1987–1993) were brought back, overhauled and sent back with the next expedition (*Army Workshop Report,* 2012). This is a proven method of building equipment expertise as part of developing major logistics capability and effect considerable savings. This has since been discontinued. Now an engineer from Pisten Bully manufacturers (Kassaboherer, Germany) visits Maitri every austral summer to train the technical personnel and to help in training and repairs. This again goes contrary to the principles of self-reliance and building institutional expertise.

8.16.5 *Food, Medicines, and Clothing*

Over the years, NCAOR arranges all the food and medicine items through indigenous sources. DRDO had developed a number of items based on initial experience (DRDO, 1997, *Special Issue of Technology Focus on DRDO in Antarctica*). However, many items of regular use, too, like boots and overalls are also imported, perhaps an easy but not good solution from country's point of view.

8.16.6 *Communications*

Communication has undergone a sea change over time. As of now, a satellite communication link between Maitri and NCAOR has been established to create an archetype in logistics and scientific operations, planning and coordination. There is 24x7 connectivity @ 1 mbps, provided to facilitate online collaboration, real-time data transfer, video conferencing, email and web browsing and live television broadcast. Like they did with Maitri in 2007, Electronics Corporation of India Ltd. (ECIL) has setup a ground station in Bharati for these facilities. With this, the Antarctic experience of only a couple of decades ago has changed dramatically for the better.

8.16.7 *Travel Time*

Travel time was reduced first in 1999 when the Antarctic team was flown from Goa to Cape Town to start expedition ships from there. With the entry of DROMLAN and ALCI on the scene, now teams fly from Goa to Mumbai (1 hour), Mumbai to Cape Town (8 hours), Cape Town to Novo (6 hours) and then go to Maitri just 15 km away in a Pisten Bully vehicle. This has made a great difference in terms of mobility, time availability for scientific activities and other aspects. However, India does not possess any air transportation capability of its own, unlike the US, Russia, UK, South Africa, Australia, Norway and some other countries.

8.16.8 *Personnel*

Details of personnel are shown at *Appendix VII*. After thirty years, a total of about two thousand people have been part of Antarctic expeditions with a third of them as winterers (Mishra, 2012).

Due to the use of helicopters from private agencies, personnel from the Indian Navy (IN) and the Indian Air Force (IAF) had stopped being a part of Indian expeditions from 1995. Till the twenty-sixth expedition in 2006, some personnel for logistics tasks were from the Indian Army. Since 2007, personnel from the Border Roads Development Organisation (BRDO) and retired army personnel have been recruited for these tasks on a contractual basis. The association with DRDO has also been reduced to a great extent.

This, like in the case of helicopters, may put total stop in building institutional expertise in Antarctic logistics in India.

8.17 Element of Self-Reliance and Logistics as a Military Operation other than War

The concept of Antarctic logistics as a military operation other than war can be effectively used in enhancing national strategic interest. Several countries due to and during their presence in Antarctica over a long period have built a self-reliant logistics network, be it aircrafts, ships, vehicles, equipment or personnel. The Antarctic presence has also been used to develop institutionalised expertise of the armed forces in polar logistics operations. Though IAP has progressed in various aspects of bureaucracy, scientific programmes and creation of a network to provide logistics support, it does not seem to have given attention to the aspects of self-reliance and using the management of Antarctic logistics as a military operation other than war. This is in stark contrast with other countries associated with Antarctic matters since decades. India is still dependent on outsourcing most of its vital logistics needs and this may not be very useful in times of crisis. A report by an expert group formed by the Government of India to study the ten-year work profile of the IAP in 1995 dwells on the long-term scientific programme and logistics facilities, but is silent on the need for self-reliance or India's strategic interests as mentioned in the NCAOR mission statement (Rao, 1995). Over the last few years, the Indian armed forces' Antarctic connection, too, seems to have been totally or severely curtailed, affecting the creation of their institutionalised expertise. The merits or otherwise of this aspect and how the Indian armed forces and the country may lose out on an excellent opportunity to train itself in polar logistics and operations if not associated with IAP, have been brought out in previous chapters.

8.18 Performance of NCAOR vis-à-vis the Mission Mandate

As pointed out earlier, the mission mandate of NCAOR and, by extension that of the IAP, is 'To plan, promote, coordinate and execute the entire gamut of polar science and logistics activities of the country in order to ensure a perceptible and influential presence of India in Antarctica and to uphold our strategic interests in the global framework of nations in the southern continents and the surrounding oceans.'

The mandate makers' probably intended for India's Antarctic presence to be perceptible (able to be seen or noticed) and influential (dominant, having great influence) laying equal emphasis on executing the scientific and logistics activities to protect the country's strategic interests. An analysis of the organisation, expedition reports and activities indicate that all its efforts and attention are concentrated on scientific activities in a disproportionate one-sided emphasis.

Leave aside the association, representation given to armed forces on the governing council is laughable. As for logistics, barring a few exceptions, India is totally dependent on outsourcing. Does this sort of IAP management, protect India's strategic interests? Does a totally dependent logistics network really protect the national strategic interests in a situation where logistics accounts for ninety per cent of resources, time and money in any Antarctic venture the world over? How will the near total outsourcing without any logistics institutional involvement from indigenous sources make the Indian Antarctic presence either perceptible or influential in any manner? Perhaps, the NCAOR or the ministry need to re-evaluate the activities and priorities on ground to see if they conform to the words of the mission mandate.

It seems obvious that in order to have a perceptible and influential presence in Antarctica and to uphold India's national strategic interest, logistics based on self-reliance and institutional expertise is critical. Towards this, an association of the armed forces or associated institutions should form an essential part of the operations, as is the case with many countries.

On this vital aspect, NCAOR seems to be unaware or undecided on its own mission mandate. Their present governing council of eleven members consists of a couple of bureaucrats, scientists and one member from the armed forces, that too from the Naval Research and Development (*NCAOR Annual Report 2013–14*). There is a special research advisory committee and finance committee but logistics does not seem to find a place for obtaining advice from real experts. So one can safely assume that either the long-term strategic interests of the country are not on the NCAOR radar or agenda or they include the successful execution of scientific programmes only as a part of IAP, notwithstanding what the mission mandate states. As a participant at the National Conference on Science and Geopolitics of Arctic and Antarctic (SaGAA 2011) Dr Rasik Ravindra, the then Director NCAOR, in an article titled 'India in Arctic and Antarctica and its Geopolitical Significance' (*Science and Geopolitics of Arctic and Antarctic: SaGAA,* 2011), states while concluding the article that, 'Considering that India is not a claimant nation in the Antarctic

Treaty regime, there is a need to define our long-term goals, apart from scientific and research aspects. India needs to develop a long-term strategy for Polar Regions, especially as many South Asian countries have started active exploration strategy both in Arctic and Antarctica.' Coming from the director of the apex body in the country, the stark contrast between mission mandate of the NCAOR and what they are doing on ground, does not need any further elaboration.

If India is really serious about the protection of its strategic interests through the NCAOR mission mandate, maybe some major rethinking is required.

8.19 Public Awareness about Polar Areas in India

Unfortunately, in India the awareness level about polar-regions is very poor. Dissemination of information on the subject via teaching, the presence of books in libraries and others means are glaringly absent. It would not be out of place to narrate a couple of the author's experiences spaced over quarter of a century. Before departing for the Antarctic expedition in 1988, the author was purchasing warm clothing in a Delhi shop. When queried as to why so particular, the author said he was on his way to Antarctica. The shopkeeper looked surprised and said that he has seen quite a bit of the world but had neither heard nor visited any such place. Years later, in 2008, the author, on one of his visits to NCAOR, went for a walk. At a spot, within a stone's throw distance from NCAOR, he met an educated person and asked him if he knew about Antarctica. The person had never heard of such a place. Antarctica may have changed but the level of awareness about it in India stands mostly unchanged after quarter of a century.

8.20 In Conclusion

In just over thirty years, IAP has come a long way with considerable progress on various aspects of the country's Antarctic quest. With the establishment of three Antarctic research stations in three out of four areas with different logistics parameters, India has gained very valuable Antarctic experience. The experience includes the running and decommissioning of a station (DG) in very harsh environmental conditions; operating a station on an inland oasis (Maitri), requiring extensive convoys running all through the year and other exacting logistics parameters; and establishing a station (Bharati) about 3000 km from Maitri on a costal promontory after successfully completing all the exacting requirements of the AT and Madrid Protocol. India also has taken effective steps so that its stations maintain and protect the pristine Antarctic environment. By association with about seventy institutions, including the armed forces and viable scientific projects, it has acquired scientific and logistics handling skills. Very valuable experience has also been gained by the DRDO through establishing Maitri.

On the bureaucratic front, it has established a nodal agency, NCAOR, to handle all Arctic and Antarctic matters on behalf of the country. As of now, it is one of the ten countries with multiple stations, a membership of the AT with CP status and important participation in all its organs in the ATS.

India, however, is overdependent on overseas organisations and resources for vital logistics needs like structures, ships, aircrafts, helicopters, vehicles, etc. The experience in building Antarctic structures and systems, gained through the successful establishment of Maitri, was a major lesson and could have been consolidated if Bharati's construction had not been outsourced. The same has happened in case of other logistics' needs. So, though expertise on some of the operational needs has been gained, dependence on outside sources may pose problems in the event of a crisis in the geopolitical situation, especially also since IAP's association with the armed forces seems to have been greatly curtailed after the few initial years. While India should continue the present system, self-reliance may be the only solution to guard the long-term strategic interests of the country. It should, therefore, examine how to utilise its Antarctic presence for enhancing national security by treating logistics as a military operation other than war while working within the Treaty parameters for scientific research. It is never too late.

Lastly, effective steps should also be instituted to enhance public awareness about polar areas in India.

Chapter 9

The Emerging Global Picture and the Year 2041

"I am hopeful that Antarctica in its symbolic robe of white will shine forth as a continent of peace as nations working together there in the cause of science set an example of international cooperation."

Admiral Richard E. Byrd (1888–1957)
(Pioneering American aviator, polar explorer and organiser of modern polar logistics)

9.1 Background

Summerland Beach on Phillip Island near Melbourne is like any other beach in Australia but for its four thousand little inhabitants, attracting droves of tourists. The smallest among the eighteen species of penguins found in the world, they form an indigenous Australian variety. Just 33 cm tall, they leave the island well before sunrise and return to their sand dune burrows after daylong fishing at sea. Day after day, this occurs exactly at the time of sunset. A hush prevails among the audience sitting at specially erected viewing stands as the little creatures make their appearance on waves, waddle to the shore and walk to their nests with a typical gait. Watching the parade with my two little Melbourne-born granddaughters during one of my visits, I and a few other spectators wondered just how ancient this ritual was. As we pondered, the penguins, oblivious of the special status and protection accorded to them and their burrows, walked to their hideaways.

Apart from the daily spotlight they enjoy on their return from fishing duties, the lives of these penguins might not have seen any change for thousands of years. But their global ethnic brethren and other animals living a few thousand kilometres away, in the icy conditions of Antarctica and the Southern Ocean, are not so lucky. Their life is tough and humans have made it tougher and more dangerous since their appearance on the scene two centuries ago.

Though the scientifically proven evolution period of Antarctica goes back well beyond 200 Ma, no human being had seen or touched the continent till Bellingshausen became the first human to sight it in 1820 and Henryk Bull made the first landing in 1895. Before that, for quite some time, voyages had mainly centred around sealing and whaling due to their commercial lure. The scientific process of mere thinking about evolution to proven theory, however, had passed through many interesting phases. It included the concept of a sphere-shaped earth before BC (Pythagoras and Aristotle), flat-earth orthodoxy of medieval Europe, hypothesis from the arguments of thinkers and scientists and Edward Suess' naming of the southern supercontinent as Gondwanaland (1885) based on early evidences after a tribe in central India. Suess, however, could not explain the mechanism and this was done by Alfred Wegener through a masterly analysis of continental drift in his book *On The Origins of Continents and Earth* (1915). Only after physical exploration started that fossils and other objects, like marsupials and a shallow-water amphibian, to support Suess' early evidence of glossopteris, could be collected to conclusively establish the existence of Gondwana, with Antarctica as the hub and Africa, South America, Australia, New Zealand and India as conjoined components. All this culminated in a widely accepted modern theory of plate tectonics, developed later in the last century, completely changing our understanding of the earth.

There was a spurt in exploration at the end of nineteenth century up until the first two decades of the last century, known as the Heroic Age. Amundsen became the first human to reach the South Pole with his team in December 1911. Many other noteworthy explorers like Scott (who reached South Pole one month after Amundsen), Shakleton, Mawson and others too made their substantial contributions. However, exploration was mostly on the periphery of the continent and the interior was largely unknown. This period also gave rise to territorial claims from seven nations, repudiation from others and a few minor conflict situations. The two World Wars (WW I and WW II) were a period of lull as they diverted attention from the continent. Byrd and a few others introduced the organised use of motorised vehicles and aircrafts, changing the old concepts and techniques of polar mobility and exploration. After the end of WW II, the US signalled the strategic importance it attached to the polar areas by organising two massive military and scientific exercises, Operations High Jump and Wind Mill during 1946–48 in Antarctica.

Eighteen months of International Geophysical Year (IGY, 1957–58), as a sequel to the International Polar Year (IPY, 1933–34), set the stage for international scientific cooperation in Antarctica, keeping aside political divides. For the first time, an overland crossing of the continent was made (Vivian Fuchs) and for the first time the South Pole was reached (Edmund Hillary), fifty years after Scott's ill-fated expedition. The thickness of the ice cap (over three kilometres deep) was measured for the first time, with sound waves using explosives, to the disbelief and then astonishment of even scientists. For the first time, massive data on scientific experiments in many research spheres was collected. Realisation also dawned that conventional colonisation would not be possible in Antarctica due to its harsh environmental conditions precluding any permanent inhabitation. This fact prompted negotiations leading to the birth of the Antarctic Treaty and its instruments, collectively known as Antarctic Treaty System (ATS).

By 1961, Antarctica entered the present system of world governance and an international presence in the form of year-round permanently maintained research stations. It also commenced the era of collaborative scientific research and an awareness of the need to protect the pristine Antarctic environment in global interests, due to its interdependence with the rest of the world. Coincidently, around the same time, the theory of plate tectonics was also proved to its finality. It was a case of exploration, geopolitics and science working in tandem.

In previous chapters, in the light of scientific progress and physical exploration, we have studied, reviewed and analysed in detail the natural and anthropogenic environmental aspects of the entire area covered by the Antarctic Treaty (AT/Treaty), that is, the area south of 60°S latitude, including the continent proper and the surrounding Southern Ocean, from both global and Indian perspectives.

It is revealed that any impact to the environment of Antarctica has repercussions on the rest of the world. For example, the ozone hole, noticed only since the late 1970s, and changes in the Antarctic ice, both continental and sea, have serious implications for worldwide climate and geography. Regarding the anthropogenic aspects, world governance in Antarctica through ATS, considered a model in international diplomacy, has its own drawbacks and hiccups, such as voluntary compliance without any legally binding liability for non-compliance or damage, enabling defaulters to get away. For example, the clause in the International Whaling Commission (IWC) states that whale hunting is banned but that whales can be killed for scientific purposes. The loophole created through this led to the complaint by Australia against Japan which had to be settled by the International Court of Justice (ICJ). Another example is the non-ratification of the liability annexure in the Madrid Protocol by all Consultative Parties, preventing its coming into force. This happens nearly to all international agreements. Environmental changes at the other end of the earth, that is, the Arctic Ocean also have the potential to adversely affect the stability of the Antarctic Treaty System.

Though all individual elements like ships, aircrafts, vehicles, communications and others have changed considerably due to technological advances, Antarctic logistics has retained its pre-eminence, in terms of resources, time and money requirements.

India's Antarctic quest and presence is recent, but it has made concerted efforts and succeeded in establishing a presence on the continent and in the system of world governance. However, the Indian Antarctic Programme (IAP), too, needs to address international complexities to attain, maintain and enhance the effect and influence of its Antarctic presence with matching logistics. IAP, however, has not made serious efforts in this regard. All this makes the study interesting and complex with considerable global ramifications.

9.2 The Emerging Antarctic Kaleidoscope

Through detailed discussion and examination of all the environmental aspects, a global picture around Antarctica emerges with both natural and anthropogenic aspects intertwined together. Sometimes the picture is clear, at other times, it is uncertain, hazy and cloudy, obscuring the future. Also the picture has continuously changed over the years and the trend may endure as new scientific findings frequently appear, mainly due to climate change. The picture seems more like an Antarctic kaleidoscope with different overlapping shades. It can be categorised into seven major shades – an unstable continent and the surrounding ocean, global warming and Antarctic climate, international presence and the ATS, Antarctic logistics and its pre-eminence, the situation in the Arctic, the Madrid Protocol and likely situation by year 2041.

It would be interesting to see what pointers these shades of the kaleidoscope, shown in Figure 9.1, offer, based as they are on all the discussion, data and analysis done earlier.

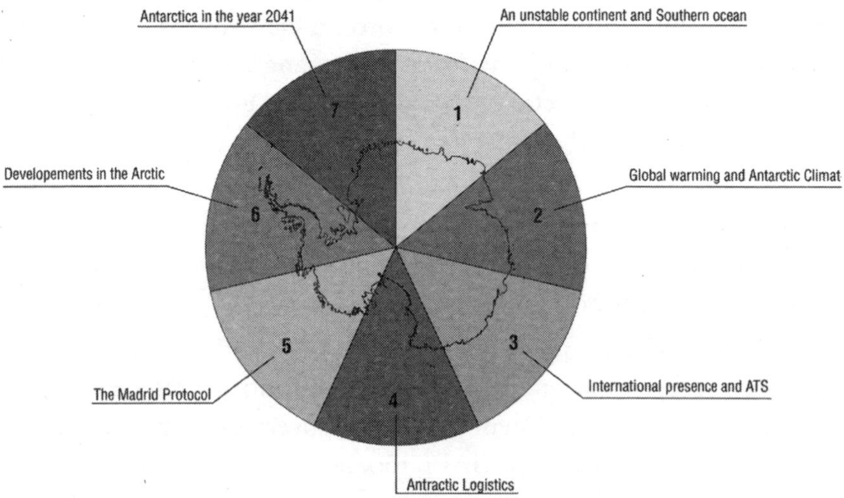

Fig. 9.1: The Antarctic Kaleidoscope

The extent of space each shade occupies at present in this pie-shaped spectrum and how that may change over the years with time and the changing geopolitical situation, is a subject of a separate study.

9.2.1 *Shade 1: An Unstable Continent and the Surrounding Southern Ocean*

Geological research in Antarctica is still limited to only a miniscule percentage of the total area of the continent, essentially to the ice-free oases strewn all over. Traces of ample mineral resources, similarly found in other Gondwana counterparts, are found here but their extent is not known. The massive ice cap makes physical exploration very difficult. It is the same with the offshore find of hydrocarbons in the Ross Sea area. Of course, with deep drilling on a commercial basis, this can be done. But the Madrid Protocol totally prohibits any commercial mining till it runs its course in 2048. Antarctica still contains many geological secrets under its ice.

The enormous ice cap, comprising about ninety per cent of the world's total ice, the thickness varying from just under a kilometre to a few kilometres deep, is a storehouse of millions of years of climatic history. This information gives details about the earth's climatic evolution, vitally important to further science and its application in forecasting. The ice cap gains from snowfall, bottom freezing and other ways and loses mainly through the calving of icebergs. Current data shows a considerable ice loss in the western Antarctica ice sheet (about twelve per cent of the total ice in Antarctica) and a slight gain in eastern Antarctica (eighty-eight per cent). Frequently, there are new articles, findings and surveys on the subject. As of now, there is a near balance between gain and loss, slightly in favour of the former. There are varying projections as to what may happen to ocean levels the world over if these ice sheets melt, with serious consequences to the many low-lying parts of the world.

While life on the continent proper is limited to plants like mosses, lichens and microscopic invertebrates, conditions in the surrounding ocean present a different picture. The marine environment is richer and abounds with life. Interdependent food chains, with nutrients, plants and animals, together form a self-sustaining giant food web with krill at the centre and whales, seals, penguins, birds and fish as other elements. The seasonal pack ice expanse over the Southern Ocean (about 16 million sq km) is between the permanent or minimum fast ice (4 million sq km) and maximum pack ice expanse at the end of winter (20 million sq km). This forms the basis for the phytoplankton bloom as the primary food source in the giant food web. It is only due to the massive ocean-freezing cycle every winter that salty heavy water flows down, enabling a vertical mixing of and circulation between world oceans, sustaining benthic life. Though krill has not lived up to its earlier expectations in terms of human taste and consumption, it is still seen as a potential food source should the world face a further spurt of population and depletion of conventional food sources. The next few years will definitely see further research and new findings on the subject. Sealing and whaling has now mostly stopped and maybe their stocks will recover in the future for a better eco-balance.

Though studies for additional geological secrets might be limited for a while, those concerning glacial ice, ice cores, sub-glacial lakes, pack ice and individual elements of the giant food web, oceanic circulations and various aspects of the Antarctic climate will remain major areas for research in the continent, mainly due to their importance and environmental interdependence with the rest of the world. The world will be wary of any changes caused by global warming and its potential to disturb the present balance in any of the aspects mentioned above.

9.2.2 *Shade 2: Global Warming and Antarctic Climate*

The warming of the earth's atmosphere, mainly due to the burning of fossil fuels and its likely consequences, has engaged serious world attention in recent times. The earlier chapters discuss the reports from Scientific Committee on Antarctic Research (SCAR), the last two Intergovernmental Panel on Climate Change (IPCC) reports in 2007 and 2014 and various climate conferences of the last few years and conferences of parties (COP 1 at Rio to COP 21 at Paris). The melting of the Arctic sea ice is already attributed to global warming. It is now widely accepted that unless checked from asttainment of the limit of a 2°C temperature rise, popularly referred to as a tipping point, consequences like the melting of ice from the western Antarctica ice sheet, glacier ice melting in the Himalayas and other places and the release of greenhouse gases from the Siberian permafrost (mostly more dangerous methane) may occur, triggering further warming. Consequences may be felt in terms of danger to low-lying areas in the world, depletion of river waters and crop production and an adverse effect on the giant food web and ocean circulations. Countries in the world are unanimous about the effects of the ozone hole and its likely adverse effects, particularly in the southern hemisphere, and a number of treaties and outside measures are in place.

Realising the gravity of the situation arising out of climate change, the world attention was focused on COP 21 held in Paris under the aegis of the United Nations Framework Convention on Climate Change (UNFCCC) from 30 November 2015 to 11 December 2015 to reach a globally acceptable and binding agreement to combat problems relating to climate change and consequent global warming. It is to the credit of over 140 heads of state/governments and over 40,000 delegates from 196 countries that COP 21 has adopted an agreement focusing on mitigation measures to keep the rise in earth's average temperature well under 2° C (called as tipping point) by the year 2100. The agreement also includes the transfer of technology for generating renewable energy resources, grants from developed to developing countries to the tune of US $100 billion per year from 2020 and other measures. The temperature has already risen nearly by 1° C since the pre-industrial base year of 1870, warranting considerable sincerity, transparency and generosity amongst the nations, rising above parochial considerations in near future, to achieve objectives of the Paris Agreement.

So in the scenario concerning global warming, its effects on the Antarctic climate, ice cap and other factors, the success or failure of COP 21 would play an important part, adding more complexity to this shade of the kaleidoscope. As pointed out in earlier chapter, entry and exit of leaders like Donald Trump and Francois Hollande respectively in near future are sure to cause global worry for sure. One can only hope that serious danger to successful implementation of the Paris accord does not occur.

9.2.3 *Shade 3: International Presence and Likely Fissures in the ATS*

There has been a global engagement with Antarctica since early days, whether it pertained with its discovery, exploration, early indiscriminate slaughter of seals and whales, territorial claims laid by various countries, local conflicts and the dawning realisation of the futility to colonise the continent in a conventional sense. The enactment of the Antarctic Treaty and its instruments were global events, wherein countries declared their intention to safeguard the pristine environment and the fragile Antarctic ecosystem and engender peaceful co-existence through permanent research stations for promoting science.

As of now, of the fifty-three acceding countries to the AT, thirty of them maintain about seventy permanent stations. About ten of them, including India, maintain two or more stations. Many international agreements for specific purposes like the protection of whales, ozone layer, dumping of waste and others, though not part of the main Treaty, contribute towards the overall environmental cause. Even with a few hiccups, the ATS is holding strong, though it needs monitoring.

Interest, influence and presence-wise, the AT countries can be divided into few distinct groups. Seven countries (UK, New Zealand, France, Australia, Norway, Chile and Argentina in chronological order) have made formal territorial claims. Though the claims are frozen while the Treaty is in force, most of these countries maintain stations in their claimed areas and try to protect their interests through active presence. Two non-claimant, but influential, countries – the US and Russia, do not recognise any claims already made and have reserved their rights to make their own claims. South Africa and Brazil have declared their zones of interests but not made claims. China and India are late comers but have established a major presence within a short span of time. China has even established a station on the interior polar plateau like the US and Russia. Countries like South Africa and Norway manifest their commercial interests with organisations such as Antarctic Logistics Company International (ALCI) and Dronning

Maud Land Air Network (DROMLAN). Most of the other countries maintain their Antarctic presence as interested parties in world affairs. These groups may influence Treaty operations at times.

Even with countries having differing long-term perceptions about their presence, it seems, in present-day Antarctica, a sort of tranquillity has been reached by the member countries of the Treaty, with the common goal of pursuing scientific research wait and watch policy.

9.2.4 *Shade 4: Antarctic Logistics*

We have seen that with technical advancements in different aspects like station structures, ability to extend life with measures like stilts and skis, life-support systems, ships, aircrafts and landing systems, snow vehicles and equipment, clothing, food, medicines and communications, Antarctic logistics has come a long way. The hardships of the early explorers is a thing of the past, travel time is greatly reduced and expeditions are safer. People can now visit the continent within a very short time. Also, the tourism industry has started and is growing by the day. However, given the continent's isolation and its harsh environment, Antarctic logistics still retains its overwhelming importance, requiring a major part of resources, effort, time and money.

Antarctica can be divided into four areas where permanent stations exist. The two types of areas, which are on oases (coastal promontories and hilly oasis areas near ice shelves), are in no danger of snow burial and the stations there are long lasting. Stations on the other two types of areas (ice shelves and interior polar plateau) are on ice and have to be abandoned and rebuilt elsewhere after a while due to snow burial. Logistics for stations near the coast is comparatively easy. Logistics for inland oasis stations is difficult and for stations on the polar plateau in the interior polar areas, many hundred kilometres from the coast, is extremely difficult requiring high level of logistics organisation and expertise.

As we have seen in earlier chapters, the strength of any country in its Antarctic presence can truly be gauged by its ability to maintain permanent year-round stations in the interior polar areas. If logistics are in place, stations can be maintained anywhere and scientific programmes will simply follow. As of now, only five stations are located on ice in the interior polar plateau and these belong to the US, Russia, France/Italy, Japan and China. Out of these only France is a claimant country. Whereas the first four have established very elaborate and competent logistics systems of their own, the Chinese station will soon join the group. These countries plus the six others which have made territorial claims, and a few others, are well-versed with matters Antarctic and have developed all-round institutional skills, including the non-military use of their armed forces, and are well-equipped for the time of reckoning as and when it approaches.

Countries like Norway, South Africa and Russia provide services on a commercial basis to those who have no logistics network of their own through recently established international organisations like ALCI and DROMLAN. This has also added a commercial angle at the governments' level along with the private sector's growing tourism industry.

9.2.5 *Shade 5: The Madrid Protocol*

The failure of the Convention for Regulation of Antarctic Mineral Resources Activities (CRAMRA) in the 1980s paved way for negotiations leading to the signing of the Madrid Protocol as a part of ATS in 1991 and its coming into force in 1998. Though the Protocol is binding till 2048, AT itself will be reviewed in 2041, so maybe both will be reviewed together. This of course depends on how strong the ATS will be by then and the future developments in the other six shades of the kaleidoscope.

The Madrid Protocol is a comprehensive and legally binding document which designates Antarctica as a natural reserve, devoted to peace and science. While prohibiting mining, it subjects all activities to a prior assessment for their environmental impact. Stringent provisions require, among other things, a Comprehensive Environmental Evaluation (CEE) before new stations are established or old stations are rebuilt or major modifications are carried out. Its Committee for Environmental Protection (CEP) advises the Antarctic

Treaty Consultative Meetings (ATCM) among other provisions. Annexures to the Protocol detail specific measures and procedures for all activities from members of the AT.

Due to this Protocol, the environment of Antarctica can be safeguarded from possible adverse environmental effects from research stations, tourism industry and others. Even organisations like the International Association of Antarctic Tour Operators (IAATO) have made guidelines for environmental protection and abide by the Committee of Managers of National Antarctic Programmes (COMNAP) directions, auguring well for the cause. The only glitch is that, even now, the last annexure of the Protocol containing the liability clause, has not been enforced due to its non-ratification by some members. However, as of now the Protocol is holding well due to wholehearted support from the CPs.

Unless disturbed, this Protocol may ensure a sort of environmental and geopolitical balance in Antarctica, at least till 2048.

9.2.6 *Shade 6: Developments in the Arctic and Antarctic Treaty System (ATS)*

WW II had just ended when the then US President Roosevelt felt that the next world battlefield might be Arctic and the country needed to be prepared. This led to massive US exercises in Antarctica, the enactment of IGY and the signing of the Antarctic Treaty in the following years. The successful execution of the AT and its instruments can be seen as holding hope in the arena of international politics.

In stark contrast, whereas Antarctica is an isolated icy continent surrounded by the Southern Ocean, Arctic is an ocean surrounded by many countries. Antarctica's mineral wealth is unknown due to its massive ice cap, while the Arctic is known to have twenty-three per cent of the world's known oil and natural gas reserves. As the Arctic Ocean water was always frozen (about four meters or so), it was not possible to exploit the resources till recently. Global warming has truly impacted the Arctic Ocean like nowhere else and it is expected to be largely ice-free for shipping in summer sometime between 2020 and 2050. This may make exploitation of natural resources and shipping in Arctic a commercial reality in the near future. Shipping through the Arctic has the potential to completely change sea-transport equations between Asia, Europe and Americas by reducing distances and rendering bottlenecks like the Malacca Strait, Suez and Panama Canals redundant.

Smelling this, five littoral (US, Canada, Russia, Norway and Denmark) and three neighbouring (Finland, Iceland and Sweden) countries formed the Arctic Council in 1996, claiming combined sovereignty, despite having conflicts among themselves. As per the constitution, observer status at the Arctic Council can be granted to countries who accept this position. Till 2013, a total of twelve countries have been granted observer status, including China and India. Surprisingly, an application from the European Union (EU) for the status has been rejected twice, though many EU countries are individual members of the Arctic Council.

So far, a sort of normalcy prevails, but as the effects of global warming are likely to become more pronounced in the Arctic, in the near future, the situation may rapidly change. Opening of Arctic sea routes, possibility of oil and gas exploitation, changes in the global trade and commercial situation and differing national perceptions, may cause serious changes in geopolitics.

The situation in the Arctic is fluid as compared to the one in Antarctica, and has the potential to affect peace in the ATS. The developments in the Arctic would keenly be watched in the near future.

9.2.7 *Shade 7: Antarctica in the Year 2041*

As of now, it can be said that, despite minor hiccups, the ATS is stable and holding well. It continues to enjoy the broad support of all member countries. The AT will be up for review in 2041 and it would be interesting to see the present position of various counties.

Most members of the AT would like to play a waiting game till 2041 for different reasons. However, they keep a keen eye on any developments contrary to their perceptions and expectations from the Treaty. Claimant countries are well established (mostly in their claimed areas) and would like the current situation to continue undisturbed and wait for the review. Countries like the US, Russia and others have gained an excellent Antarctic foothold and may like to give more attention to the situation in the Arctic and would, therefore, welcome a status quo in Antarctica and wait and watch the developments. Some countries like Norway, South Africa and Russia would like to utilise their advanced Antarctic skills for commercial purposes and have given an indication for doing so by establishing organisations like ALCI. Countries like Japan would be keen to safeguard their commercial fishing interests. Eleven countries, including India, have signalled coordinated logistics actions by establishing consortium like DROMLAN. Countries like China are making concerted efforts to join the select group of countries with stations on the interior polar ice and create highly developed Antarctic logistics. Other countries wish some sort of Antarctic presence and would try to consolidate their positions within Treaty parameters. All the countries wish to participate in scientific activities in Antarctica with a view to advance their scientific research capabilities and world standing.

However, aspects like global warming and the fluid situation in the Arctic may add new twists and dimensions to the situation in Antarctica. As of now, the effects of global warming are more pronounced in terms of ice melting in the Arctic Ocean and Greenland. The rate of ice melting in West Antarctica is more than in East Antarctica but, overall, the ice gain is marginally more than the loss. Paradoxically, contrary to the Arctic, pack ice in Antarctica in the last few years shows a slight increase. The picture here is uncertain and hazy and, perhaps, in the next few years, definitive trends will be available from scientific findings and the situation on ground. IPCC and future climate summits will be viewed with seriousness. We have also seen a fluid situation in the Arctic so far as natural resources and Arctic sea routes are concerned.

All these and other unpredictable developments in geopolitics have the potential to create fissures in the ATS. Add to that the growing world population, food and crops situation, the giant food web dynamics, effects of global warming on low-lying countries and the position of natural resources in the world, status of renewable energy sources like solar, wind and wave, making the variables in the Antarctic kaleidoscope open to dramatic changes.

9.3 In Conclusion

As the shades of the Antarctic kaleidoscope keep changing, we are sure to witness new developments in the next few years leading towards 2041. A change in every shade has the potential to create a disturbance in the overall situation, thereby, affecting the Treaty and its instruments. Every country will be keeping a close watch on the situation in Antarctica, developments in the Arctic and the combined effects of all these as they will have an impact on the entire world.

Along with its scientific programmes and effective participation in the ATS, India may do well to lay down an appropriate game plan for becoming self-reliant in all its Antarctic logistics operations and execute the same to perfection by creating its own excellent logistics network. Only then, like many leading nations, will it also be ready for an effective and influential role in matters Antarctic and in the process ensure that its strategic interests are fully protected.

While the world will keep a keen watch on the execution of the recently concluded climate-change agreement at COP 21 in Paris (December 2015), the year 2041, like the years 1957–58 (IGY), 1961 (Antarctic Treaty) and 1998 (Madrid Protocol), may become another watershed year in Antarctic governance.

CHAPTER 10

India and the Antarctic Kaleidoscope
(Preparing for the Future)

"Nations have no permanent friends and no permanent enemies, only permanent interests."

(Old British adage)

The previous chapters have studied, examined and analyzed the various aspects of Antarctica and the Southern Ocean south of 60° S latitude - the area covered by the Antarctic Treaty (AT) - and their relevance to India. It would be pertinent to see what India needs to do to safeguard its global strategic interests in relation to the seven shades of the kaleidoscope as discussed in the earlier chapter in view of the constantly changing geopolitical situation.

10.1 Global Importance of Antarctica and the Southern Ocean and Year 2041

If there is one subject which can truly be called global in every sense, undoubtedly, it would be the continent of Antarctica. The seven shades of the Antarctic kaleidoscope have amply demonstrated the powerful impact that Antarctica and the Southern Ocean exert on the entire planet earth, primarily from an environmental point of view. The nature and intensity of this impact would, of course, depend upon the success or otherwise of the mitigation measures the world takes to implement the recently concluded Paris Agreement.

Data, analysis and predictions gathered by various international bodies like the Scientific Committee on Antarctic Research (SCAR), the Intergovernmental Panel on Climate Change (IPCC) and others, pinpoint the consequences of global warming on the enormous ice cap in Antarctica that contains ninety per cent of the world's ice. They also bring out the impact on the giant food web in the Southern Ocean and the oceanic circulation between oceans of the world, endangering marine life and food security, and likely fissures in the Antarctic Treaty System (ATS) due to the changing situation in the Arctic.

On the favourable side, the studies rate as highly successful the ATS and the continued use of the continent as a world laboratory for the development of science, monitoring of phenomena like the ozone hole and for the unravelling of the many secrets Antarctica still holds for the benefit of mankind.

Successful implementation of thetee agreement at the Conference of Parties in Paris (COP 21) in December 2015, under the aegis of the United Nations Framework Convention on Climate Change (UNFCCC), will hold the key in restricting global warming within limits to obviate or lessen the impact of catastrophic effects.

Presently, though the situation in Antarctica seems tranquil due to the presence of ATS and the Madrid Protocol, conditions can change should there be any disturbance in any unseen environmental and geopolitical dynamics. The global environment may witness a battle between hope and apprehension in the coming years, but whatever the case may be, the entire gamut of ATS will be up for review in the year 2041.

All countries will surely endeavour to create and maintain an effective and influential presence in Antarctica in the intervening period, demonstrating their power by the number of stations, ensuring self-reliance in logistics and through advanced scientific programmes. While developed countries, in the forefront since the beginning of Antarctic exploration, will strive to maintain and cement their existing strong positions further, latecomers to the Antarctic, like India and China, will endeavour to catch up with them and many others will make efforts to establish an effective foothold.

10.2 Position of India's Antarctic Presence and Major Recommendations for National Consideration

While the world views Antarctica and the Southern Ocean from angles that suit their purpose, India's concerns focus on the need for enhanced food security for its ever-increasing population, depleting stocks of non-renewable energy resources and the development of new ones, containment of global warming to minimise environmental disasters at home and its urge to attain global prominence based on self-reliance in the twenty-first century. And in each of these aspects, closely connected with the national strategic interests, self reliance in the seventh continent may play a significant role.

Antarctica, a subject of such significance, however, receives less than scanty attention in India in stark contrast to other leading countries. If India is to assume its rightful place in the comity of nations, it needs to make its Antarctic presence assume prominence and priority. This can happen only through the creation of a polar vision and doctrine, a long-term plan and its sustained execution over a period of time.

The following points analyze the current situation and delineate a game plan with a view to consolidate India's position in the continent.

10.2.1 *Public Awareness and Participation*

There is hardly any literature on Antarctica by Indians. Any other reading material on the subject, too, is either non-existent or sparsely included in the Indian education curriculum. Ignorance of the subject, even among the educated, in a sense, is societal myopia and is reflected in a very low public awareness level. To bring Antarctica and its strategic importance in to mainstream national thinking, it needs to be included in the education system at all levels, from school to university, at a doctoral and post-doctoral research level. Those writing articles and books on the subject should be given special grants. Libraries need to be stocked with books on polar areas in various languages. Polar-related adventure, tourism and other measures should be given a boost. All these measures will go a long way in increasing general awareness levels. Once some of them are implemented, the media can do the rest. Lately, there are a few signs of change but are very slow in coming. With the national vision and polar doctrine in place, public interest and involvement at all levels in matters Antarctic will not lag behind.

10.2.2 *Determining National Vision and Polar Doctrine*

The mission mandate of the apex body in India, directly functioning under the Union Ministry of Earth Sciences, the National Centre for Antarctic and Ocean Research (NCAOR), talks about ensuring a perceptible and influential presence in Antarctica with a focus on polar science and logistics in order to uphold India's strategic interests. But the ground situation seems different. In a recent speech, the then director, NCAOR, lamented on the failure to define our long-term Antarctic goals apart from those in science and research. He must have had logistics and strategic interests in mind when he said this. Certainly, there is an urgent need for a wider debate to determine a national vision towards Antarctica and its relevance to the strategic interests of the country. It should be followed by a white paper and a long-term plan for its execution, irrespective of the party in power. Along with this, other countries' actions, the role of armed forces and the need to build institutional expertise within the country should be carefully considered. We have seen earlier how Antarctic logistics can effectively be used as a military operation other than war. Other vital additions may be made with informed debate on the subject involving experts from various fields. Plans devised with these aspects in mind together with their consistent execution over a period of time may ensure India's perceptible and influential presence in Antarctica and safeguard its strategic interests. A thorough analysis of the Indian Antarctic Programme (IAP) and the role of NCAOR on this basis may be useful. However, pending the national vision and the doctrine as mentioned above, the present mission mandate may be used as the yardstick for analysis of the IAP.

10.2.3 *Indian Antarctic Programme (IAP) – Positive Aspects*

Since it commenced in 1981, the progress of the IAP is very impressive. Over thirty expeditions, continuous continental presence, establishment of three permanent scientific research stations with different environmental parameters and viable research programmes have all contributed to a valuable Antarctic experience.

India has taken effective steps towards maintaining the pristine Antarctic environment at its existing stations Maitri and Bharati with a duly developed Environment Management Plan (EMP), following the procedures and measures stipulated by the Madrid Protocol.

India has established a matching bureaucratic structure with NCAOR as a nodal agency for IAP, functioning directly under the Union Ministry of Earth Sciences. It is an active member of the Antarctic Treaty (AT/Treaty) and its instruments. It enjoys an important position in the Scientific Committee on Antarctic Research (SCAR), Committee of Managers for National Antarctic Programmes (COMNAP) and Committee on Environmental Protection under the Madrid Protocol (CEP). India's keen interest and its recognition can be seen from the fact that within a month of its acceding, it was granted the membership of the Treaty with Consultative Party Status (CP). Its viable scientific research programmes from Maitri and Bharati, located over 3000 km from each other on the Antarctic coast, in cooperation with other countries' stations in the vicinity, augur well for the IAP. Enhanced public awareness and wider institutional and universities' participation would be immensely helpful to take the polar areas in to national conscience.

10.2.4 *Indian Antarctic Programme (IAP) – Worrying Aspects*

While all the achievements of the IAP make good reading, a major cause for worry, from a national strategic point of view, is its inability to follow the principle of self-reliance. India totally depends on outsourcing and lacks its own logistics network support in the Antarctic, concentrating all its efforts on the development and execution of scientific programmes. For India's first station Dakshin Gangotri (DG), it imported the structure and life-support systems, but built it with trained army personnel. Subsequent maintenance till decommissioning was done by the army with technical support from the Defence Research and Development Organisation (DRDO). DRDO designed and fabricated the second station Maitri and its systems and provided maintenance requirements. Well past its designed life, the ageing Maitri is still working but awaits rebuilding. The third station, Bharati, was completely outsourced, from design, construction to subsequent maintenance, thereby depriving vital opportunity to DRDO or any other indigenous organisation to build on the Maitri experience and acquire further polar experience and expertise. As of now, in every major element of Antarctic logistics, India is ill equipped at self-reliance.It charters ships for expeditions, imports all its vehicles and equipment, which until a few years' ago used to be overhauled by the army for reuse, but even that successful practice has since been discontinued. Earlier, for some years, helicopters from the Indian Air Force and the Indian Navy were used, but that too has been discontinued. For many years, personnel from three branches of the Armed Forces used to be an integral part of the expeditions, a practice that has either been minimised or eliminated. Employing retired personnel from the Army or Border Roads Organisation will not build institutional expertise desirable to protect national strategic interests. While the consistent association of the armed forces with Antarctic expeditions will enable them to build institutional knowledge, skills, polar expertise, outsourcing of vital logistics needs will deny them the experience. India uses services of organisations like Antarctic Logistics Company International (ALCI) and Dronning Maud Land Air Network (DROMLAN) for air and sea transportation of goods and personnel. Comparatively small countries like South Africa and Norway use such organisations to advance their commercial interests also. India's total dependence on outsourcing, strikes a sore note on this score. In short, India's capability in all vital Antarctic logistics requirements, based on indigenous resources, is next to zero. Barring few exceptions, it is same case with other equipment and services.

10.2.5 *Role of the National Centre for Antarctic and Ocean Research (NCAOR)*

NCAOR, while running the IAP, has been able to fulfil its mission mandate on developing viable scientific programmes consistent with the norms of the ATS, including the Madrid Protocol. In the process, the country has become one of the ten prominent members of the ATS, maintaining two or more permanent year-round research stations in Antarctica. However, at the same time, it has fallen far short of fulfilling its aims in Antarctic logistics, leaving a vacuum in the area of securing the strategic interests of the country.

With the above in mind, the following actions consistent with the existing mission mandate of NCAOR, are suggested:

- A sincere appraisal of the mission mandate without any parochial considerations to determine where NCAOR stands today with reference to the above analysis of the IAP. With their present organisational structure, NCAOR may require competent help to understand and act in the strategic interests of the country and develop long term logistics network based on self-reliance.
- An analysis of India's thirty-five years of Antarctic presence with a view to assess the need for more stations, ongoing scientific research, strategic interests, cost and benefit examination and the like.
- Based on the above, initiation of a national debate to determine polar vision and doctrine.
- Continued active participation in the functioning of various ATS fora.
- Total support and adherence to the ATS in ensuring peaceful and non-military use of the continent, transparent and collaborative scientific research, free exchange of information, ban on commercial mining, minimising pollution from humans, including tourism, exploitation of renewable resources within laid down limits and stringent measures for environmental protection without any interference.
- The effective maintenance of the two permanent research stations, Maitri and Bharati, in accordance with provisions of the Treaty and the Madrid Protocol to preserve the pristine Antarctic environment.
- Development of a logistics network based on self-reliance to maintain the two stations and future ones, based on national policy, as suggested above.

- Establishment of an impressive tourist centre in Goa (like the Antarctic Centre in Christchurch, New Zealand) as a subsidiary of NCAOR, to promote IAP and public awareness. It may also include models of Indian Antarctic stations, a world-class library and a knowledge centre for polar studies.
- Continuance with the Arctic expeditions and keeping a vigilant watch on the happenings in the Arctic region, Arctic Council and effects on ATS.
- Development of an excellent website to serve as a one-point reference for any polar student, adventurer or reader on any aspect of polar studies.

10.3 In Conclusion

India has attained a good position as an important member of the Antarctic Treaty and its instruments, and has a major Antarctic presence, with two permanent year-round stations and well-organised scientific programmes. But to make that presence effective and influential, the country needs to develop its own logistics network based on self-reliance. Unless that is accomplished, our Antarctic presence would be meaningless, unlike that of many countries who are currently in the forefront. Fortunately, there seems to be a tranquil situation till the 2041 review, unless conditions take a turn for the worse in any of the shades of the kaleidoscope we have seen earlier. Important advantages, too, may accrue with the development of a self-reliant polar logistics network and the country will be ready should a disturbing situation arise anywhere in the world. The association and participation of armed forces personnel in polar expeditions should be viewed with the angle of using them in a military operation other than war to build institutional expertise. There, indeed, is a need for consideration as to what polar expeditions should attain for the country in the long run and how they can become an instrument in the protection of the country's strategic interests. Public awareness and national debate leading to polar vision and devising a polar doctrine may go a long way in that direction. A critical appraisal of the mission mandate and the actual ground situation, concerning the

NCAOR, after thirty-five years of Antarctic experience, may serve the country well in preparing for future involvement and the year 2041. As an apex body on polar regions, NCAOR may do well to take a lead with a broad national outlook.

There needs to be a higher level of awareness about polar areas in the country. Understandably, a study on a subject like Antarctica, which is still mostly uncharted and hardly debated in India, will raise more questions than answers, but that is to be expected as India started its Antarctic quest only in 1981, unlike many countries with polar involvement, experience and resultant expertise over centuries of involvement.

In this book, an effort has been made to present various aspects of Antarctica, accompanied with available data and evidence and some conclusions, recommendations and suggestions have been made for India's future association with Antarctica and its northern counterpart, the Arctic.

It is sincerely hoped that the book is found worthwhile in generating further interest in the Antarctic for the initiation of the recommended actions and for inspiration for future work. Should that be the case, the resultant actions will go a long way in achieving a meaningful, effective and influential Indian presence that will safeguard India's strategic interests in the world through its presence in Antarctica.

Epilogue

International perspectives on Antarctica have undergone a sea change since it was first sighted by Admiral Bellingshausen in 1820. From it being viewed as an arena of indiscriminate slaughter of seals and whales, leading to their near extinction, to its gradual geographical discovery, leading to a race to reach the South Pole during the Age of the Heroes and laying of territorial claims, Antarctica has come a long way today as an arena of global cooperation and peace. As the realisation of the impossibility of conventional colonisation dawned, nations got together to conduct the highly successful and landmark International Geophysical Year (IGY). Subsequent enactment of the Antarctic Treaty, its instruments and the Madrid Protocol, resulted in Antarctica becoming the world's scientific laboratory, signifying international cooperation despite political differences, with the aim of protecting its pristine environment. The highly successful current system of international governance has resulted in a sort of a geopolitical tranquillity in the continent. There are factors, however, which can seriously disturb the present situation, such as the effects of global warming, ice melting in the Arctic, issues hotly debated at the recently concluded Paris Agreement on climate change and upheavals in the global political environment. As of now, however, though the massive Antarctic ice cap still hides many secrets, we know almost everything the continent is likely to offer.

The twenty-first session of the Conference of Parties (COP 21) help in Paris came to an end in December 2015 with an agreement on climate change. Journalists from all over the world have widely reported about how a change of the word 'shall' to 'should' in the sentence, 'Developed country parties *shall* continue taking the lead by undertaking economy-wide absolute emission reduction targets', paved the way for the text to pass the final scrutiny of 195 countries plus the European Union. In fact, it changed the nature of the UNFCCC (United Nations Framework Convention on Climate Change, under whose aegis the COP took place), as it mandated the Common but Differentiated Responsibilities (CBDRs) from a binding obligation into a guiding principle, freeing them to act at will. But this is not new in international diplomacy. Ambiguity leaves room for individual interpretation, obviates punitive actions and puts the onus on voluntary execution. In the absence of this, ratification, like in the case of Annexure 6 (the liability clause) of the Madrid Protocol, does not occur and execution is halted. Only loose language, accommodating differing views and perceptions, can lead to wider acceptability and agreements. This reality will continue in the international arena.

In the Paris Agreement, the main issues revolved around the execution and enhancement of voluntary emission reduction targets, known as Intended Nationally Determined Contributions (INDCs), action to transfer renewable energy technology mainly by the 'should' countries to needy ones, effective mechanism for the creation and utilisation of a green climate fund and other concerns. Statistically, as per IPCC estimates, 2900 giga tonnes (GT) of CO_2 can be emitted from pre-industrialisation level (1870) till the year 2100 if global warming is to be kept within the threshold limit of 2°C, known as the tipping point. Unless preventive actions are taken, the average rise in global temperature by year 2100 will be 2.7°C with likely catastrophic effects. On ground, by the end of year 2011, the world has already emitted 1900 GT, leaving 1000 GT as the remaining global carbon emission budget till the year 2100. Another 748.5 GT is likely to be emitted by year 2030, finishing seventy-five per cent of its carbon budget till the end of the current century. This will leave hardly any carbon space for developing countries to grow.

Surveys, studies and predictions by leading world organisations like UNFCCC, IPCC, NASA, SCAR, WMO, World Bank and others also make disturbing observations on climate change, like 2015 being the hottest year on record since 1850, a sharp increase in extreme heat, rainfall, wind and ice events, disruptive agriculture, spread of diseases, rising seas, draught, over a hundred millions more pushed in poverty and other predictions. On this front, as far as the Indian subcontinent is concerned, the IPCC has warned about the likely depleted river-water situation due to receding Himalayan glaciers, setback to agriculture produce, increased natural disasters, resultant political instability and likely conflict situations in the Indian subcontinent if global warming is not contained.

All these complications continue to cause worry and it will be interesting to see how they are tackled once the implementation of the Paris Agreement commences in 2020.The world is sure to be very attentive towards research and indicators in these areas. Even partial success to arrest global warming will positively impact conditions in Antarctica, the Southern Ocean and its giant food web. Failure is not an option.

India's strategic interests are interwoven with the vital stakes in the outcome of every major aspect under discussion. Success of the Paris Agreement at keeping global warming within permissible limits has to be seen from this angle. It is vital not only for individual countries, but also for preserving the Antarctic environment and the stability of the Antarctic Treaty and its instruments and, in particular, the Madrid Protocol.

India was once a part of the supercontinent of Gondwanaland along with Antarctica, a fact known to very few people in India except few scholars. In a country beset with gigantic problems, with extremely poor polar awareness, this talk of establishing a very strong Antarctic presence as explained in the book, is easier said than done. But if done, it may serve a great deal in India's aspirations to secure a pride of place among the comity of nations in the twenty-first century. In this case, India may do well to settle on the expression 'shall' and not 'should' in its attainment of goals on the world stage. India's polar doctrine to safeguard its strategic interests and maintain an influential presence in Antarctica, based on logistics self-reliance, should be viewed from this perspective rather than from parochial institutional considerations. The alternative will not be pleasing to future generations.

Appendixes

APPENDIX I: Timeline of Antarctic Exploration

(Landis, 2001 and additions from the author for 1946, 1991 to 2048 and Note 4)

600–330 BC : Greek philosophers develop theories that earth was a sphere with equally distributed land masses and climatic zones, including the North and South Polar Regions.

AD 150 : Claudius Ptolemy writes the influential *Guide to Geography* which was translated from Greek to Arabic in the ninth century, and subsequently from Arabic to Latin by Spanish translators sometime before 1300.

1519–22 : Ferdinand Magellan undertakes the first circumnavigation of the world; Magellan sails through the strait that now bears his name in November 1520.

1577–80 : Sir Francis Drake discovers his namesake sea passage in September 1578.

1719–20 : George Shelvocke sails to 61°30' South on a voyage that inspires Samuel Taylor Coleridge's *Rime of the Ancient Mariner*.

1738–39 : Jean-Baptiste Charles Bouvet de Lozier discovers Cape Circumcision, Bouvetoya Island in January 1739.

1768–71 : Captain James Cook's *Endeavour* expedition explores Tierra del Fuego in January 1769 and searches for *Terra Australis Incognita*.

1771–72 : Yves-Joseph de Kerguelen-Tremarec discovers Iles Kerguelen.

1772–75 : James Cook's *Resolution* and *Adventure* reach 71°10' South; Cook claims South Georgia and South Sandwich Islands for England.

1777 : James Cook publishes his two-volume narrative *Resolution.*

1819 (Feb. 19): William Smith discovers South Shetland Islands.

1819 (Sept.) : *San Telmo* wrecks in the Drake Passage with loss of all hands.

1819 (Oct. 16): William Smith returns and claims South Shetland Islands.

1819–21 : Admiral Thaddeus von Bellingshausen circles Antarctica with the *Vostok* and *Mirni;* sights the continent on 27 January, 1820.

1820 (Jan. 30): Edward Bransfield and William Smith sight Antarctic Peninsula.

1820 : Nathaniel Palmer explores Deception Island with the *Hero;* sights the Antarctic Peninsula on 17 November 1820.

1821 (Jan. 7) : John Davis sails the *Cecilia* to Hughes Bay on the Antarctic Peninsula and lands on the continent.

1822–24 : James Weddell travels to Antarctica with *Jane* and *Beaufoy*; reaches 74°15' South on 20 February 1823.

1830–32 : John Biscoe expedition with the *Tula* and *Lively* sights Enderby Land and discovers Adelaide Island.

1837–40 : The French Antarctic Expedition, led by Jules-Sebastien-Cesar Dumont d'Urville, sails the *Astrolabe* and the *Zelee* to the Antarctic Peninsula and eastern Antarctica; discovers Adelie Land.

1838–39 : John Balleny discovers Balleny Islands with the *Eliza Scott* and *Sabrina.*

1838–42 : Charles Wilkes heads the United States Exploring Expedition, sailing with the *Vincennes, Porpoise, Peacock, Flying Fish, Sea Gall* (Peninsula only), and *Relief* (Tierra del Fuego only) to the Antarctic Peninsula and eastern Antarctica.

1839–43 : James Clark Ross discovers the Ross Sea, Ross Ice Shelf, Victoria Land, and many other Antarctic landmarks with the *Erebus* and *Terror*.

1892–93 : The Dundee whaling expedition, with the *Balaena, Active, Diana* and *Polar Star*, sails to the Weddell Sea in search of whales.

1893–95 : Henryk Bull sails the *Antarctic* to the Ross Sea to hunt for whales.

1897–99 : Adrien Victor Joseph de Gerlache heads the *Belgica* expedition, exploring the Antarctic Peninsula and the Bellingshausen Sea.

1898–1900 : Carsten Borchgrevink sails with the *Southern Cross* to Cape Adare and spends the winter on the continent.

1901–03 : Erich Dagoberth von Drygalski spends the winter in eastern Antarctica on the *Gauss*.

1901–04 : Otto Nordenskjold heads the Swedish Antarctic Expedition on the *Antarctic* to the Weddell Sea and Snow Hill Island.

1901–04 : Robert F Scott heads the *Discovery* expedition, spends two winters on Ross Island, and treks toward the South Pole.

1902–04 : William S. Bruce heads the Scottish National Antarctic Expedition and sails on the *Scotia* to the Weddell Sea and the South Orkney Islands.

1903–05 : Jean-Baptiste Charcot, heading the French Antarctic Expedition, sails the *Francais* to the Antarctic Peninsula and spends the winter.

1907–09 : Ernest Shackleton heads the *Nimrod* expedition, sails to Ross Island, and sledges toward the South Pole.

1908–10 : Jean-Baptiste Charcot returns to the Antarctic Peninsula on the *Pourquoi-Pas* and spends the winter.

1910–12 : Roald Amundsen sails to the Bay of Whales on the *Fram* to trek to the South Pole.

1910–13 : Robert F. Scott heads the *Terra Nova* expedition, sailing to Ross Island for a sledge journey to the South Pole.

1911–12 : Wilhelm Filchner sails the Deutschland into the Weddell Sea pack ice.

1911–14 : Douglas Mawson heads the Australian Antarctic Expedition, sailing on the *Aurora* and spends two winters at Commonwealth Bay.

1914–16 : Members of Ernest Shackleton's expedition sail the *Endurance* in to the Weddell Sea pack ice.

1914–17 : The other half of Ernest Shackleton's expedition, the Ross Sea Party, sails on the *Aurora* to Ross Island.

1921–22 : Ernest Shackleton sails to South Georgia on the *Quest*.

1928–30 : Members of the Richard E Byrd's party spend the winter at Little America on the Bay of Whales, to fly to the South Pole in November 1929.

1928–30 : Members of the Richard E Byrd's second expedition to Little America II; Byrd stays alone at Bolling Advance Station.

1933–39 : Lincoln Ellsworth leads expeditions that explore four Antarctic regions by air.

1939–41 : Richard E Byrd heads the United States Antarctic Services Expedition to Little America III.

1946 (Dec. 2): International Whaling Commission (IWC) signed.

1946–47 : Richard E Byrd undertakes Operation High Jump.

1957–58 : The International Geophysical Year (IGY), an eighteen-month year of Antarctic science, is held.

1956–58 : Vivian Fuchs of the Commonwealth Trans-Antarctic Expedition crosses the continent from the Weddell Sea to the Ross Sea, with help from Edmund Hillary.

1959 (Dec 1) : Antarctic Treaty signed.

1964 : The Agreed Measures for the Conservation of Antarctic Fauna and Flora (AMCAFF) is enacted.[1]

1965 (Dec.) : The last whaling station on South Georgia closes.

1967 : Researcher from Ohio State University discovers jaw fragments of a four-legged, lizard-like amphibian, final proof that Antarctica was indeed part of Gondwana two hundred million years ago.

1978	: Convention for the Conservation of Antarctic Seals (CCAS) is enacted.[2]
1979	: Fatal Air New Zealand crash on Mount Erebus kills all 257 passengers and crew.
1980	: Convention on the Conservation of Antarctic Marine Living Resources (CCAMLR) is enacted.[3]
1982	: Fossil of a rat-size marsupial is found on Seymour Island, near the tip of the Antarctic Peninsula, evidence that marsupials migrated from South America to Australia via Antarctica.
1988	: Convention for the Regulation of the Antarctic Mineral Resources (CRAMRA) signed.[4]
1991	: Protocol for Environmental Protection to the Antarctic Treaty (Madrid Protocol) signed.
1998	: Madrid Protocol comes into force after ratification by all the Consultative Parties (CP).
2041	: Review of the Antarctic Treaty (Planned).
2048	: Review of the Madrid Protocol (Planned).

Notes

1. The Agreed Measures for the Conservation of Antarctic Fauna and Flora (AMCAFF) protects Antarctic animals and plants from exploitation and allows ecologically sensitive areas to be out of bounds to everyone but scientists.
2. Convention for the Conservation of Antarctic Seals (CCAS) ensures that seals will not be hunted for commercial gain.
3. The Convention on the Conservation of Antarctic Marine Living Resources CCAMLR) protects krill (the foundation of the Antarctic food chain) and other marine resources from excessive harvesting.
4. Convention for the Regulation of the Antarctic Mineral Resources (CRAMRA) has not come into force due to non-ratification by some countries, notably Australia and New Zealand.

APPENDIX II: The Complete Text of the Antarctic Treaty

(ATS website ats.aq and end notes from the author)

The Governments of Argentina, Australia, Belgium, Chile, the French Republic, Japan, New Zealand, Norway, the Union of South Africa, the Union of Soviet Socialist Republics, the United Kingdom of Great Britain and Northern Ireland, and the United States of America;

Recognizing that it is in the interest of all mankind that Antarctica shall continue forever to be used exclusively for peaceful purposes and shall not become the scene or object of international discord;

Acknowledging the substantial contributions to scientific knowledge resulting from international co-operation in scientific investigation in Antarctica;

Convinced that the establishment of a firm foundation for the continuation and development of such co-operation on the basis of freedom of scientific investigation in Antarctica as applied during the International Geophysical Year accords with the interest of science and the progress of all mankind;

Convinced also that a treaty ensuring the use of Antarctica for peaceful purpose only and the continuance of international harmony in Antarctica will further the purposes and principles embodied in the Charter of the United Nations;

Have agreed as follows:

Article I

1. Antarctica shall be used for peaceful purposes only. There shall be prohibited, inter alia, any measure of a military nature, such as the establishment of military bases and fortifications, the carrying out of military manoeuvres, as well as the testing of any type of weapon.
2. The present Treaty shall not prevent the use of military personnel or equipment for scientific research or for any other peaceful purpose.

Article II

Freedom of scientific investigation in Antarctica and co-operation toward that end, as applied during the International Geophysical Year, shall continue, subject to the provisions of the present Treaty.

Article III

1. In order to promote international co-operation in scientific investigation in Antarctica, as provided for in Article II of the present Treaty, the Contracting Parties agree that, to the greatest extent feasible and practicable:
   (a) Information regarding plans for scientific programs in Antarctica shall be exchanged to permit maximum economy of and efficiency of operations;
   (b) Scientific personnel shall be exchanged in Antarctica between expeditions and stations;
   (c) Scientific observations and results from Antarctica shall be exchanged and made freely available.
2. In implementing this Article, every encouragement shall be given to the establishment of co-operative working relations with those Specialized Agencies of the United Nations and other international organizations having a scientific or technical interest in Antarctica.

Article IV

1. Nothing contained in the present Treaty shall be interpreted as:
   (a) a renunciation by any Contracting Party of previously asserted rights of or claims to territorial sovereignty in Antarctica;
   (b) a renunciation or diminution by any Contracting Party of any basis of claim to territorial sovereignty in Antarctica which it may have whether as a result of its activities or those of its nationals in Antarctica, or otherwise;
   (c) prejudicing position of any Contracting Party as regard its recognition or non-recognition of any other State's rights of or claim or basis of claim to territorial sovereignty in Antarctica.

2. No acts or activities taking place while the present Treaty is in force shall constitute a basis for asserting, supporting or denying a claim to territorial sovereignty in Antarctica or create any rights of sovereignty in Antarctica. No new claim, or enlargement of an existing claim, to territorial sovereignty in Antarctica shall be asserted while the present Treaty is in force.

Article V

1. Any nuclear explosions in Antarctica and the disposal there of radioactive waste material shall be prohibited.
2. In the event of the conclusion of international agreements concerning the use of nuclear energy, including nuclear explosions and the disposal of radioactive waste material, to which all of the Contracting Parties whose representatives are entitled to participate in the meetings provided for under Article IX are parties, the rules established under such agreements shall apply in Antarctica.

Article VI

The provisions of the present treaty shall apply to the area south of 60° South Latitude, including all ice shelves, but nothing in the present Treaty shall prejudice or in any way affect the rights, or the exercise of the rights, of any State under international law with regard to the high seas within that area.

Article VII

1. In order to promote the objectives and ensure the observance of the provisions of present Treaty, each Contracting Party whose representatives are entitled to participate in the meetings referred to in Article IX of the Treaty shall have the right to designate observers to carry out any inspection provided for by the present Article. Observers shall be nationals of the Contracting Parties which designate them. The names of observers shall be communicated to every other Contracting Party having the right to designate observers, and like notice shall be given of the termination of their appointment.

2. Each observer designated in accordance with the provisions of paragraph 1 of this Article shall have complete freedom of access at any time to any or all areas of Antarctica.
3. All areas of Antarctica, including all stations, installations and equipment within those areas, and all ships and aircraft at points of discharging or embarking cargoes or personnel in Antarctica, shall be open at all times to inspection by any observers designated in accordance with paragraph 1 of this Article.
4. Aerial observation may be carried out at any time over any or all areas of Antarctica by any of the Contracting Parties having the right to designate observers.
5. Each Contracting Party shall, at the time when the present Treaty enters into force for it, inform the other Contracting Parties, and thereafter shall give them notice in advance, of
   (a) all expeditions to and within Antarctica, on the part of its ships or nationals, and all expeditions to Antarctica organized in or proceeding from its territory.
   (b) all stations in Antarctica occupied by its nationals; and
   (c) any military personnel or equipment intended to be introduced by it into Antarctica subject to the conditions prescribed in paragraph 2 of Article I of the present Treaty.

Article VIII

1. In order to facilitate the exercise of their functions under the present Treaty, and without prejudice to the respective positions of the Contracting Parties relating to jurisdiction over all other persons in Antarctica, observers designated under paragraph 1 of Article VII and scientific personnel exchanged under sub-paragraph 1 (b) of Article III of the Treaty, and members of the staffs accompanying any such persons, shall be subject only to the jurisdiction of the Contracting Party of which they are nationals in respect of all acts or omissions occurring while they are in Antarctica for the purpose of exercising their functions.
2. Without prejudice to the provisions of paragraph 1 of this Article, and pending the adoption of measures in pursuance of sub-paragraph 1(e) of Article IX, the Contracting Parties Concerned

in any case of dispute with regard to the exercise of jurisdiction in Antarctica shall immediately consult together with a view to reaching a mutually acceptable solution.

Article IX

1. Representatives of the Contracting Parties named in the preamble to the present Treaty shall meet at the City of Canberra within two months after the date of entry into force of the Treaty, and thereafter at suitable intervals and places, for the purpose of exchanging information, consulting together on matters of common interest pertaining to Antarctica, and formulating and considering, and recommending to their Governments, measures in furtherance of the principles and objectives of the Treaty, including measures regarding:-
   (a) Use of Antarctica for peaceful purposes only;
   (b) Facilitation of scientific research in Antarctica;
   (c) Facilitation of International scientific co-operation in Antarctica;
   (d) Facilitation of the exercise of the rights of inspection provided for in Article VII of the Treaty;
   (e) Questions relating to the exercise of jurisdiction in Antarctica;
   (f) Preservation and conservation of living resources in Antarctica.
2. Each contracting party which has become a party to the present Treaty by accession under Article XIII entitled to appoint representatives to participate in the meetings referred to in paragraph 1 of the present Article, during such times as that Contracting Party demonstrates its interest in Antarctica by conducting substantial scientific research activities there, such as the establishment of a scientific station or the dispatch of a scientific expedition.
3. Reports from the observers referred to in Article VII of the present Treaty shall be transmitted to the representatives of the Contracting Parties participating in the meetings referred to in paragraph 1 of the present Article.

4. The measures referred to in paragraph 1 of this Article shall become effective when approved by all the Contracting Parties whose representative were entitled to participate in the meetings held to consider those measures.
5. Any or all of the rights established in the present Treaty may be exercised as from the date of entry into force of the Treaty whether or not any measures facilitating the exercise of such rights have been proposed, considered or approved as provided in this Article.

Article X

Each of the contracting parties undertakes to exert appropriate efforts, consistent with the charter of the United Nations, to the end that no one engages in any activity in Antarctica contrary to the principles or purposes of the present Treaty.

Article XI

1. If any dispute arises between two or more of the Contracting Parties concerning the interpretation or application of the present Treaty, those Contracting Parties shall consult among themselves with a view to having the dispute resolved by negotiation, inquiry, mediation, conciliation, arbitration, judicial settlement or other peaceful means of their own choice.
2. Any dispute of this not so resolved shall, with the consent, in each case, of all parties to the dispute, be referred to the International Court of Justice for settlement; but failure to reach agreement on reference to the International Court shall not absolve parties to the dispute from the responsibility of continuing to seek to resolve it by any of the various peaceful means referred to in paragraph 1 of this Article.

Article XII

1. (a) The present Treaty may be modified or amended at any time by unanimous agreement of the Contracting parties whose representatives are entitled to participate in the meetings

provided for under Article IX. Any such modification or amendment shall enter into force when the depository Government has received notice from all such Contracting parties that they have ratified it.

(b) Such modification or amendment shall thereafter enter into force as to any other Contracting Party when notice of ratification by it has been received by the depository Government. Any such contracting Party from which no notice or ratification is received within a period of two years from the date of entry into force of the modification or amendment in accordance with the provision of sub paragraph 1(a) of this article shall be deemed to have withdrawn from the present Treaty on the date of the expiration of such period.

2. (a) If after the expiration of thirty years from the date of entry into force of the present Treaty, any of the Contracting Parties whose representatives are entitled to participate in the meetings provided for under Article XI so requests by a communication addressed to the depository Government, a Conference of all the Contracting Parties shall be held as soon as practicable to review the operation of the Treaty.

(b) Any modification or amendment to the present Treaty which is approved at such a Conference by a majority of the Contracting Parties there represented, including a majority of those whose representatives are entitled to participate in the meeting provided for under Article IX, shall be communicated by the depositary Government to all Contracting Parties immediately after the termination of the Conference and shall enter into force in accordance with the provisions of paragraph 1 of the present Article.

(c) If any such modification or amendment has not entered into force in accordance with the provisions of sub-paragraph 1 (a) of this Article within a period of two years after the date of its communication to all the Contracting Parties, any Contracting Party may at any time after the expiration of the period give notice to the depositary Government of its withdrawal from the present Treaty; and such withdrawal shall take effect two years after the receipt of the notice by the depositary Government.

Article XIII

1. The present Treaty shall be subject to ratification by the signatory States. It shall be open for accession by any State which is a Member of the United Nations, or by any other State which may be invited to accede to the Treaty with the consent of all the Contracting Parties whose representative are entitled to participate in the meetings provided for under Article IX of the Treaty.
2. Ratification of or accession to the present Treaty shall be effected by each State in accordance with its constitutional processes.
3. Instruments of ratification and instruments of accession shall be deposited with the Government of the United States of America, hereby designated as the depositary Government.
4. The deposited Government shall inform all signatory and acceding States of the date of each deposit of an instrument of ratification or accession, and the date of entry into force of the Treaty and of any modification or amendment thereto.
5. Upon the deposit of instruments of ratification by all signatory States, the present Treaty shall enter into force for those States and for States which have deposited instruments of accessions. Thereafter the Treaty shall enter into force for any acceding State upon the deposit of its instruments of accessions.
6. The present Treaty shall be registered by the depositary Government pursuant to Article 201 of the Charter of the United Nations.

Article XIV

The present Treaty, done in the English, French, Russian and Spanish languages, each version being equally authentic, shall be deposited in the archives of the Government of the United States of America, which shall transmit duly certified copies thereof to the Governments of the signatory and acceding States.

Important Notes

1. Antarctic Treaty was signed on 01 December. After due ratification from all signatories, it came in to effect on 23 June, 1061.
2. Presently there are 53 parties including the 12 who have signed originally.
3. A total of 29 parties including the original 12 enjoy Consultative Party (CP) status.
4. As per present calendar, Antarctic Treaty will be up for review in 2041.
5. The Secretariat of the Antarctic Treaty headquarters is located at Buenos Aires (Argentina), since 2004.
6. Antarctic Treaty and its various instruments like AMCAFF, CCAS, CCAMLR and the Madrid Protocol are collectively called the Antarctic Treaty System (ATS).

APPENDIX III: Details of the Current Status of Various Countries in the Antarctic Treaty

(Refers to information from various sources, COMNAP Information Chart, Wikipedia, and Cool Antarctica 2015)

Sr. No	*Country*	*Antarctic Treaty*			*Notes*
	Name	*Signature*	*Accession to Treaty*	*Consultative Status*	
1.	Argentina (claim)	1 December 1959	23 June 1961	23 June 1961	OS/CN/CP
2	Australia (claim)	1 December 1959	23 June 1961	23 June 1961	OS/CN/CP
3.	Austria	No	25 August 1987	–	AS
4.	Belarus	No	27 December2006	–	AS
5.	Belgium	1 December 1959	26 July 1960	23 June 1961	OS/CP
6.	Brazil	No	16 May 1975	27 September 1983	AS/CP
7.	Bulgaria	No	11 September 1978	5 June 1998	AS/CP
8.	Canada	No	4 May 1988	–	AS
9.	Chile (claim)	1 December 1959	23 June 1961	23 June 1961	OS/CN/CP
10.	China	No	8 June 1983	7 October 1985	AS/CP
11.	Colombia	No	31 January 1989	–	AS
12.	Cuba	No	August 16, 1984	–	AS
13.	Czech Republic	No	1 January 1993	1 April 2014	AS (Succession from Czechoslovakia, which acceded on 14 June 1962)

Sr.	*Country*	*Antarctic Treaty*			*Notes*
No	*Name*	*Signature*	*Accession to Treaty*	*Consultative Status*	
14.	Denmark	No	20 May 1965	–	AS
15.	Ecuador	No	15 September 1987	19 November 1990	AS/CP
16.	Estonia	No	17 May 2001	–	AS
17.	Finland	No	15 May 1984	20 October 1989	AS/CP
18.	France (claim)	1 December 1959	16 September 1960	23 June 1961	OS/CN/CP
19.	Germany (claim) (rests since 1945)	No	5 February 1979	3 March 1981	AS/CP (Ratified as West Germany. East Germany also acceded on 19 November 1974 and received consultative status on 5 October 1987 prior to its reunification with West Germany.)
20.	Greece	No	8 January 1987	–	AS
21.	Guatemala	No	31 July 1991	–	AS
22.	Hungary	No	27 January 1984	–	AS
23.	Iceland	No	13 October 2015	–	AS
24.	India	No	19 August 1983	12 September 1983	AS/CP
25.	Italy	No	18 March 1981	5 October 1987	AS/CP
26.	Japan	1 December 1959	4 August 1960	23 June 1961	OS/CP
27.	Kazakhstan	No	27 January 2015	–	AS
28.	Malaysia	No	31 October 2011	–	AS

Sr.	Country	Antarctic Treaty			Notes
No	Name	Signature	Accession to Treaty	Consultative Status	
29.	Monaco	No	31 May 2008	–	AS
30.	Mongolia	No	23 March 2015	–	AS
31.	Netherlands	No	30 March 1967	19 November 1990	AS/CP
32.	New Zealand (claim)	1 December 1959	1 November 1960	23 June 1961	OS/CN/CP
33.	North Korea	No	21 January 1987	–	AS
34.	Norway (claim)	1 December 1959	24 August 1960	23 June 1961	OS/CN/CP
35.	Pakistan	No	1 March 2012	–	AS
36.	Papua New Guinea	No	16 March 1981	–	AS (Succession from Australia Effective from their independence on 16 September 1975)
37.	Peru	No	10 April 1981	9 October 1989	AS/CP
38.	Poland	No	8 June 1961	29 July 1977	AS/CP
39.	Portugal	No	29 January 2010	–	AS
40.	Romania	No	15 September 1971		AS
41.	Russia	December1, 1959	2 November 1960	23 June 1961	AS/CP (Ratified as the Soviet Union)
42.	Slovakia	No	1 January 1993	–	AS Succession from Czechoslovakia, which acceded on 14 June 1962
43.	South Africa	December1, 1959	21 June 1960	23 June 1961	OS/CP

Sr. No	Country Name	Antarctic Treaty Signature	Accession to Treaty	Consultative Status	Notes
44.	South Korea	No	28 November 1986	9 October 1989	AS/CP
45.	Spain	No	31 March 1982	21 September 1988	AS/CP
46.	Sweden	No	24 April 1984	21 September 1988	AS/CP
47.	Switzerland	No	15 November 1990	–	AS
48.	Turkey	No	24 January 1996	–	AS
49.	Ukraine	No	28 October 1992	4 June 2004	AS/CP
50.	United Kingdom (claim)	1 December 1959	31 May 1960	23 June 1961	OS/CN/CP
51.	United States	1 December 1959	18 August 1960	23 June 1961	OS/CP
52.	Uruguay	No	11 January 1980	7 October 1985	AS/CP
53.	Venezuela	No	24 May 1999	–	AS

OS: Original Signatories; AS: Acceding State; CN: Claimant Nation; CP: Consultative Party.

Important Notes

1. Czechoslovakia acceded to the Treaty on 14 June 1962, separated into the Czech Republic and Slovakia on 1 January 1993.
2. German Democratic Republic acceded to the Treaty on 19 November 1974 and gained Consultative Status 5 October 1987. It united with the Federal Republic of Germany on 3 October 1990, and is now known as Germany.
3. Papua New Guinea succeeded to the Treaty after independence from Australia.
4. Russian Federation succeeded to the Treaty as part of the Soviet Union.
5. The Hydrographic Committee on Antarctica (HCA). A Registered Hydrographic Commission of the International Hydrographic Organisation (IHO) is responsible for hydrographic surveying and nautical charting matters in the 'International Region M', which corresponds to the Antarctic Treaty area. The Antarctic Treaty Secretariat, COMNAP and SCAR have observer status at the HCA and contribute to its work. The work of the HCA is essential for the safety of ship operations in the Antarctic.
6. The Secretarial of the Antarctic Treaty System (ATS) is in Buenos Aires, Argentina. Internet address: http://www.ats.aq
7. The Secretariat of the Council of Managers of National Antarctic Programs (COMNAP) is in Hobart, Australia. Internet address: http://www.comnap.aq
8. The Secretariat of the Commission for the Conservation of Antarctic Marine Living Resources (CCAMLR) is in Hobart, Australia. Internet address: http://www.ccamlr.org
9. The Secretariat of the Scientific Committee on Antarctic Research (SCAR) is in Cambridge United Kingdom. Internet Address: http://www.scar.org

APPENDIX IV: Details of Currently Occupied Scientific Research Stations in Antarctica

(Refers to information from various sources, COMNAP Information Chart, Wikipedia, 2012, Research Stations in Antarctica and Cool Antarctica, 2015)

Sr. No.	*Station Name*	*Nationality*	*Location*	*Opened*	*Status*
1.	Aboa	Finland	Queen Maud Land	1989	Seasonal
2.	Amundsen-Scott	USA	South Pole	1956	Year round
3.	Arctowski	Poland	King George Island Peninsula	1977	Year round
4.	Artigas	Uruguay	King George Island Peninsula	1984	Year round
5.	Arturo Parodi	Chile	Patriot Hills		
6.	Arturo Prat	Chile	South Shetland Islands	1947	Seasonal
7.	Belgrano II (1)	Argentina	Coats Land	1955	Year round
8.	Bellingshausen	Russia	King George Island Peninsula	1968	Year round
9.	Bharati	India	Larsemann Hills	2012	Year round
10.	Browning Pass	Italy	Victoria Land		
11.	Casey	Australia	Vincennes Bay Wilkes Land	1969	Year round
12.	ComandanteFerraz	Brazil	King George Island Peninsula	1984	Year round
13.	Concordia (2)	France & Italy	Dome C-Antarctic Plateau	1997	Year round

Sr. No.	Station Name	Nationality	Location	Opened	Status
14.	D 10 skiway	France	Adelie Land		
15.	D 85 skiway	France	Adelie Land		
16.	Davis	Australia	Princess Elizabeth Land	1957	Year round
17.	Dome Fuji	Japan	Queen Maud Land	1995	Seasonal
18.	Druzhnaya 4	Russia	Prydz Bay	1987	Seasonal
19.	Dumont d'Urville	France	Adelie Land	1956	Year round
20.	Enigma Lake	Italy	Victoria Land		
21.	Escudero	Chile	King George Island Peninsula	1994	Year round
22.	Esperanza	Argentina	Hope Bay	1952	Year round
23.	Fossil Bluff	United Kingdom	Alexander Island Peninsula		
24.	Frei	Chile	King George Island Peninsula	1969	Year round
25.	Gabriel de Castilla	Spain	Deception Island	1990	Seasonal
26.	Great Wall	China	King George Island Peninsula	1985	Year round
27.	Halley	United Kingdom	Brunt Ice Shelf	1956	Year round
28.	Juan Carlos Primero	Spain	Livingston Island Peninsula	1989	Seasonal
29.	Jubany	Argentina	King George Island Peninsula	1982	Year round
30.	King Sejong	Korea	King George Island Peninsula	1988	Year round
31.	Kohnen	Germany	Dronning Maud Land	2001	Seasonal
32.	Law Racovita	Romania	Princess Elizabeth Land	1987	Seasonal

Sr. No.	Station Name	Nationality	Location	Opened	Status
33.	Macchu Picchu	Peru	King George Island Peninsula	1989	Seasonal
34.	Maitri	India	Schirmacher region	1989	Year round
35.	Maldonado	Ecuador	Greenwich Island South Shetland Islands	1990	Seasonal
36.	Marambio	Argentina	Seymour-Marambio Island	1969	Year round
37.	Mario Zucchelli	Italy	Terra Nova Bay Ross Sea	1986	Seasonal
38.	Mawson	Australia	Mac Robertson Land	1954	Year round
39.	McMurdo	USA	Ross Island	1955	Year round
40.	Mid-Point	Italy	Ross Sea Area		
41.	Mirny	Russia	Davis Sea	1956	Year round
42.	Molodezhnaya	Russia	Alasheyev Bay, East Antarctica		
43.	Neumayer	Germany	Atka Bay	1981	Year round
44.	Novolazarevskaya	Russia	Queen Maud Land	1961	Year round
45.	Bernardo O'Higgins	Chile	Antarctic Peninsula	1948	Year round
46.	Odell Glacier	USA	Allan Hills, Ross Sea Area		
47.	Ohridiski	Bulgaria	Livingston Island, South Shetland Islands	1988	Seasonal
48.	Orcadas	Argentina	South Orkney Islands	1904	Year round
49.	Palmer	United States.	Anvers Island Peninsula	1965	Year round
50.	Princes Elisabeth	Belgium	Dronning Maud Land	2009	Seasonal

Sr. No.	Station Name	Nationality	Location	Opened	Status
51.	Progress 2	Russia	Prydz Bay	1989	Year round
52.	Refugio Ecuador	Ecuador	Antarctic Peninsula	1990	Refuge
53.	Rothera	United Kingdom	Adelaide Island Peninsula	1976	Year round
54.	S17	Japan	30 km E of Syowa station		
55.	San Martin	Argentina	Antarctic Peninsula	1951	Year round
56.	Sanae IV (3)	South Africa	Queen Maud Land	1962	Year round
57.	Scott Base	New Zealand	Ross Island	1957	Year round
58.	Signy	United Kingdom	South Orkney Islands	1947	Seasonal
59.	Sitry (Zucchelli)	Italy	Terra	Enlarged in 2010	Seasonal
60.	Sky Blu	United Kingdom	Eastern Ellsworth Land	1995	Seasonal
61.	Syowa/Showa	Japan	East Ongul Island	1957	Year round
62.	Tor	Norway	Dronning Maud Land	1985	Seasonal
63.	Troll (4)	Norway	Dronning Maud Land	1990	Year round
64.	Vernadsky	Ukraine	Galindez Island	1996	Year round
65.	Vostok	Russia	Antarctic polar plateau	1957	Year round
66.	Wasa	Sweden	Dronning Maud Land	1989	Seasonal
67.	Zsongshan	China	Prydz Bay	1989	Year round

Important Notes

1. The original Belgrano Station opened 1955. It was replaced by Belgrano II in 1979.
2. Concordia Station opened December 1997 for summer-only operation. It opened for year-round operations in December 2005.
3. The original Sanae Station opened in 1962; Sanae IV opened in 1997 at a new location, 200 km South of Sanae I to III.
4. Troll Station opened in February 1990 for summer-only operations. It opened for year-round operations in February 2005.
5. Detailed information on airfields is available in the Antarctic Flight Information Manual (AFIM). Copies can be purchased from COMNAP. Internet address: (http://comnap.aq/publications/afim)
6. Details given above keep changing frequently.

The Antarctic Spatial Data is from the following source:

The Topographic data, Antarctic Polar Front and Maximum Sea Ice Extent data is supplied by the Australian Antarctic Division. Internet address: www.aad.gov.au

- The Maximum Sea Ice Extent is based on passive microwave data between 1989 and 1999.
- Topographical data is derived from the Antarctic Digital Database (Version 4). Features include Coastline, Contours, Glacier Flow Lines and Exposed Rock.
- Scientific Committee on Antarctic Research (SCAR).

All Other Information is from the following sources:

Antarctic Treaty (AT) membership data is supplied by the Antarctic Treaty Secretariat. Internet address: http://www.ats.aq

Committee for Environmental Protection (CEP) membership data is supplied by the CEP. Internet address: http://www.cep.aq

Council of Managers of National Antarctic Programs (COMNAP) membership data is supplied by the COMNAP. Internet address: http://www.comnap.aq

Commission for the Conservation of Antarctic Marine Living Resources (CCAMLR) membership supplied by the CCAMLR. Internet address: http://www.ccamlr.org

Scientific Committee on Antarctic Research (SCAR) membership data is supplied by the SCAR. Internet address: http://www.scar.org

Hydrographic Committee on Antarctic (HCA) membership data is supplied by the International Hydrographic Organisation (IHO). Internet address: http://www.iho.shom.fr

The data for the Main Antarctic Facilities operated by national Antarctic programmes in the Antarctic Treaty area is supplied by the Council of Managers of National Antarctic Programs (COMNAP). Internet address: http://www.comnap.aq

The World and Bathymetric data is derived from the Vector Map Level. V Map is supplied by the US NIMA (National Imagery and Mapping Agency).

APPENDIX V: Details of Ships Used for Indian Antarctic Programme (IAP) Expeditions

(Mishra, Shubhrata and NCAOR Technical Publications)

Sr. No of Expedition	*Year*	*Name of The Ship*	*Agency and Country*
1.	1981	MV Polar Circle	Reber Shipping, A S Norway
2.	1982	MV Polar Circle	Reber Shipping, A S Norway
3.	1983	MV Fin Polaris	O Y Finlands, Finland
4.	1984	MV Fin Polaris	O Y Finlands, Finland
5.	1985	MV Thuleland	Bostroms Rediar, AB Sweden
6.	1986	MV Thuleland	Bostroms Rediar, AB Sweden
7.	1987	MV Thuleland	Bostroms Rediar, AB Sweden
8.	1988	MV Thuleland	Bostroms Rediar, AB Sweden
9.	1989	MV Thuleland	Bostroms Rediar, AB Sweden
10.	1990	MV Thuleland	Bostroms Rediar, AB Sweden
11.	1991	MV Thuleland	Bostroms Rediar, AB Sweden
12.	1992	MV Thuleland	Bostroms Rediar, AB Sweden
13.	1993	MV Stephen Kashnekov	Russia
14.	1994	MV Polar Bird	Reber Shipping, A S Norway
15.	1995	MV Briknes	Reber Shipping, A S Norway
16.	1996	MV Polar Bird	Reber Shipping, A S Norway
17.	1997	MV Polar Bird	Reber Shipping, A S Norway
18.	1998	MV Polar Bird	Reber Shipping, A S Norway
19.	1999	MV Merdelina Oldendorf	Germany
20.	2000	MV Merdelina Oldendorf	Germany

Sr. No of Expedition	*Year*	*Name of The Ship*	*Agency and Country*
21.	2001	MV Merdelina Oldendorf	Germany
22.	2002	MV Merdelina Oldendorf	Germany
23.	2003	MV Emerald C	MS Elmira Shipping and Trading, Greece
24.	2004	MV Emerald C	MS Elmira Shipping and Trading, Greece
25.	2005	MV Pardebeng	MS Dynamic Shipping Service, Israel
26.	2006	MV Emerald C	Ms Eastland Finance Company, Liberia
27.	2007	MV Emerald C	Ms Eastland Finance Company, Liberia
28.	2008	MV Emerald C	Ms Eastland Finance Company, Liberia
29.	2009	MV Ivan Papanin and MV Vladimir Ignatyuk	Murmansk Shipping Company, Russia
30.	2010	MV Ivan Papanin & MV Vladmir Ignatyuk	Murmansk Shipping Company, Russia
31.	2011	Same as year 2010	Same as year 2010
32.	2012	Same as year 2010	Same as year 2010
33.	2013	MV Ivan Papanin	Murmansk Shipping Company, Russia
34.	2014	Same as year 2013	Same as year 2013
35.	2015	Same as year 2013	Same as year 2013

Note: Details for years 2011 to 2015 are taken from an email dated 27 April, 2016 from Ms. Reena S. Naik of NCAOR to the author. The author is thankful to her and Mr. M.J. Beg, Director (Logistics), NCAOR for the information.

APPENDIX VI: Details of Helicopters Used for Indian Antarctic Programme (IAP) Expeditions

(Mishra, Shubhrata and NCAOR Technical Publications)

Sr. No of Expedition	*Year*	*Name of Helicopter*	*Agency and Country*
1.	1981	–	–
2.	1982	–	–
3.	1983	MI-8 & Elotis	Indian Air force (IAF)
4.	1984	MI-8 & CHETAK	Indian Air force (IAF)
5.	1985	MI-8 & CHETAK	Indian Air force (IAF)
6.	1986	MI-8 & CHETAK	Indian Air force (IAF)
7.	1987	MI-8 & CHETAK	Indian Air force (IAF)
8.	1988	MI-8 & CHETAK	Indian Air force (IAF)
9.	1989	MI-8 & CHETAK	Indian Air force (IAF)
10.	1990	MI-8 & CHETAK	Indian Air force (IAF)
11.	1991	MI-8 & CHETAK	Indian Air force (IAF)
12.	1992	MI-8 & CHETAK	Indian Air force (IAF)
13.	1993	CHETAK	Indian Navy (IN)
14.	1994	CHETAK	Indian Navy (IN)
15.	1995	Norwegian	Norwegian Helicopters, Norway
16.	1996	JERO	JERO Helicopter, Australia
17.	1997	JERO	JERO Helicopter, Australia
18.	1998	Pawan Hans	Pawan Hans Helicopters Ltd, New Delhi
19.	1999	Pawan Hans	Pawan Hans Helicopter Ltd, New Delhi
20.	2000	NZ	New Zealand Helicopter Ltd, New Zealand

Sr. No of Expedition	*Year*	*Name of Helicopter*	*Agency and Country*
21.	2001	NZ	New Zealand Helicopter Ltd, New Zealand
22.	2002	NZ	New Zealand Helicopter Ltd, New Zealand
23.	2003	NZ	New Zealand Helicopter Ltd, New Zealand
24.	2004	AS 350 A Aerospatiale	Ms Heliserve Pvt Ltd, Australia
25.	2005	V0 105 CVS 4/5	Ms Helicopter Service, GMBH, Germany
26.	2006	NZ	New Zealand Helicopter Ltd, New Zealand.
27.	2007	NZ	New Zealand Helicopter Ltd, New Zealand.
28.	2008	NZ	New Zealand Helicopter Ltd, New Zealand.
29.	2009	KAMOV KA 32 A and AS 350 B2	LG International Corporation, Norway & AS Airlift, Norway
30.	2010	Same as year 2009	
31.	2011	KAMOV KA 32 A and Bell 407	LG International Corporation, Norway and Ms. Heli Korea, Korea
32.	2012	KA 32 C and Bell 407	Ms JSC Avialift, Vladivostok (Russia) and Ms. Heli Korea, Korea
33.	2013	Same as year 2012	Same as year 2012
34.	2014	KA 32 C and AS 350 B3	Ms JSC Avialift, Vladivostok (Russia) and Ms. Ultimate Heli, Johannesburg (South Africa)
35.	2015	Same as year 2014	Ms. Artic Group Ltd, Ukraine and Ms. Ultimate Heli, Johannesburg (South Africa)

Note: Details for years 2011 to 2015 are taken from an email dated 27 April, 2016 from Ms. Reena S. Naik of NCAOR to the author. The author is thankful to her and Mr. M.J. Beg, Director (Logistics), NCAOR for the information.

APPENDIX VII: Details of Personnel Participating in Indian Antarctic Programme (IAP) Expeditions

(Mishra, Shubhrata and NCAOR Technical Publications)

Sr. No of Expedition	*Year*	*Summer Team*	*Winter Team*	*Total*
1.	1981	21	–	21
2.	1982	28	–	28
3.	1983	69	12	81
4.	1984	69	13	82
5.	1985	74	14	88
6.	1986	73	17	90
7.	1987	76	15	91
8.	1988	58	42	100
9.	1989	55	25	80
10.	1990	75	25	100
11.	1991	72	26	98
12.	1992	31	25	56
13.	1993	33	25	58
14.	1994	38	25	63
15.	1995	23	26	49
16.	1996	36	25	61
17.	1997	29	24	53
18.	1998	30	26	56
19.	1999	24	24	48
20.	2000	28	25	53
21.	2001	26	25	51
22.	2002	25	23	48
23.	2003	30	25	55
24.	2004	29	25	54
25.	2005	27	25	52
26.	2006	33	24	57
27.	2007	36	25	61
28.	2008	41	25	66
29.	2009	31	23	54
30.	2010	21	23	44
Total		**1241**	**657**	**1898**

Note: Details from NCAOR after year 2010 could not be ascertained.

References

Books/Articles

Alam, S.I., Singh, L., Dube, S., Reddy, G.S.N. and Shivaji, S., "Psychrophilic Planococcus Maitriensis", *Antarctica, Systematic and Applied Microbiology*, Vol. 26 (2003).

Antarctic Archives, Research and Development Organisation. *RDEE,* Pune (1991).

Antarctic Climate Change and the Environment, Report by SCAR (UK) (2008).

Antarctic Environment Officer Network (AEOW), Workshop on Waste Management in Antarctica. COMNAP, Hobart, Australia (2006).

Antarctica: Great Stories from the Frozen Continent, Reader's Digest Services Pty Limited, Australia (1985).

Antarctica New Zealand Profile, International Antarctic Centre, New Zealand (2009).

Antarctica: 25 Years Research by German Democratic Republic, GDR (1984).

Army Workshop Report (2012). 512 Army Base Workshop Letter No 70101/ Gen/TC/PP&C (20 July 2012) addressed to Lt Col J.P. Khadilkar.

Bhave, Prashant and Dixit Medha, "Environmental Status at Maitri Antarctic Station". NCAOR TP No. 20, NCAOR (2008).

Bolten, Doug' "Warming may lead to 30% drop in polar bear population", Times of India (21 November, 2015).

Buildings and Structures in Antarctica, Books LLC, Memphis, USA (2010).

Chaturvedi, A. and Ravindra, R., "Conspicuous Evidence of Global Warning from Schirmacher Range: Antarctica in Role of Earth Sciences", In Integrated Development and Related Social Issues, GSI-NR, Lucknow (2001).

Chaturvedi, S., Khare, N. and Pandey, P.C., India in the Antarctic: Scientific and Geopolitical Perspective. South Asian Publishers, New Delhi (2005).

Chaturvedi, A., Antarctica: The Frozen Continent. Geological Survey of India (2009).

Copeland, S., Global Warning. Earth Ware Edition, UK (2007).

Crane, D., Scott of the Antarctic: A Life of Courage and Tragedy in the Extreme South, Harper Collins, London (2005).

Gaikwad, S.S., "Impact of Antarctic Climate on Structural Components and Life Support Systems of Maitri Station," *NCAOR TP* No. 11, NCAOR (1996).

"Geology of Schirmacher and Wolthot Region, Central Dronning Maud Land, Antarctica (scale 1:250000)". Government of India, GSI (1991).

"Geology of Schirmacher Oasis, Central Dronning Maud Land, East Antarctica (Scale 1:250000)". Government of India, GSI (1999).

"Geological Map of Orvinfjella Central Dronning Maud Land, East Antarctica, (Scale 1:250000)". Government of India, GSI (2006).

"Geomorphologic Map of Schirmacher Oasis, Central Dronning Maud Land, East Antarctica (scale 1:250000)". Government of India, GSI (2006).

Gorodotskaya, Irina V., "The role of atmospheric rivers in anomalous snow accumulation in East Antarctica", *Geoohysical Research letters* (2014).

Gupta, P., Reddy, G., Delittle, D. and Shivaji, S., "Arthobacter Gangotriensis and Arthobacter Kerguelensis from Antarctica", *International Journal of Systematic and Evolutionary Microbiology,* Vol. 54 (2004).

HIMVAT (Magazine from Dakshin Gangotri), DOD (1990).

Hudson, S. and Brandt, R., PPT on Surface Based Temperature Inversion on the Antarctic Plateau. University of Washington (2008). (http://www.atmos.washington.edu/hudson/CMDL_meeting/hudson_CMDL_talk.ppt# 256).

Ice Age: Celebrating 50 years of New Zealand in Antarctica, Scott Base 1957–2007. New Zealand Memories (2007).

Jain, S.L., "Ozone Hole over Poles Current Station." *Science and Geopolitics of Arctic and Antarctic' Sa GAA* (2011).

Jain, S.L., Ghude, S. and Arya, B.C., "Signature of Early Ozone Hole Recovery During 2002", *Current Science,* Vol. 86 (2004).

Jenouvrier, Stephanie (2012). "Emperor Penguins Feel the Heat", Times of India, 24 June 2012.

Jonathan, Amos, "Gigantic chasm discovered under Antarctic Ice", *BBC Science and Environment* (24 January, 2016).

Khadilkar, Lt Col J.P., Report of the Sixth Wintering Team at Dakshin Gangotri. DOD (1990).

Koltz, Gen Frank, "America on the Ice". Antarctic Policy Issues, National Defence University Press, USA (1990).

Koppar, A.L., India Antarctic Research: A Station Review on Meteorology, National Workshop on India Antarctic Research at NCAOR, 27–29 October 2004. NCAOR (2004).

Kriwoken, K. and Rootes, D., "Tourism on Ice", Environmental Assessment of Antarctic Tourism, *Impact Assessment and Project Appraisal*, Vol. 18, No. 2, June 2000.

Lal, R.P., "Science and Geopolitics of Arctic and Antarctic", *Review of Meteorological Observation Programme of IMD in Antarctica*, SaGAA (2011).

Landis, Marilyn J., Antarctica, Exploring the Extreme: 400 years of Adventure, Chicago Review Press (2001).

Launching of First National South Pole Expedition, NCAOR, Goa (2010).

Larson, Edward J., "An Empire on Ice (Scott, Shackleton and the Heroic Age of Antarctic Science", Yale University Press (2011).

Mawson, Douglas, The Home of the Blizzard: Story of Australian Antarctic Expedition 1911–14. Wakefield Press, South Australia (1996).

May, John, Green Peace Book of Antarctica: A New View of the Seventh Continent, Dorling Kindersley Limited, London (1989).

McGonigal, D., Antarctica: Secrets of the Southern Continent, Firefly Books Ltd., USA (2008).

Meridian, Discover Ross Island: The World's Coolest Wind Farm, Meridian, New Zealand, meridian.co.nz (2009).

Mishra, Dr Shubhrata, Bharatiya Antarctic Sambhar Tantra (Trs Indian Antarctic Logistics).

NCAOR, Goa (2012).

Murray T and others, "Reverse glacier motion during ice berg calving and causes of glacier earthquake", *Science Express* (2015).

Nair, Sushma R., "Maitri: Environmental Perspective". *NCAOR TP* No. 12 (1998).

Outposts of Antarctica: Research Stations, Books LLC Memphis, USA (2010).

Reports of Indian Expeditions to Antarctica (IEA), Technical Publications (TP) NCAOR, Department of Ocean Development (DOD) /Ministry of Earth Sciences:

1st IEA (1981–82), TP No. 1 (1983), 2nd IEA (1982–83), TP No. 2 (1985),

3rd IEA (1983–84), TP No. 3 (1986), 4th IEA (1984–85), TP No. 4 (1987),

5th IEA (1985–86), TP No. 5 (1988), 7th IEA (1987–88), TP No. 7 (1990),

9th IEA (1989–90), TP No. 6 (1994), 11th IEA (1991–92), TP No. 9 (1995),

12th IEA (1992–93), TP No. 10 (1996), 13th IEA (1993–94), TP No. 11 (1997),

14th IEA (1994–95), TP No. 12 (1998), 15th IEA (1995–96), TP No. 13 (1999),

16th IEA (1996–97), TP No. 14 (2000), 17th IEA (1997–98), TP No. 15 (2000),

18th IEA (1998–99), TP No. 16 (2002), 19th IEA (1999–2000), TP No. 17 (2005),

20th IEA (2000–2001), TP No. 18 (2006), 21st IEA (2001–02), TP No. 19 (2010),

22nd IEA (2002–03), TP No. 20 (2011), 23rd IEA (2003–04), TP No. 21 (2007),

24th IEA (2003–04), TP No. 22 (2012).

(*Note:* Numbers of IEA, TP and years are not consistent as in some years reports are not published as also years of publication are different).

National Geographic (April 1987).

National Geographic (April 1990).

NCAOR Annual Report (2009–10).

NCAOR Annual Report (2010–11).

NCAOR Laboratory Facilities, NCAOR, Goa (2010).

NCAOR Research Activities, NCAOR, Goa (2010).

NCAOR Through the Years, NCAOR, Goa (2010).

NCAOR Final Comprehensive Environmental Evaluation of New Indian Research Base at Larsemann Hills, Antarctica (CEE), NCAOR, Goa (2010).

Nizampurkar, V.N., Rao, D.K., Clausen, H.B., Kaul, M.K. and Chaturvedi. A "Records of Climatic Change and Volcanic Events in an Ice Core from Central Maud Land (East Antarctica) During Past Century". Indian Academy of Sciences (Earth Planet Science), Vol. III, No. 1 (2002).

Pandey, P.C. "India: Antarctic Programme". Encyclopaedia of the Antarctic, Riffenberg, Abingdon and New York (2007).

Pathak, Brig R.C., Antarctic Panorama: A Geographical and Geotechnical Account, Diamond Publications, Pune (2010).

Peat, Neville, Antarctica the Big Ice: Celebrating 50 Years of New Zealand in Antarctica. Otago Museum, New Zealand (2007).

Technology Focus, Special Issue on DRDO in Antarctica, Research and Development Organisation, RDEE (1997).

Maitri Manual, Research and Development Organisation, RDEE, Pune (1988).

Raghava, G. and Murthy, S.G.N., "Structural Assessment of Maitri Building of Antarctica", NCAOR TP No21, NCAOR (2010).

Raina, V.K., Reminiscences, NCAOR, Goa (2006).

Rajan, S. and Pandey, P., Antarctic Geoscience, Ocean Atmosphere Interaction and Paleoclimatology, NCAOR, Goa (2004).

Rajendran, Girija and Reddy, B.M., Atmosphere Research from Antarctica: The Indian Contribution (1993).

Rao, U.R. Report of Expert Group on Ten Years Work Profile of Indian Antarctic Programme. Government of India, Department of Ocean Development (DOD), New Delhi (1995).

Ravi, Kumar N. and Malaimani, E.C., "Recent Results of the Strain Accumulation between Antarctica and Southern Indian Peninsula by GPS-Geodesy" (2000).

(http://www.tu-dresden.de/ipg/polenet/program/004-kumar.rn.pdf)

Reddy, G.S.N., Prakash, J.S.S., Matsumoto, G.I., Stackebrandt, E. and Shivaji, S. "Arthobacter Roseus, A Psychrophilic Bacterium, Isolated from an Antarctic Cynobacterial Mat Sample", International Journal of Systematic and Evolutionary Microbiology, Vol. 52, (2002 a).

Reddy, G.S.N., Prakash, J.S.S., Vairamani, M., Prabhakar, S., Matsumoto, G.I. and Shivaji, S., "Planococcus Antarcticus and Planococcus Psychrophiles, Isolated from Cynobacterial Mat Sample Collected from Ponds in Antarctica, Extremophiles", International Journal of Systematic and Evolutionary Microbiology, Vol. 6(3) (2002 b).

Reed, Margaret, Antarctic Logistics Defined as Military Operations other than War. Naval War College, New Port (US) (1995).

Samson, Lou, Annual Antarctic Conference, University of Otago, New Zealand (2008).

Saran, Shyam, "India's Stake in Arctic Cold War". The Hindu, Chennai (01/2/2012).

Scott, K., Geographic Book of Antarctica, Australian Geographic Pty Ltd, Australia (1993).

Sehra, P.S., A Visit to South Pole and Explore the Antarctic Continent. HKT Publication, New Delhi (2003).

Sethi, Nitin, "Firm India Forces Climate Breakthrough at Durban". Times of India, Pune, (12/12/2011).

Sharma, Maj Gen S.S., Breaking the Ice in Antarctica. New Age International Publishers, New Delhi (2001).

Shivaji, S., Rao, N.S., Aisree, L., Sheth, V., Reddy, G.S. and Bhargava, P.M., "Isolation and Identification of Pseudomonas, from Schirmacher Oasis, Antarctica", Applied and Environmental Microbiology, Vol. 55, No. 3, (1989).

Singh, Deepak K., "West Has Failed at Copenhagen". Banking Services Chronicle, Vol. 17, No. 5 (February 2010).

Sinha, Anurag, "Structural Assessment of Maitri Station at Antarctica". NCAOR TP No. 21, NCAOR (2010).

Operating and Maintenance Instructions for Dakshin Gangotri. M/s Structa Ply Ltd, UK (1983).

Swan, R., Antarctic 2041, New York (2009).

Thamban, M., Chaturvedi, A., Naik, S., Dsouza, W., Singh, A., Rajan, S. and Rajakumar, A., "Aerosol Perturbations Related to Volcanic Eruptions during Past Few Centuries as Recorded in an Ice Core from CDML", Antarctica Current Science, Vol. 91 (2006).

Tiwari, Anju, Krishnan, K.P. and Ravindra, R., The Story of Antarctica. NCAOR, Goa and Geological Survey of India, Bangalore (2008).

Tiwari, Anoop and Kulkarni, Sunil, "Environmental Status at Indian Polar Station Maitri: A Comprehensive Study", NCAOR TP No. NCAOR (2004).

Trewby, Mary, Antarctica: An Encyclopaedia from Abbott Ice Shelf to Zooplankton, Firefly Books Ltd, New Zealand (2002).

Triggs, Gillian D., The Antarctic Treaty Regime, Law, Environment and Resources, Press Syndicate, University of Cambridge, UK (1987).

Veerbhadraiah, G. and Jain, A. "Environmental Management Services at Maitri, Antarctica", NCAOR TP No. 21, NCAOR (2010).

Walton, D.W.H. and Doake, C.S.M., "Keystone to Gondwana", Chapter 13, Antarctic Science. Press Syndicate of the University of Cambridge, UK (1987).

Walton, D.W.H. and Dudeney, J.R., "Climate of Extreme", Chapter 15, Antarctic Science, Press Syndicate of the University of Cambridge, UK (1987).

Walton, D.W.H. and Everson, I., "Biogeography and Ecological Niches", Chapter 5, Antarctic Science. Press Syndicate of the University of Cambridge, UK (1987).

Public Domain Sites Visited

AAD Seals and Sea Lions (2012): http://www.antarctica.gov.au/about-antarctica/wildlife/animals/seals-and-sealions *(seen on 15/06/2012).*

AAD History of Australian Antarctic Stations (2010): http://www.aad.gov.an/default.asp *(seen on 29/06/2010).*

ALCI (2012): http://www.antartic_company.com/alci.htm *(seen on 22/03/2012).*

Antarctic Treaty System (2012): http://www.ats.aq *(seen in 2009, 2012, 2015).*

Australian Antarctic Aviation (2012): http://www.antarctica.gov.an/living-and-working/travel-and-logistics/aviation *(seen on 04/07/2012).*

Australian Antarctic Data Centre, Antarctic Facilities: http://www.comnap.aq/facilities *(seen in 2012).*

cnn.com, Health (2009) & Cnn.com, Report (2010): http://www.edition.cnn.com/2009/health/06/23/obitjerrinielsen *(seen on 24/06/2009 & 01/06/2010 respectively).*

Cool Antarctica, Human Impacts (2012): http://www.coolantarctica.com/human_inpacts_and_threats_to_the_enviroment_sealing_and_whaling *(seen on 29/02/2012).*

Cool Antarctica, The Madrid Protocol (2012): http://www.coolantarctica.com/madrid/protocol *(seen on 29/02/2012).*

Dronning Maud Land Air Network DROMLAN Information Book (2011) referred from: www.http.google.com *(seen in 2011 and 2012).*

Google, IPCC (2007) Fourth Assessment Report & IPCC (2013–14) Fifth Assessment Report *(seen in 2010 & 2012; 2014 & 2015 respectively).*

Google, Structure of the Earth: http://www/133.5.170.64/museum/museum-e/parts5-e/PSI-e/earth_structre-e.htm *(seen on 08/07/2012).*

IAATO-Home (2012), News from, Antarctic Tourism Factsheet: http://www.iaato.org *(seen on 07/08/2012).*

Jones, Nicola (2012), Nature News and Comment: http://www.nature.com/news/russians_celebrate_vosok_victory *(seen on 06/04/2012)*

MSN *Green News* (2011): http://www.green.in.msn.com/greennews/article.aspx?cp_documentid=5651715*(seen on 16/12/2011).*

NSF, 4.0 Antarctica Past and Present (1997): http://www.nsf.gov/pubs/1997/antpane/4past.htm *(seen on 05/08/2009).*

Russian Antarctic Station Overview (2012): http://www.aari.aq/stations/list_en.html *(seen on 24/3/2012).*

SCAR, Antarctic Research Stations (2010): http://www.scar.org/information *(seen on 24/02/2010 and 24/03/2012).*

Symposium on Antarctic logistics (2012): http://www.forumforum.com/3-9/showthread.php *(seen on 22/03/2012).*

The Age, "China Widens Antarctic Operations" (2012): http://www.theage.com.au/national/china_widens_antarctic_operations *(seen on 01/04/2012)*

The Antarctic Sun (2008): http://antarcticsun.usap.gov/science/contenthandler.cfm?id=1505 *(seen on 05/06/2012).*

The Economist, "Arctic Politics Cosy Amid the Thaw" (2012): http://www.economist.com/node/21551029 *(seen on 06/07/2012).*

Wikipedia, Antarctic Treaty System (2010): http://www.en.wikipedia.org/wiki/antarctic_treaty-system *(seen on 23/2/2010).*

Wikipedia, Antarctica (2009): http://www.en.wikipedia.org/wiki/antarctica *(seen on 28/11/2009).*

Wikipedia, Arctic and Antarctic Research Institute (2012): http://www.en.wikipedia.org/wiki/arctic_and_antarctic_research_institute *(seen on 06/04/2012).*

Wikipedia, Arctic Council (2012): http://www.en.wikipedia.org/wiki/arctic_council *(seen on 06/07/2012).*

Wikipedia, Argentina Antarctic Program (2012): http://www.en.wikipedia.org/wiki/argentina_antarctic_program *(seen on 30/3/2012).*

Wikipedia, British Antarctic Survey (2012): http://www.en.wikipedia.org/wiki/british_antarctic_survey *(seen on 24/6/2012).*

Wikipedia, Chinese Arctic and Antarctic Administration: http://www.en.wikipedia.org/wiki/chinese_arctic_and_antarctic_administration *(seen on 01/4/2012).*

Wikipedia, Global warming (2011): http://www.en.wikipedia.org/wiki/global_warming *(seen on15/12/2011).*

Wikipedia, Global Warming (2011): http://www.en.wikipedia.org/wiki/global_warning *(seen on 15/12/2011).*

Wikipedia, Indian Antarctic Programme (2012): http://www.en.wikipedia.org/wiki/inidian_antarctica_program *(seen on 03/06/2012)*

Wikipedia, Lithosphere (2010): http://www.en.wikipedia.org/wiki/lithosphere *(seen on 02/10/2010).*

Wikipedia, Queen Maud Land (2012): http://www.en.wikipedia.org/wiki/queen_maud_land *(seen on 31/03/2012).*

Wikipedia, Southern Ocean (2010): http://www.en.wikipedia.org/wiki/sonthernocean *(seen on 02/03/2010).*

Important Websites Referred

Antarctic Cooperative Research Centre in Australia: www.acecrc.org.au

International Association of Antarctic Tour Operators (IAATO): http://iaato.org

Delivery of the latest news and comment about Antarctic life, South Polar endeavours, world of the ice and Southern Ocean: www.antarctican.com

Australian Antarctic Division: www.aad.gov.au

British Antarctica Survey (BAS): www.antarctica.ac.uk

Byrd Polar Research Centre in USA: http://bprc.osu.edu

Council of Managers for National Antarctic Programmes (COMNAP): www.comnap.aq

Facts and statistics concerning NAPs: www.scar.org

Information on stamps and postal history, south polar expeditions and vast collection of photo illustrations: www.south-pole.com

New Zealand Antarctic Institute: www.antarcticanz.govt.nz

Quark Expeditions: www.quarkexpeditions.com

Rice University in Texas: www.glacier.rice.edu

Scott Polar Research Institute: www.spri.cam.ac.uk

United States National Science Foundation (NSF): www.nsf.gov

University of Canterbury in New Zealand: www.anta.canterbury.ac.nz

Selected Bibliography

Antarctic Adventure: Scott's Northern Party (Reprinted). London: Hurst and Company, 1974.

Antarctic Climate Change and Environment: A Contribution to the IPY 2007–08. SCAR Publication, 2008.

Antarctica: An Encyclopaedia from Abbot Ice-shelf to Zooplankton edited by Trewby M. Auckland: A Firefly Book, 2002.

Antarctica: The Extraordinary History of Man's Conquest of the Frozen Continent. Sydney: Reader's Digest, 1988.

Antarctica: Great Stories from the Frozen Continent (Sydney: Reader's Digest Services Pty Ltd, 1985).

Antarctic Treaty System: An Assessment. Washington DC: National Academy Press, 1986.

A Voyage of Discovery and Research in the Southern and Antarctic Regions during Years 1839–43. (Reprinted). London: Latimer, Trend and Company Limited, 1969.

Borchgrevink, C.E. *First on the Antarctic Continent.* Montreal: McGill-Queen's University Press, 1980.

Byrd, Richard E. *Alone.* New York: G P Putnam's Sons, 1938.

Chaturvedi, S., Khare, N. and Pande, P. *India in Antarctica: A Scientific and Geopolitical Perspective* by New Delhi: South Asian Publishers, 2005.

Cherry-Garrard, Apsley. *The Worst Journey in the World* (Second Edition). New York: Carrol and Graham Publishing, Inc., 1997.

Chipman, Elizabeth. *Women on the Ice: A History of Women in the Far South.* Melbourne University Press, 1986.

Crane, David. *Scott of the Antarctic: A Life of Courage and Tragedy in the Extreme South.* London: HarperCollins, 2005.

Cold Recall: Roald Amundsen's Reflections from the South Pole (Oslo: The Fram Museum, 2011).

Fuchs, Sir Vivian, and Hillary Sir Edmund. *The Crossing of Antarctica* (New York: Macmillan Publishing, 1958).

Hadoram, Shirihai. *A Complete Guide to Antarctic Wildlife: The Birds and Marine Mammals of the Antarctic Continent and Southern Ocean.*Helsinki: Alula Press, 2002.

Landis, Marilyn J. *Antarctica: Exploring the Extreme: 400 Years of Adventure.* Chicago: Chicago Review Press, 2001.

Larson, Edward J. *An Empire on Fire (Scott, Shackleton and the Heroic Age of Antarctic Science).* New Heaven and London: Yale University Press, 2002.

Lyttle, Richard B. *Polar Frontiers: A Background Book on the Arctic, the Antarctic, and Mankind.* New York: Parent Magazine Press, 1972.

Mawson, Douglas. *The Home of the Blizzard: The Story of the Australasian Antarctic Expedition, 1911–14.* Kent Town, Wakefield Press, South Australia, 1996.

May, John. *The Green Peace Book of Antarctica: A New View of the Continent.* London: Dorling Kindersley, 1988, reprinted with revisions 1989.

McGonigal, David. *Antarctica: Secrets of the Southern Continent.* Sydney: Firefly Books Ltd, 2008.

Mishra, Dr Shubhrata. *Bharatiya Antarctic Sambhar Tantra (*trs Indian Antarctic Logistics). Goa, NCAOR, 2012.

Ponting, Herbert. *The Great White South: Traveling with Robert F. Scott's Doomed South Pole Expedition.* London: Duckworth, 1921, Reprinted in 1932.

Prahalad, C.K. and Ramaswamy, Venkat. *The Future of Competition: Co-Creating Unique Value with Customers.* New Delhi: Penguin Books, India Pvt. Ltd., 2004.

Quigg, Philip W. *A Pole Apart: The Emerging Issue of Antarctica.* New York: McGraw Hill Book Company, 1983.

Rajaram, Girija and Reddy, B.M. *Atmospheric Research from Antarctica: The Indian Contribution.* Mumbai: A monograph published by IIG and NPL, 1993.

Sale, Richard. *To the Ends of the Earth: The History of Polar Exploration.* London: HarperCollins, 2002.

Sebastian, Copeland. *Antarctica: The Global Warning.* Earth Aware edition, 2007.

Sehra, P.S. *A Visit to the South Pole.* Chandigarh: HKT Publication, 2003.

Swan, Robert. *Antarctica 2041: My Quest to Save the Earth's Last Wilderness.* New York: Broadway Books, 2009.

Riffenburgh, Beau. *Racing with Death: Douglas Mawson, Antarctic Explorer.* London: Bloomsbury, 2008.

Triggs, Gillian, ed: *The Antarctic Treaty Regime: Law, Environment and Resources*. Cambridge University Press, 1987.

Walton, D.W.H., ed., with contributions from C.S.M. Doake, J.R. Dudley, I. Everson and R.M. Laws. *Antarctic Science*. Cambridge University Press, 1987.

Wheeler, Sara. *Terra Incognita: Travels in Antarctica.* New York: The Modern Library, 1999.

Worsley, Frank Arthur. *Shackleton's Boat Journey*. New York: W Norton and Company, Inc., 1977.

Index

About the Author

Dr Jagadish Khadilkar (Born 17 September, 1944) has been a keen scholar throughout his life. He has to his credit two graduations in engineering, a master's in defence studies and a doctorate in environmental studies. After a short teaching stint in an engineering college, he joined the Indian Army. A graduate from the Indian Military Academy (IMA), Dehradun, and a postgraduate from the Defence Services Staff College (DSSC), Wellington, he has enjoyed prestigious field and staff appointments in active service, during war and insurgency. He has also completed numerous academic, military and professional courses in India and abroad over half a century.

Dr Khadilkar led a major army expedition in the Thar Desert. Later, as a Lieutenant Colonel, he served as the leader and station commander of the Indian Antarctic Station, Dakshin Gangotri, during its dangerous and thrilling last wintering before it had to be decommissioned due to snow burial.

After taking premature retirement from the army, Dr Khadilkar served in senior corporate positions in many countries for a decade and half. After he returned to India, he served with couple of educational institutions as Dean and Executive Director. Currently, settled in Pune, he travels widely, writes and speaks on a variety of subjects in the media, academic and professional fora.